THE SCIENCE OF PHOTOTHERAPY: AN INTRODUCTION

# The Science of Phototherapy: An Introduction

*by*


Leonard I. Grossweiner†
*Illinois Institute of Technology,*
*Chicago, IL, U.S.A.*


*edited by*


Linda R. Jones
*College of Charleston,*
*Charleston, SC, U.S.A.*


*co-authors*


James B. Grossweiner
*Edward Hospital,*
*Naperville, IL, U.S.A.*

*and*

B.H. Gerald Rogers
*University of Chicago Pritzker School of Medicine,*
*Chicago, IL, U.S.A.*


Springer

A C.I.P. Catalogue record for this book is available from the Library of Congress.

ISBN 1-4020-2883-0 (HB)
ISBN 1-4020-2885-7 (e-book)

---

Published by Springer,
P.O. Box 17, 3300 AA Dordrecht, The Netherlands.

Sold and distributed in North, Central and South America
by Springer,
101 Philip Drive, Norwell, MA 02061, U.S.A.

In all other countries, sold and distributed
by Springer,
P.O. Box 322, 3300 AH Dordrecht, The Netherlands.

*Printed on acid-free paper*

springeronline.com



Printed in the Netherlands.

# TABLE OF CONTENTS

# FOREWORD

Leonard Grossweiner was proud of his career as a scientist. He once told me that he had accomplished more than he had dreamed, as a person who finished college on the GI bill after World War II and who obtained the doctorate in physics in night school while working full-time as a chemical engineer. He found and opened "windows of opportunity" in the most unlikely places. How would a stranger assess his accomplishments? Well … when he came to Illinois Institute of Technology [IIT], the Physics department had a limited variety of basic research activities, as well as traditional undergraduate and graduate programs. He developed new courses in statistical mechanics, quantum mechanics, and solid state physics, while maintaining an active experimental research program using flash photolysis to study organic molecules. He studied the hydrated electron, which plays a central role in producing radiation damage in biological materials, and he began an intensive and thorough study of biological physics. He was a leader in founding the American Society for Photobiology over thirty years ago. His research in photobiology broadened and deepened, and he became a leading world expert in tissue optics. His analysis of this subject is characterized by its breadth. He explored and understood a broad range of the aspects of tissue optics at a fundamental level. He made theoretical calculations of heat flow, comparing them with data taken on phantom samples -- as well as potatoes, hamburger, and beefsteak. He studied and understood enzyme migration and dosimetry in the body. He participated in clinical trials, and actually counseled terminally ill patients on their treatment protocol. Others may have explored one aspect or another of photophysics in more detail, but very few understood it with such breadth.

Leonard Grossweiner was my role model, for being able to concentrate upon the goals of a project, and to avoid being side-tracked by tangential issues. He understood that one must focus upon important questions, and ignore matters that would ultimately turn out to be extraneous and inconsequential. He understood that our most important resource, and perhaps the most irreplaceable resource, was time itself. By his example we learned how to focus upon important issues, and not to dwell upon things that could not be changed. He always knew instinctively what the "next step" should be, and refused to be discouraged by disappointments. He showed us that there is a great distinction between a "lack of resources" to develop a project and a "lack of ideas". While my colleagues and I found his disciplined approach to be difficult at times, we understood that he always was supportive of the person who worked hard to discover something new. He supported his students, his colleagues, and his collaborators, and set an example that we could all hope to emulate. We miss him.

Porter W Johnson
IIT, Chicago IL 2004

# PREFACE

The beneficial effects of sunlight are a ubiquitous concept in folk medicine originating from sun worship. Indian medical literature dating to 1500 BC describes a treatment of non-pigmented areas of skin by applying black seeds of the plant Bavachee or Vasuchika followed by exposure of the skin to sunlight. This plant was identified as *Psoralea corylifolia* which contains photosensitizing *psoralen* compounds. Similar references appear in Buddhist literature from about 200 AD and in 10th century Chinese documents from the Sung period. The clearing of psoriasis by exposing the skin to sunlight has been known for at least a century. Sunlight also has deleterious biological effects. Exposure of horses, sheep, and cattle to sunlight after ingestion of certain plants leads to ulceration, and infection, and in severe cases, convulsions and death. In some regions of Arabia white horses were not used and even non-pigmented patches on colored horses were darkened with henna or tobacco juice. In the Tarentine fields of Italy only black sheep were grown because the white ones grew ill after exposure to sunlight. The connection between plants of the genus *Hypericum* and the sunlight sensitivity of grazing animals was made in the 19th century, if not earlier. The active photosensitizer was identified with the red pigment hypericin which is found in some insects and plants including St. John's wort. The cell killing properties of ultraviolet light are well known. The use of germidical lamps for this purpose is receiving new consideration in connection with the increasing incidence of contaminated food products. The possible link between skin cancer in humans and sunlight exposure had been suspected for many years based on statistical correlations between skin cancer incidence and the average intensity of sunlight. Recent basic research provides essentially incontrovertible evidence that frequent sunlight exposure generates DNA lesions which can initiate skin cancer. This anecdotal history provides ample evidence that light can act as a powerful drug, either alone or in combination with chemical agents. The objective in a phototherapy is to utilize the beneficial action of light for treatment of disease while minimizing the harmful effects.

The era of controlled phototherapy started at the turn of the 20th century. Niels Finsen, the father of modern phototherapy, discovered that natural sunlight and ultraviolet radiation from a carbon arc are beneficial treatments for cutaneous tuberculosis. Finsen received the Nobel Prize in medicine for his pioneering work in 1903. The slow acceptance of phototherapy in the

U.S. may have resulted from a residual distrust of a discredited blue light therapy for arthritis and other afflictions proposed Augustus J. Pleasonton, a Civil War general. Regulatory approvals of clinical phototherapy started in the 1960s. Although the currently approved procedures are relatively few in number, each of them is directed to a major affliction that affects a large patient population. One of the first approved phototherapies was the treatment of severe plaque-type psoriasis with crude coal tar followed by ultraviolet irradiation of the diseased skin. This approach has been largely superceded by whole-body ultraviolet irradiations and by exposing the skin to "black light" after oral administration of a psoralen drug. This photochemotherapy termed "PUVA" was motivated by knowledge of the ancient writings. The treatment of jaundice in pre-term and very young infants by exposing the skin to blue light is a widely used phototherapy. Photodynamic therapy of malignant tumors (PDT) utilizes the combined action of normally benign red light and a photosensitizing drug that localizes in solid tumors after intravenous administration. PDT is approved in the U.S. for cancer of the esophagus and early-stage lung cancer and for other cancers elsewhere, including malignancies of the skin, urinary bladder, and gastrointestinal tract. The PDT approach has been extended to non-cancer diseases, including new light treatments for "wet" age-related macular degeneration and pre-cancerous skin disease, actinic keratosis. PDT motivated significant advances in related areas including fluorescence diagnosis of cancer and non-invasive optical imaging. The recent laser treatments of port wine stain and other vascular lesions follow directly from modeling of skin interactions with light based on tissue optics.

Phototherapy exemplifies scientific medicine in which basic research and new technologies developed in parallel with the clinical applications. The author's principal objective was to provide a consistent and logical treatment of the entire field for readers coming from different professional backgrounds. Accordingly, some of the content may be very familiar to some readers and quite new to others. *Chapter 1* provides an overview of the entire field of phototherapy which may be useful for a reader who wishes to concentrate on specific topics. *Chapter 2* though *Chapter 4* emphasize the underlying basic science of phototherapy and the relevant optical-electronics technology. Ancillary topics relevant to the action mechanisms are covered in the next four chapters. *Chapter 5* is an introductory treatment of tissue optics, including the basic elements of radiative transfer theory and Monte Carlo simulation. The biological effects of optical radiations at the level of biological molecules and cells are described in *Chapter 6 including* repair mechanisms and target theory. *Chapter 7* describes optical methods of diagnosis and tomographic imaging. *Chapter 8* comprises a comprehensive description of light dosimetry modeling for phototherapy based on the principles in *Chapter 5.* The photothermal interactions of light are reviewed in *Chapter 9* emphasizing applications to laser surgery. The final four chapters review the current status of the major phototherapy procedures. *Chapter 10* and *Chapter 11* are devoted to photodynamic therapy. The clinical applications are exemplified by detailed case histories of esophageal PDT reported by B. H. Gerald Rogers, M.D. The phototherapy of skin diseases is described in *Chapter 12* including case histories provided by James B. Grossweiner, M.D. *Chapter 13* reviews the current status of the phototherapy of neonatal jaundice. Many worked numerical examples are included as a guide to applications of the theoretical concepts. Each chapter provides an annotated listing of textbook and symposium volumes for additional information. The citations to the journal articles are limited to those deemed especially relevant to the text. A comprehensive glossary is provided with definitions of the more frequently employed technical terns, jargon, and abbreviations.

*Leonard I. Grossweiner*
*Chicago IL 2001*

# ACKNOWLEDGEMENTS

It has been an honor to participate in the completion of this book. On behalf of Professor Grossweiner, I wish to express sincere appreciation to Bess Grossweiner for her untiring help with the formatting and proofreading of this text. Also on his behalf I would like to express appreciation to B. H. Gerald Rogers, M.D. and James Grossweiner, M.D. for their interesting and informative chapters on clinical phototherapy.

*Linda R. Jones*

## IN MEMORIAM

**Leonard I. Grossweiner**
**1924 – 2001**

Professor Leonard I. Grossweiner, a native of New York, began his career in Chicago at the Argonne National Laboratory in 1947. He earned a Ph.D. in Physics at the Illinois Institute of Technology in 1955 and joined the physics faculty of IIT as Associate Professor in 1957. His major research interests at that time were fast reaction photochemistry, color centers in alkali halides and solid state photosensitization. This unlikely combination motivated him to invite other kindred souls to the first International Conference on Photosensitization in Solids at IIT in 1963. In the 1970's, he organized the Division in Biological Physics of the American Physical Society, serving as the first secretary-treasurer and subsequently as the chairperson. He helped to organize a joint meeting between the Biophysical Society and APS in Washington DC in 1977.

Prof. Grossweiner's research interests moved towards photobiology in the 1980's and he became active in the American Society for Photobiology, serving as its president in 1988. In the same period he was the last chairman of the U.S. National Committee on Photobiology. He joked that the reason the committee was terminated by the National Research Council is because in the words of the NRC administrator "all we ever did at the meetings was talk science." He also served as Chair of the IIT Physics Department from 1970-1980.

Prof. Grossweiner's earlier work involved molecules of biological interest. Eventually he appreciated that the work was actually not very interesting to biologists because biology takes place in tissues and not dilute aqueous solutions. Later work involved more realistic models of biological photosensitizations in light-scattering environments. He became interested in tissue optics, laser-tissue interactions and light dosimetry for photodynamic therapy. After his retirement from IIT in 1996 he became the research director of the Wenske Laser Center of the Ravenswood Hospital Medical Center and continued directing research and participating in clinical applications of photodynamic therapy until his death.

Prof. Grossweiner viewed his research career as typical of a science professor at a technical school or a small private university. He described himself as a "sergeant-major" of research and science teaching, working on research projects that are scientifically interesting, potentially useful and within [his] intellectual and physical resources." His research philosophy was to ignore artificial boundaries and continue learning. This book is a wonderful tribute to his career and will be useful to scientists, clinicians and students. Those who knew him will recognize his voice, his reverence for science and even his sense of humor throughout these pages.

*Linda R. Jones*
*College of Charleston, Charleston SC 2004*

# CHAPTER 1

# AN OVERVIEW OF PHOTOTHERAPY

The beneficial effects of sunlight are a ubiquitous concept in folk medicine originating from sun worship. Indian medical literature dating to 1500 BC describes a treatment of non-pigmented areas of skin by applying black seeds of the plant Bavachee or Vasuchika followed by exposure of the skin to sunlight. This plant was identified as *Psoralea corylifolia* which contains photosensitizing psoralen compounds. Similar references appear in Buddhist literature from about 200 AD and in 10th century Chinese documents from the Sung period. The clearing of psoriasis by exposing the skin to sunlight has been known for at least a century. Sunlight also has deleterious biological effects. Exposure of horses, sheep, and cattle to sunlight after ingestion of certain plants leads to ulceration, and infection, and in severe cases, convulsions and death. In some regions of Arabia white horses were not used and even non-pigmented patches on colored horses were darkened with henna or tobacco juice. In the Tarentine fields of Italy only black sheep were grown because the white ones grew ill after exposure to sunlight. The connection between plants of the genus *Hypericum* and the sunlight sensitivity of grazing animals was made in the 19th century, if not earlier. The active photosensitizer was identified with the red pigment hypericin which is found in some insects and plants including St. John's wort. The cell killing properties of ultraviolet light are well known. The use of germidical lamps for this purpose is receiving new consideration in connection with the increasing incidence of contaminated food products. The possible link between skin cancer in humans and sunlight exposure had been suspected for many years based on statistical correlations between skin cancer incidence and the average intensity of sunlight. Recent basic research provides essentially incontrovertible evidence that frequent sunlight exposure generates DNA lesions which can initiate skin cancer. This history provides ample evidence that light can act as a powerful drug, either alone or in combination with chemical agents. The objective in a phototherapy is to utilize the beneficial action of light for treatment of disease while minimizing the harmful effects.

The study of light occupies the central position in the history of science (Table 1.1). Virtually every discovery cited in the table plays a direct or indirect role in the delivery or mechanism of phototherapy. The central principle of photochemistry, first stated by Grotthus and Draper in 1818, is "only absorbed

**Table 1.1 Chronology of Discoveries about Light**

In the beginning God created the heaven and the earth. Now the earth was unformed and void, and darkness was upon the face of the deep; and the spirit of God appeared over the face of the waters. And God said: "Let there be light", and there was light. *Genesis 1*

Atomistic theory of nature; *430 BC*, Democritus of Abdera
Discovery of fluorescence; *1577*, Nicholas Monardes
Discovery of phosphorescence; *1603*, Vincenzo Cascariolo
Laws of light refraction; *1637*, Rene Descartes
Explanation of the rainbow; *1665*, Isaac Newton
Discovery of double refraction in calcite; *1670*, Erasmus Bartholin
Measurement of the speed of light. *1676*, Øle Roemer
Wave theory of light; *1690*, Christiaan Huygens
Discovery of diffuse light; *1760*, Johann Lambert
Discovery of infrared radiation; *1800*, William Herschel
Discovery of ultraviolet radiation; *1801*, Johann Ritter
Discovery of absorption and emission spectra; *1802*, William Wollaston
Three color theory of vision; *1802*, Thomas Young and Ludwig von Helmholtz
The first artificial light source; *1826*, Humphry Davy
Invention of photography; *1826*, Joseph Niépce
Rotation of light polarization by a magnetic field; *1845*, Michael Faraday
Theory of electromagnetic radiation; *1865*, James Clerk Maxwell
Theory of light scattering; *1871*, Lord Rayleigh
Discovery of the photoelectric effect; *1887*, Heinrich Hertz
Invention of the optical interferometer; *1881*, Albert Michelson
First controlled phototherapy; *1893*, Niels Finsen
Quantum theory of black body radiation; *1900*, Max Planck
Discovery of biological photosensitization; *1900*, Oscar Raab
Quantum theory of light; *1905*, Albert Einstein
Quantum theory of the hydrogen atom; *1913*, Niels Bohr
Wave mechanics; *1924-1925*, L. de Broglie, W, Heisenberg, E. Schroedinger and others
Theory of the chemical bond; *1927*; W. Heitler, F. London, R. S. Mulliken, F. Hund
Theory of radiative transfer; *1950*, S. Chandresekhar
Invention of the laser; *1960*, Theodore Maiman

light can produce a chemical change". This deceptively simple concept is the basis of all photobiology. Every process in photobiology has the common feature that the first light-activated step involves very fast events at the molecular level.

The photophysical act of light absorption initiates a sequence of actions and reactions that can lead to a remarkable diversity of physiological endpoints, for example, plant growth, animal vision, circadian rhythms, and sunburn. The unique feature of phototherapy is that light acts as a powerful drug. This property of light differs from the historic view that light is a benign entity that is required for vision. However, it is now well known that excess exposure to sunlight is responsible for many disorders of the skin including a risk of skin cancer. The objective in phototherapy is to utilize light to cure disease while minimizing the adverse effects. All interactions of light with biological systems utilize unique light-absorbing molecular units or *chromophores* located within the illuminated tissues. In natural photobiology the chromophores are normally present in the tissue matrix, for example, chlorophyll molecules in green plants and retinal pigments in vertebrate retina. The tissue structure is an important determinant of the overall reaction pathway. The matrix may modify the "quality" of the incident light, establish structural relationships between chromophores and reacting molecules and store intermediate products. The photosynthetic reaction center in a plant leaf performs all of these functions. Evolution has optimized the structural conditions required for efficient utilization of sunlight. In phototherapy, however, the chromophores are exogenous chemical agents or metabolic products. The optimal treatment parameters must be determined by experimentation and clinical experience. The usual research sequence starts with photochemical studies on the relevant biological molecules, followed by *in vitro* cell cultures, experimental animal models, and clinical trials on human patients. This protracted and frequently disjointed program may be expedited by utilizing the methods of tissue optics. Recent developments in this field have made it possible to model key aspects of phototherapy by using a combination of analytical techniques and experiments with *tissue phantoms*, which are inert systems that mimic the optical properties of biological tissues.

The era of controlled phototherapy started at the turn of the 20th century. Niels Finsen, the father of modern phototherapy, discovered that natural sunlight and ultraviolet radiation from a carbon arc are beneficial treatments for cutaneous tuberculosis. Finsen received the Nobel Prize in medicine for his pioneering work in 1903. The slow acceptance of phototherapy in the U.S. may have resulted from a residual distrust of a discredited blue light therapy for arthritis and other afflictions proposed A. J. Pleasonton, a Civil War general. Regulatory approvals of clinical phototherapy started in the 1960s. Although the currently approved procedures are relatively few in number, each of them is directed to a major affliction that affects a large patient population. Clinical reports demonstrating the anti-psoriatic action of mid-range ultraviolet (UV-B) from mercury-arc lamps first appeared in the 1920s. W. H. Goeckerman found that the effectiveness of the mercury arc was enhanced by first applying and then washing off crude coal tar from the psoriatic skin. The Goeckerman regimen has been a standard phototherapy for severe plaque-type psoriasis for many decades. Remarkably, this combination of two carcinogenic agents, coal tar and UV-B, is not known to induce skin cancer in humans. The modern *photochemotherapy* of psoriasis originated from the ancient literature. This procedure is based on the combined action of an oral psoralen drug and UV-A ("black light"). It is referred to as *PUVA*. The extension of ancient sunlight treatments for vitiligo using topical plant extracts to modern PUVA therapy of psoriasis was an exceptionally insightful accomplishment. Basic research on psoralen photochemistry established the molecular action mechanism for PUVA even before the clinical procedure was developed. The dimensions of linear psoralen derivatives span the pyrimidine bases across the DNA double helix. Absorption of UV-A by the localized psoralen molecules generates covalent cross-links across the DNA chains that block DNA synthesis. PUVA exemplifies a *photosensitization* in which light absorbed by a "colored" agent present in small amounts activates physical, chemical, or biological effects in a "colorless" substrate. *Extracorporeal photophoresis* (ECP) is an extension of the PUVA concept which is being used to treat cutaneous T-cell lymphoma and inhibit acute rejection of heart transplants. ECP is performed by irradiating a small amount of blood taken from the patient with UV-A after administration of the psoralen drug and then re-infusing the treated blood. The photochemically altered white cells initiate immunological responses, the details of which are the subject of active research.

*Photodynamic therapy* (PDT) is a light treatment for localized tumors utilizing the combined action of light and a chemical drug targeted to malignant tissues. In general, a *photodynamic action* is defined as a biological photo-sensitization that requires the participation of molecular oxygen. PDT was first attempted in 1903 by H. von Tappenier and A. Jesionek. who treated skin cancers with eosin dye followed by exposure of the lesions to sunlight. The lack of success probably resulted from an insufficient dye concentration in the tumors and an inadequate light dose. The first effective PDT drug was synthesized by R.L. Lipson in 1960 by treating impure hematoporphyin with strong acids. The resultant brown powder termed *hematoporphyrin derivative* (HPD) is a mixture of porphyrins, including a unique active component that localizes in and is retained by tumors after intravenous injection. This active component termed *dihematoporphyrin ether* (DHE) can be identified by a reddish fluorescence excited by blue or near-ultraviolet light. HPD was used sparingly for tumor identification and a few cancer treatments in the 1960s and 1970s. In 1978 T. J. Dougherty and co-workers reported controlled clinical trials indicating that HPD is effective for eradication of skin cancers. The procedure was first referred to as "photoradiation therapy" which was then changed to "photodynamic therapy" to avoid confusion with cancer treatments that use ionizing radiation. Subsequent studies worldwide showed that HPD is effective for PDT of other localized cancers, including tumors of the head and neck, esophagus, urinary bladder, and bronchus. PDT has received regulatory approvals in 11 countries for specific malignancies, including esophageal cancer and early-stage bronchogenic lung cancer in the U.S. The putative action mechanism in PDT involves the generation of *singlet molecular oxygen* ("singlet oxygen") by the light-activated HPD and its attack on endothelial membranes of the tumor microvasculature. The shutdown of the tumor oxygen supply induces necrosis and killing of cancer cells. Recent research indicates that the vascular pathway is augmented by direct killing of cancer cells by *apoptosis*. This process is a form of regulated cell death that is normally involved in tissue hemostasis, and regulation of the immune system. The promising results with HPD motivated a concerted search for new and possibly more effective PDT drugs. Much attention has been given to the topical PDT agent δ-aminolevulinic acid (ALA) which is converted to photosensitizing protoporphyrin after uptake by cancer cells. The ability of PDT to inactivate abnormal endothelial cells has led to experimental extensions of the procedure to light treatments for non-cancer disorders, including Barrett's esophagus, endometriosis, and atherosclerotic vascular disease. Inhibition of age-related macular degeneration is a recently approved PDT application in the U.S.

The reduction of jaundice in newborn infants by exposure to blue-light was discovered by accident. In 1957 a nursing sister in the clinic of R. J. Cremer in Essex U.K. reported that neonatal jaundice in the sunlit parts of the nursery faded after a short period of exposure. A significant fraction of healthy term newborns now receive the blue-light phototherapy. Jaundice originates from the accumulation of unbound bilirubin in the infant's serum, which is a normal breakdown product of heme generated by turnover of red blood cells. Non-pathologic hyperbilirubinemia results from delayed development of liver enzymes that conjugate bilirubin with albumin and permit its elimination. The action mechanism in this phototherapy involves the photochemical conversions of bilirubin to a different molecular shape that is more readily eliminated. The process of *photoisomerization* is relatively uncommon in natural photobiology, except in animal vision in which it is the primary photochemical reaction of the retinal pigments.

The recent developments in phototherapy, and especially PDT, motivated far-reaching interest in relevant pre-clinical areas. The rapidly-developing field of *tissue optics* exemplifies this phenomenon. Crystalline and amorphous substances, such as pure liquids, glasses, and crystals, are optically homogeneous materials in which light propagation is accurately described by the laws of geometrical and physical optics Classical optics is inapplicable for most biological tissues which are "cloudy" or *turbid* substances. Tissues contain microscopic entities that scatter light rays into many different directions. This effect is responsible for much of the attenuation of an entering light beam, in addition to the loss resulting from absorption by chromophores. The most general approach to light propagation in turbid media was developed by S. Chandrasekhar in the 1930s and 1940s. This theory termed *radiative transfer* was originally developed to explain light propagation between stars and planets, and subsequently extended to such diverse applications as diffusion of neutrons in nuclear reactors, atmospheric scattering of radar, underwater acoustics, and most recently, to tissue optics. The basic objective in tissue optics is to calculate the distribution of optical energy within tissue for specified irradiation conditions. The complex mathematics of the original theory has been augmented by new digital computer techniques. Tissue optics has been used for modeling PDT light dosimetry since about 1980. Optical imaging is a more recent application of radiative transfer directed towards non-invasive diagnosis of tissue abnormalities. The resolution achieved in optical imaging has improved considerably since the mid-1990s by utilization of tomographic techniques and image reconstruction algorithms borrowed from x-ray computed tomography (CT) and magnetic resonance imaging (MRI). *Monte Carlo* simulation is an alternative and very powerful analytical technique based on statistical methods. Another imaging approach utilizes the *in situ* fluorescence of a localized PDT drug excited by external light. This technique is referred to *photodynamic diagnosis* (PDD). The light sources used for phototherapy are either strong conventional lamps or relatively low- power lasers. The apparently different modality of laser surgery requires high-power lasers in order to coagulate, vaporize, and ablate tissues. However, the initial stages in laser surgery and phototherapy are very similar. In both modalities incident light must penetrate tissue and be absorbed by the relevant chromophores. The subsequent events in phototherapy involve photochemical reactions, while in laser surgery the absorbed energy is converted to heat or photomechanical effects. Recent analytical modeling techniques combine tissue optics with heat transfer theory. The applications include modeling of cancer therapy by hyperthermia and laser treatments of port wine stain and other vascular lesions.

Radiation dosimetry is an essential aspect of modern radiotherapy. In conjunction with new imaging techniques, improvements in radiation dosimetry are responsible for the improved efficacy of radiotherapy since the 1950s. It is generally accepted that clinicians utilizing these techniques should have a basic understanding of the underlying science and a detailed knowledge of the relevant technology. The same criteria are applicable for phototherapy. The apparatus used for phototherapy is an integral part of each procedure. Testing of a new PDT drug frequently requires the development of a new light source and delivery system. Basic information about the properties of light sources, optical fibers, and electro-optical devices can assist the phototherapy clinician in the planning and delivery of the procedure. The most up-to-date information is more likely to appear in commercial literature than in general textbooks. In principle, light dosimetry for phototherapy parallels radiation dosimetry for radiotherapy. However, the phototherapy case is more complicated owing to the weaker penetration of light compared to x-rays and gamma-rays and the greater variability of tissue optical properties. The problems are further complicated by the use of *photometric units* in the photography and lighting industries and *radiometric units* in photobiology and photomedicine. The recent advances in tissue optics and optical imaging should provide a good basis for eventual rational treatment planning based on accurate light dosimetry. Except for the light treatment of neonatal jaundice, phototherapy is not widely used at present. PUVA has an established "niche" in dermatology. The recent use of ECP to prevent acute rejection of organ transplants will almost certainly expand during the next decade. The anti-cancer applications of PDT have increased steadily since 1980. Major tasks for the future will be to determine the most advantageous PDT procedure for individual patients, and evaluate the role of PDT as a treatment of first choice instead of a last resort when all other modalities have failed. It is interesting to note that cancer treatments with x-rays and radium started almost immediately after their discovery some 100 years ago. However, the major advances in radiotherapy were made after 1950 with the availability of new radiation sources, quantitative radiation dosimetry, and advances in the knowledge of radiation damage at the cellular level. PDT is entering a comparable phase of rapid growth driven by patient interest, increased receptiveness of the medical community, and significant investments of resources by the private sector. The term "PDT" is now being used to describe virtually any phototherapy utilizing an exogenous photosensitizing agent, whether or not photodynamic action has been demonstrated. The extension of PDT to the treatment of pre-cancerous and non-cancer diseases is a recent development. Inhibition of "wet" age-related macular degeneration by PDT is a recent application which has received regulatory approvals. ALA is approved in the US for PDT of actinic keratosis. The clinical potential of PDT has engendered a significant "spin-off" in related fields. Basic research on the PDT action mechanism led to new information about cellular and tissue responses to photosensitization. Photosensitizing agents which may not be useful for PDT are being tested for other applications, including photochemical detoxification of viral and bacterial contamination of stored blood products and photosensitization of oral pathogenic dental bacteria. PDT motivated the synthesis of a new class of "pseudo-porphyrin" photosensitizing compounds, the metallo-texaphyrins. Lutetium texaphyrin is an effective drug activated by tissue-penetrating near-infrared light. The gadolinium texaphyrin derivative shows promise as an anoxic sensitizer for radiotherapy of brain tumors, with the added feature that it localizes in tumors and can be visualized by MRI. Recent advances in tissue optics were motivated by applications to modeling of PDT light dosimetry and laser surgery.

# CHAPTER 2

# OPTICAL PHYSICS AND BIOTECHNOLOGY

This chapter reviews the underlying optical physics relevant for phototherapy, including the properties of light, light sources, and the devices used to deliver, control, and measure light.

## 2.1 LIGHT WAVES

Light is a wave motion. In general, a wave is an oscillatory disturbance moving through space. The nature of the disturbance depends on the type of wave. The disturbance in a mechanical wave is a transient dislocation of atoms in a medium. For example, a hydrodynamic surface wave is generated by dropping a pebble into water. This motion is *transverse* because the direction of the disturbance is perpendicular to the propagation direction of the wave. The disturbance in a sound wave consists of alternating regions of high and low pressure. Sound waves are *longitudinal* because the pressure variations oscillate in the same direction as the wave motion. A *standing wave* is a wave motion in which stationary regions or nodes are fixed in space. The displacement of a violin string exemplifies transverse standing waves. Sound waves in an organ pipe are longitudinal standing waves. As the name implies, the disturbance carried by a *continuous wave* (CW) is repetitive over a relatively long time period. An ordinary light bulb emits a mixture of CW light waves. The mathematical technique of *Fourier analysis* shows that virtually any wave can be decomposed into a group of infinitely long sinusoidal waves.[1] A pure musical tone generated by a tuning fork corresponds to an approximately single-wavelength or *monochromatic* sound wave. A musical chord consists of a mixture of pure tones. An ideal monochromatic wave is described by its amplitude, frequency, and wavelength (Figure 2.1). The amplitude is the maximum strength of the disturbance. The corresponding power carried by a wave is proportional to the square of the amplitude. The wave frequency ($f$) is the rate of oscillation at any point along the motion. The wavelength ($\lambda$) is the shortest distance between maximum amplitudes. The parameters of an idealized sinusoidal wave are related by

$$v=\lambda f \tag{2.1}$$

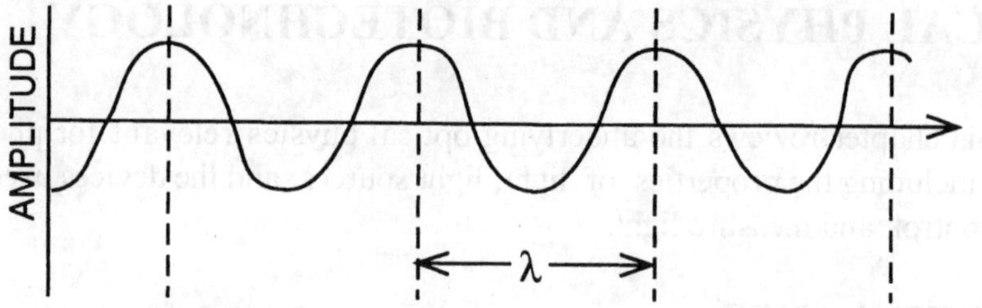


**Figure 2.1** An ideal monochromatic wave is depicted as a sinusoidal disturbance. In a "running" wave the peaks and valleys propagate along the *Z*-axis with the wave velocity *V*. In a "standing" wave the points of zero amplitude or "nodes" are fixed in space and the peaks and valleys oscillate with the wave frequency *f*.

where $v$ is the wave velocity. The *bandwidth* specifies the range of frequencies or wavelengths in a more general type of wave.

Light waves are *electromagnetic waves* (EMW). Other EMW include radio waves, microwaves, x-rays, and gamma-rays. The different forms of EMW are characterized by their range of wavelengths or frequencies. An EMW transports an electric field (E-field) and a magnetic field (B-field) at the speed of light (*c*) which equals 3.00 x $10^8$ m/s in free space (Figure 2.2). The speed of an EMW in a medium depends on the type of wave and the properties of the material. The radiant power in an EMW is proportional to the square of the E-field (or B-field) amplitude. Typical optical power levels range from microwatts for small sources to gigawatts for powerful lasers.[2]

The wavelength (or frequency) is the principal determinant of how an EMW interacts with matter. The wavelength of visible light ranges from 400 nm (blue light) to 760 nm (far-red light). White light refers to a mixture of visible wavelengths that is perceived by the human eye as "white". The human eye can distinguish the order of ten million visible colors formed by adding the primary colors, blue, green, and red, in different ratios. A substance appears colored in white light because it either reflects or absorbs visible wavelengths. For example, a plant leaf is green in reflected sunlight because chlorophyll absorbs blue light and red light.

Radio waves have much longer wavelengths than light waves. Heating of biological tissues by radio waves is caused by interactions of the oscillating E-field with the dipolar water molecules in tissues.[3] Rapid back and forth rotations of the water molecules generates friction that heats the material. Similar

a

b

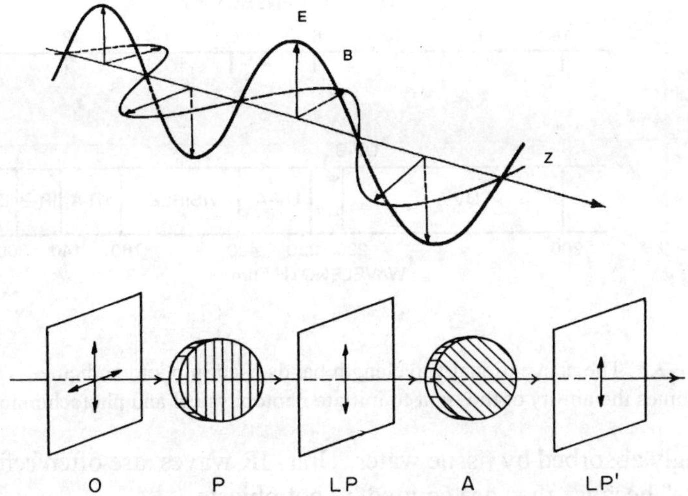


**Figure 2.2** (a) The **E**-vector and **B**-vector of an electromagnetic wave are mutually perpendicular in a plane transverse to the direction of wave motions and oscillate with the wave frequency *f*. (b) The **E**-field of non-polarized light has a random orientation, as depicted by perpendicular projections of equal amplitude (O). Linear polarized light (LP) is transmitted by a polarizer (P). A second polarizer or "analyzer" (A) transmits 0-100% of the incident light depending on its relative angle to the polarizer.

*Dielectric* interactions are involved in tissue heating by microwaves. Gamma-rays are very short- wavelength EMW. Gamma-rays interact with the atomic electrons in biological matter. This process may lead to ejection of electrons (*photoelectric effect*). Optical radiations having shorter wavelengths than visible light are termed *ultraviolet* (UV) and those optical radiations having longer wavelengths than visible light are *infrared* (IR). The UV and IR bands are divided into sub-bands which are indicative of their interactions with biological macromolecules (Figure 2.3). UV-C or "far-UV" (200 nm - 290 nm) is strongly absorbed by all cellular constituents including proteins, nucleic acids, carbohydrates, and lipids. This band includes "germicidal" radiations strongly absorbed by DNA. UV-B or "mid-range UV" (290 nm - 320 nm) is absorbed mostly by aromatic amino acids and nucleic acids. The responses of skin to UV-B include vitamin-$D_3$ production, erythema, hyper-pigmentation, and skin cancer induction. UV-A or "black light" (320 nm - 400 nm) is weakly absorbed by "colorless" cellular constituents. Similarly, the IR spectrum is divided into sub-bands. IR-A or "near-IR" (0.76 μm - 4 mm) is an optical "window" with minimal attenuation in tissues. IR-B or "mid-range IR" (1.4 mm - 3 mm) is absorbed mostly by water and bone. IR-C or "far-IR" (3 mm - 1000 mm) is very

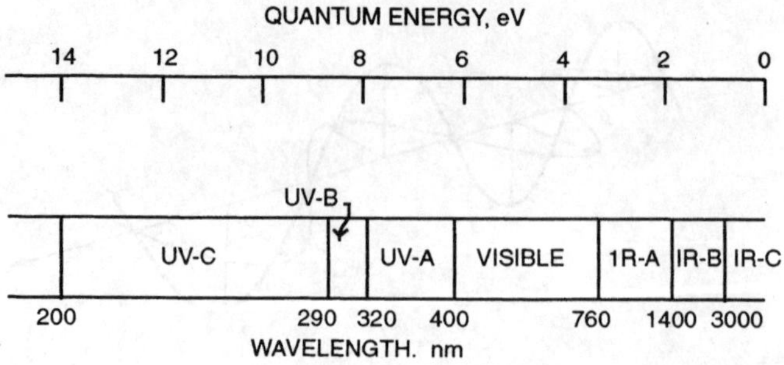


**Figure 2.3** The conventional wavelength bands used in photomedicine. The quantum energy determines the ability of the light to initiate photophysical and photochemical processes.

strongly absorbed by tissue water. Far- IR waves are often referred to as "heat waves" because they are emitted by hot objects.

In addition to wavelength, frequency, and speed, EMW are characterized by the property of *polarization*. The EMW theory of Maxwell shows that E-field and B-field are located in a plane perpendicular to the direction of the wave motion or "ray" and are oriented perpendicular to each other. The vectors representing these fields oscillate with the wave frequency *f*. The orientation of the E-field in space defines the polarization. Ordinary lamps emit *non-polarized light* in which the time-averaged E-field has no preferred direction. The E-field of *linearly polarized light* maintains a fixed direction in space as the wave propagates. Linear polarization affects light absorption by long-chain molecules. This effect is utilized by Polaroid® to convert non-polarized light to linearly polarized light. This phenomenon is illustrated in Figure 2.4. The "polarizer" transmits preferentially in the direction determined by the orientation of the molecules in the film. The light intensity transmitted by the "analyzer" varies from almost zero to 100% depending on the relative orientation of the two films. In *circularly polarized light* the E-vector rotates with frequency *f* in its plane. Molecules having regions of spiral structure such as proteins and DNA interact differently with right-handed and left-handed circularly polarized light.

## 2.2 LIGHT QUANTA

The wave picture provides satisfactory explanations for many light properties, including mirror reflection, imaging by a lens, and light scattering by small particles. The interaction of light with a transparent object has a minute effect on the object itself. For example, a glass prism separates white light into

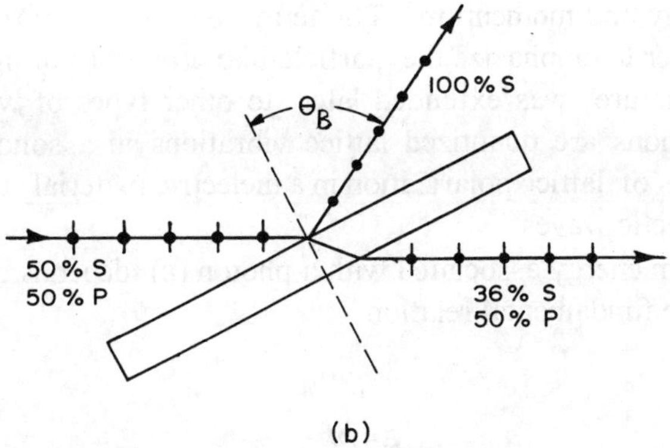


**Figure 2.4** Polarization of ordinary light by a glass plate tilted at the Brewster angle. From Grossweiner, L.I., *The Science of Photobiology*, 2nd ed., Plenum Press, New York, 1989. With permission.

its constituent colors and the only effect on the prism is an extremely weak *radiation pressure* induced by reflection of light at the air-glass interface. However, light absorption may involve large interactive effects. The key role of light absorption in photobiology is expressed by extending the Grotthus and Draper central principle: *Only absorbed light can produce a physical, chemical, or physiological change in a biological system.* A proper description of light absorption and its consequences requires a quantum theory based on the molecular structure of matter.

Historically, the spectrum of thermal radiation emitted by hot objects was the first indication that classical electromagnetic theory led to drastically incorrect results. All matter emits radiation generated by the thermal motions of the constituent atoms. The red color emitted by the heating element on an electric stove is a familiar example. The classical theory of radiation predicts that thermal radiation should be increasingly intense into the UV region. However, measurements show that the thermal radiation spectrum of extra-terrestrial sunlight increases from low intensity in the IR region to a maximum near 550 nm (green light) and then decreases into the UV region. The 1901 radiation theory of Planck removed the "ultraviolet catastrophe" by discarding the classical concept that thermal radiation has a continuous spectrum of frequencies. Planck's revolutionary assumption was that thermal radiation is *quantized* in energy packets, each of which corresponds to a specific frequency. Quantization of light was extended by Einstein in 1905 to explain the *photoelectric effect* in metals and by Bohr in 1913 to explain the emission spectrum of atomic hydrogen. According to quantum theory, light and other EMW can be considered as consisting of hypothetical zero-mass particles that move with the speed of light and carry energy and momentum. The term *photon* was introduced by G. N. Lewis much later to emphasize the particle-like properties of light quanta. The particle-like picture was extended later to other types of wave motion, for example, phonons are quantized lattice vibrations in a solid, polarons are quantized waves of lattice polarization in a dielectric material and magnons are quantized magnetic waves.

The quantum energy associated with a photon ($\varepsilon$) depends on its frequency according to the fundamental relation

$$\boxed{\epsilon = hf} \qquad (2.2)$$

where $h$ is Planck's constant equal to $6.62 \times 10^{-34}$ J-s. The energy equivalent of a photon is usually stated in units of the electron-volt (eV) where 1 eV = $1.60 \times 10^{-19}$ J.[4] In thermochemical units, 1 eV = 23.06 kcal/mol. Figure 2.3 indicates the quantum energy associated with the optical radiations. Light interactions with matter are most efficient when the photon energy is comparable to the difference between the final and initial energy states of the constituent atoms or molecules. UV photons can break chemical bonds and eject electrons from atoms and molecules. Photons of visible light "excite" valence electrons in atoms and molecules to higher energy states. IR photons interact with the vibrational motions in molecules and solids.

---

**Example 2.1** Conversion of optical power to photons

A laser emits an optical power of 1 W at 500 nm in a 1 mm diameter collimated beam. How many photons are contained in a 1 m length of the beam?

The power per unit area in the laser beam is: 1 W/[$(\pi/4) \times (1 \times 10^{-3}\ \text{m})^2$ ] = $1.27 \times 10^6$ (J/s-m$^2$ ). The total energy in 1 m of beam length is: $1.27 \times 10^6$ (J/s-m$^2$) x 1 m / $3.00 \times 10^8$ (m/s) = $4.24 \times 10^{-3}$ J. The photon energy at 500 nm from Equation (2.1) and Equation (2.2) is: $6.62 \times 10^{-34}$ (J-s) x $3.00 \times 10^8$ (m/s) / ($500 \times 10^{-9}$ m) = $3.97 \times 10^{-19}$ J. The number of photons in a 1 m beam length is: $4.24 \times 10^3$ J/ $3.97 \times 10^{-19}$ J/photon = $1.07 \times 10^{16}$ photons.

---

It should be noted that not all radiations used in medicine are EMW. The alpha-rays and beta-rays, emitted by radioactive elements are energetic helium nuclei and electrons, respectively. *Ultrasound* is a longitudinal pressure wave, similar to ordinary sound waves but having a higher frequency and power level. Propagation of ultrasound requires a medium such as air, water, or tissue. The wavelength and frequency of ultrasound are related by Equation (2.1) with $v$ equal to the velocity of sound in the medium. The frequencies of ultrasound used for medical diagnosis range from 1 to 10 MHz with wavelengths in tissues about 0.15-1.5 mm. This range limits the resolution that can be achieved by ultrasound imaging. Recent interest in optical imaging was motivated by the much greater resolution feasible at shorter light wavelengths compared to ultrasound (see Chapter 7).

## 2.3 GEOMETRICAL AND PHYSICAL OPTICS

The properties of virtually all optical devices are based on the laws of classical optics. *Geometrical optics* provides an adequate description of those phenomena in which light rays encounter objects having dimensions much larger than the light wavelength including mirror reflection, imaging by lenses, and light transmission by optical fibers. The light rays are termed "directed" because their directions are pre-determined by the geometry of the source and interacting devices. Light interactions with objects comparable in size to the wavelength require the more complex concepts of *physical optics*, for example, lens aberrations, the spread of a laser beam, light scattering by small particles, and holographic imaging. *Interference* and *diffraction* are significant effects in physical optics. Interference occurs when two or more light waves occupy the same region of space. This effect leads to cancellation or reenforcement of intensity depending on how the waves "match up". Diffraction occurs when a physical object such a narrow slit blocks part of a propagating wave. Interference between those waves propagating past different regions of the object leads to intensity variations referred to as diffraction patterns. This section reviews the basic aspects of geometrical and physical optics which are involved in the properties of common optical devices.

### 2.3.1 Reflection and Refraction of Light

The speed of light is slower in a medium compared to free space owing to interactions of the E-field with atomic electrons. The velocity of light in a medium is expressed as the *refractive index* ($n$) according to

$$v = \frac{c}{n} \tag{2.3}$$

where $v$ is the wave velocity in the medium. Refractive index depends on both the medium and the light wavelength. The variation of refractive index with wavelength is termed *dispersion* after the separation of white light into its constituent colors by a glass prism. Approximate values of $n$ for visible light are 1.0 in air, 1.3 in water, 1.5 in glass, and 1.4 in biological tissues.

---

**Example 2.2** Light waves in a medium

(a) Calculate the frequency of 500 nm green light in air. (b) Calculate the frequency, wavelength, and velocity of the same light in a glass of refractive index equal to 1.50.

(a) The light frequency in air from Equation (2.1) is: $3.00 \times 10^8$ (m/s)/500 $\times 10^{-9}$ m = $6.0 \times 10^{14}$ $s^{-1}$ = 600 THz.

(b) The velocity in the glass from Equation (2.3) is: $3.00 \times 10^8$ (m/s)/1.5 = $2.0 \times 10^8$ (m/s). The corresponding wavelength in the glass from Equation (2.1) is: $2.0 \times 10^8$ (m/s)/$6.0 \times 10^{14}$ $s^{-1}$ = $3.333 \times 10^{-7}$ m or 333 nm.[5]

---

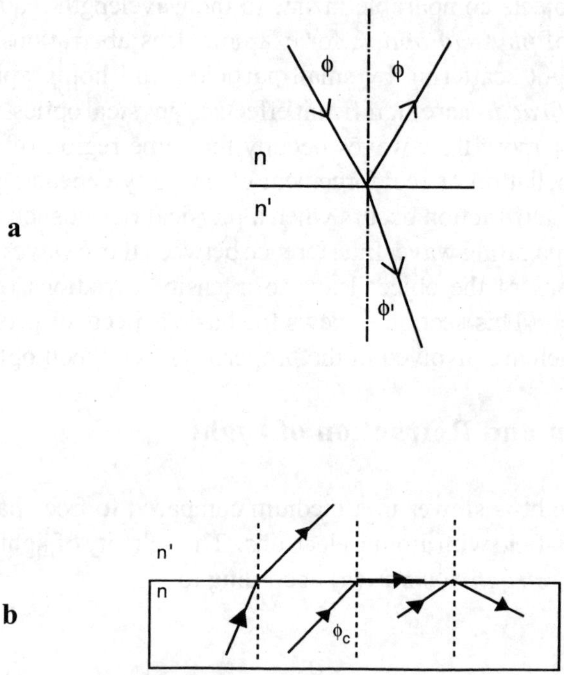


**Figure 2.5** (a) Reflection and refraction of a collimated light beam at the interface between two media of refractive indices $n$ and $n'$ where $n<n'$. (b) Total internal reflection takes place for $n>n'$ and the angle of incidence exceeds the critical angle defined by Equation (2.5).

Figure 2.5a depicts a collimated beam of monochromatic light in air incident on a polished glass surface at an incidence angle $\phi$ from the perpendicular. A
small fraction of the incident intensity is reflected at angle $\phi$, typically 3% - 5%. Mirror reflection is referred to as *specular* (originally from shiny speculum metal) to distinguish it from *diffuse reflection* by a rough surface. The wave that enters the medium is bent or *refracted* at a less oblique angle $\phi'$. The angle between the incident and refracted rays is determined by *Snell's law*

$$n \sin\phi = n' \sin\phi' \tag{2.4}$$

where $n$ and $n'$ are the refractive indices of the first medium and the second medium, respectively. The *relative refractive index* is defined as: $n_r = n'/n$. Refraction always takes place according to Equation (2.4) when a light ray enters from a medium of higher refractive index ($n' > n$). However, when a light ray encounters a medium of lower refractive index ($n' < n$) the value of $\sin\phi'$ can exceed unity for a sufficiently oblique incidence angle $\phi$. This condition cannot satisfy Equation (2.4). *Total internal reflection* (TIR) is the physical consequence of this apparent paradox in which the incident ray is reflected back into the initial medium. A familiar example of TIR is the disappearance of a swimming pool bottom when viewed from outside at an oblique angle. The *Snell's law critical angle* ($\phi_c$) for TIR is given by

$$\phi_c = \arcsin(n'/n) \tag{2.5}$$

where $n$ and $n'$ refer to the higher and lower refractive index media, respectively ( Figure 2.5b ). Optical fibers utilize TIR to transport light over long distances (Section 2.6.3.)

---

**Example 2.3** Snell's law

---

A light beam in air enters glass at 45°. The refractive index of the glass is 1.50. Calculate: (a) the refraction angle within the glass; (b) the maximum angle at which a light ray propagating within the glass can exit to air.

(a) Refraction at the air-glass interface obeys Snell's law. From Equation (2.4.): $\sin\phi' = (1/1.50) \sin 45° = 0.471$; $\phi' = \arcsin 0.471 = 28.1°$.

(b) Total internal reflection occurs for incidence angles exceeding: $\phi_c$ = arc sin(1.0/1.5) = 41.8°.

---

Snell's law does not predict the distribution of intensity between the reflected and refracted waves. This information is provided by the *Fresnel relations* which are derived from EMW theory. The special case of perpendicular incidence leads to the following expression

$$R(\phi=0)=\frac{(n_r-1)^2}{(n_r+1)^2} \tag{2.6}$$

where $R$ is the relative intensity of the reflected light or *reflection coefficient.*[6] For a typical glass with $n'_r$ = 1.5 the value of R($\varphi$ = 0) is 0.04.

Snell's law applies for "ordinary" or non-polarized light. More generally, $R$ depends on the polarization of the incident light relative to the interface. The E-vector of linearly polarized light can be resolved into two components, which are perpendicular and parallel to the *plane of incidence*. In Figure 2.4a the plane of incidence is the plane of the page. For the component of the electric vector perpendicular to the plane of incidence (referred to as S-polarization), the reflection coefficient $R$ starts out at the value given by Equation 2.6 and increases with increasing angle of incidence to $R$ = 1 at 90° or "grazing" incidence (Figure 2.6). For the component of the electric vector parallel to the plane of incidence (referred to as P-polarization), $R$ starts out at the same value given by Equation (2.6), falls to zero at an angle of incidence referred to as the *Brewster angle*($\varphi_B$), and then increases to unity at grazing incidence. The Brewster angle depends on the relative refractive index according to

$$\tan \varphi_B=n_r \tag{2.7}$$

A typical value of ($\varphi_B$ ) is 56.3° for an optical glass with $n_r$ = 1.50. The dashed line in Figure 2.6 shows the total reflection coefficient for non-polarized light with equal contributions from each polarization component. The reflection and refraction of polarized light have important practical applications It is shown in elementary optics that *the reflected ray and transmitted ray are perpendicular to each other at the Brewster angle*. When ordinary light is incident at $\varphi_B$ , the reflected ray is 100% linearly polarized in the S-direction and the refracted ray is partially polarized in the P-direction (Figure 2.6). A high degree of polarization can obtained by passing a light beam through a stack of plates tilted at $\varphi_B$ . Each successive plate increases the P-polarization of the transmitted beam without any reflection losses. The end widows of a laser cavity are tilted at $j_B$. Multiple traverses of the laser light through the Brewster window are equivalent to a single passage through a stack of plates.

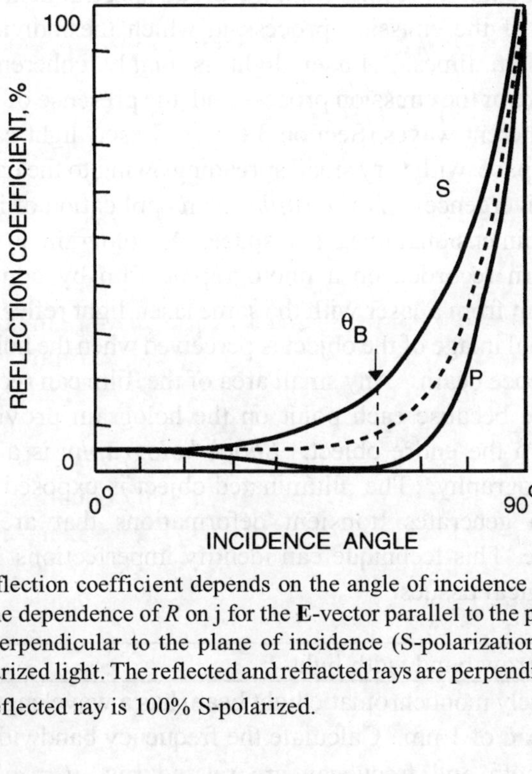


**Figure 2.6** The reflection coefficient depends on the angle of incidence and the polarization. The figure shows the dependence of *R* on j for the **E**-vector parallel to the plane of incidence (P-polarization) and perpendicular to the plane of incidence (S-polarization). The dashed line applies for non-polarized light. The reflected and refracted rays are perpendicular at the Brewster angle $\varphi_B$ and the reflected ray is 100% S-polarized.

## 2.3.2 Interference

If an approximately monochromatic light source is divided into two beams which are then brought together after traveling different optical paths, the optical image of the combined beam consists of an array of bright and dark fringes. This effect is *interference*. The fringes originate because the amplitudes in different regions of the two beams either add or subtract. Interference is possible only if the two beams are at least partially *coherent*. In coherent light the individual waves move together in space and time. Coherent light is analogous to a platoon of soldiers marching together in the same direction at exactly the same speed. However, if there are small variations in either speed or direction, the individual marchers will separate after some distance. Ideal coherence can never occur because the output of even the most highly monochromatic light source has a finite spectral bandwidth and beam divergence. The maximum path difference over which two light beams can form an interference pattern is the *coherence length*. Some typical values of coherence length are 1000 m for a helium-neon laser, 6 mm for a low-pressure sodium arc, and 0.001 mm for sunlight. The weak coherence of sunlight results from the broad spectral distribution and also from the nature of the emission process in which the individual atoms emit radiation at random times. Laser light is highly coherent owing to the synchronous nature of the emission process and the presence of an optical cavity that reinforces coherent waves (Section 3.6.5). Laser light can propagate for long distances in space with very small spreading owing to the narrow bandwidth and small beam divergence. [7] *Holography* is an application of interference used to generate three-dimensional images in space. A hologram is two-dimensional interference pattern recorded on a photographic film by combining a highly coherent light beam from a laser with the same laser light reflected by an object. A three-dimensional image of the object is perceived when the hologram is viewed by the same reference beam. Any small area of the film can recreate a complete holographic image because each point on the hologram provides interference contributions from the entire object. *Shock-holography* is a recent practical application of holography. The illuminated object is exposed to simultaneous ultrasound which generates transient deformations that are visible on the holographic image. This technique can identify imperfections in manufactured objects and biological tissues.

---

**Example 2.4** Narrow bandwidth light

An approximately monochromatic light beam has a wavelength of 500 nm and a wavelength spread of 1 nm. Calculate the frequency bandwidth.

The wavelength and frequency are related by: $f = c/\lambda$. Elementary differentiation leads to: $\delta f = (c/\lambda^2)\delta\lambda$, where $\delta f$ and $\delta\lambda$ are the frequency bandwidth and wavelength bandwidth respectively. Substituting the quantities gives: $\delta f = [3 \times 10^8 \text{ m/s}/ (500 \times 10^{-9} \text{ m})^2] \times (1 \times 10^{-9} \text{ m}) = 1.2 \times 10^{12} \text{ s}^{-1} = 1{,}200$ GHz. Note that an approximately monochromatic light wave has a very large frequency bandwidth.

---

### 2.3.2.1 More Details About Interference

Figure 2.7 depicts interference between two ideally monochromatic waves having the same wavelength, frequency, and amplitude (dashed lines). The relative displacement of two waves in distance or time is referred to as a *phase shift* because one wave reaches its maximum amplitude before or after the other

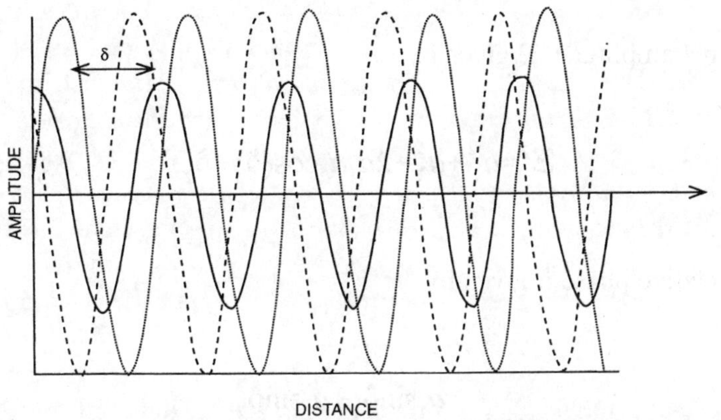


**Figure 2.7** Superposition of two sinusoidal waves having the same wavelength and amplitudes. The resultant wave (solid line) has the same wavelength and an amplitude that depends on the relative phase difference between the two waves.

wave.The combined wave in Figure 2.7 (solid line) is sinusoidal at the same frequency, with an amplitude ranging from zero to twice the individual amplitudes depending on the relative phase shift. Phase shift is frequently expressed in units of the wavelength. The maximum reinforcement occurs when the waves are "in phase" by a multiple of the wavelength ($\Delta\lambda = 0, \lambda, 2\lambda$, etc.) and maximum cancellation occurs when the waves are "out of phase" by a multiple of one half-wavelength: $\Delta\lambda = \lambda/2, 3\lambda/2$, etc. The effect of phase on interference can be analyzed by considering the simpler case of two oscillations fixed in space having the same frequency *f*, with different maximum amplitudes $\psi_1$ and $\psi_2$ and different phases $\delta_1$ and $\delta_2$ . The instantaneous amplitudes $\psi_1$ and $\psi_2$ are given by

$$\psi_1 = a_1 \sin(2\pi f t - \delta_1)$$

$$\psi_2 = a_2 \sin(2\pi f t - \delta_2)$$

For $\delta_2 > \delta_1$, amplitude $\psi_1$ lags behind amplitude $\psi_2$ by a time difference $\Delta t = (\delta_2 - \delta_1)/2\pi$ The algebraic sum of the two amplitudes leads to the combined oscillation:

$$\psi_1 + \psi_2 = \Xi\cos(2\pi ft - \delta) \tag{2.8}$$

with the net amplitude $\Xi$ given by

$$\Xi^2 = a_1^2 + a_2^2 + 2a_1a_2\cos(\delta_1 - \delta_2)$$

and the relative phase $\delta$ given by

$$\tan\delta = \frac{a_1\sin\delta_1 + a_2\sin\delta_2}{a_1\cos\delta_1 + a_2\cos\delta_2}$$

Equation (2.8) shows that the combined oscillation is sinusoidal at the same frequency as the constituent oscillations with a resultant amplitude $\Xi$ and phase $\delta$. The equivalent analysis is applicable for light waves occupying the same region of space. The radiant power carried by the resultant wave is proportional to the $\Xi^2$ in Equation (2.8). The key results are most easily seen for equal initial amplitudes $a_1 = a_2 \equiv a$.

Case (1) The combining waves are in phase. In this case $\delta_1 = \delta_2$ and $\Xi^2 = 4a^2$. Constructive interference results. The resultant intensity is twice as high as the sum of the intensities of the two constituent waves.

Case (2) The combining waves are 180° out of phase: In this case $\delta_2 - \delta_1 = 180°$ and $\Xi = 0$. Destructive interference results. There is no resultant wave.

Case (3) The combining waves are incoherent: In this case there is a random relationship between $\delta_1$ and $\delta_2$. The value of $\cos(\delta_2 - \delta_1)$ averages to zero over many oscillations and $\Xi^2 = 2a^2$. In this case the total intensity equals the sum of the individual intensities.

Coherent light can attain extremely high power levels in regions of space where the individual waves are in phase. For example, if 10 equal monochromatic waves of unit amplitude are added incoherently the total power is 10 x $1^2$ = 10. This situation applies to the combination of ordinary light beams. However, if the waves are added coherently the total power is $(1+1+1+1+1+1+1+1+1+1)^2$ = 100. The high optical power levels in a laser cavity result from the superposition of coherent waves.

#### 2.3.2.2 Interference Filters

Interference filters are widely used for wavelength selection. The basic operation of an interference filter is illustrated in Figure 2.8 which depicts a thin glass film illuminated with white light. Constructive interference takes place when the phase difference between successive reflected rays (a,b,c, ..) or successive transmitted rays (a',b',c', ..) equals 0, λ , 2λ , 3λ, etc. Destructive interference takes place for phase differences of λ/2, 3λ /2, 5λ /2, etc. The phase shift induced by a thin film originates from two effects. Parallel rays that travel different distances through the film before combining are phase-shifted by the differences in their optical paths. The other source of phase shift is less obvious. Electromagnetic theory shows that reflection of light at a medium of higher refractive index (the top face in Figure 2.8) leads to an additional λ/2 phase shift that does not occur for reflection at a lower refractive index medium (the bottom face in Figure 2.8).

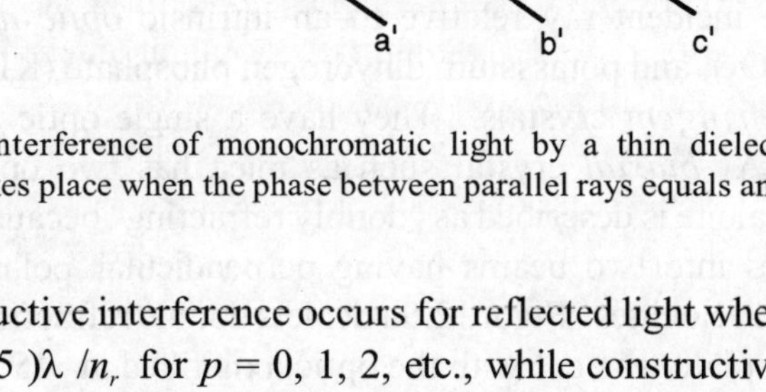


**Figure 2.8** Interference of monochromatic light by a thin dielectric film. Constructive interference takes place when the phase between parallel rays equals an integral multiple of the wavelength.

Thus, constructive interference occurs for reflected light when the path difference equals $(p + .5)\lambda / n_r$ for $p = 0, 1, 2$, etc., while constructive interference of the transmitted rays occurs when the path difference is $p\lambda/n_r$. The minimum path difference for perpendicular incidence is 2d. Therefore, if a thin film is illuminated by white light and viewed by reflection, bright fringes are observed only for those wavelengths that satisfy the reinforcement condition. An additional requirement for fringe formation is that the film thickness must be less than the coherence length of the incident light. Thin-film interference of sunlight requires a very thin film owing to the short coherence length. Interference of laser light is feasible with relatively thick films.

---

**Example 2.5.** Thin film interference

Consider a 0.3 μm film of soapy water in air, for example, as might be formed by dipping a wire ring. If the film is illuminated with white light at perpendicular incidence, which colors of fringes are seen by reflection and transmission?

Bright reflection fringes will be seen for wavelengths that satisfy the reinforcement condition: $2d = (p + 0.5)\,\lambda / n_r$. Substituting $d = 350$ nm and $n_r = 1.33$ for water gives: $\lambda = (2 \times 350 \times 1.33)/(p + 0.5)$. Setting $p = 0, 1, 2$, etc. leads to $\lambda = 1862$ nm, 621 nm, 372 nm, etc. Only the red light fringes at 621 nm fringes will be observed by the human eye. For the same soap film, interference of transmitted light satisfies: $2d = p\lambda / n_r$. The wavelengths of reinforcement are: $2 \times 350 \times 1.33/p = 931$ nm, 466 nm, 310 nm, etc. This film can act as a selective transmission filter for blue light at 466 nm. However, the spectral bandwidth would be broad because only a few waves can combine owing to the low reflection coefficient of the soap-air interface.

---

### 2.3.2.3 Optical Activity of Crystals

The combination of interference and polarization has been used in passive optical devices for more than a century. Some natural crystals have the property of *optical activity* that results from an anisotropic arrangement of the atoms. The refractive index in an optically active crystal varies with the polarization and the direction of the incident ray relative to an intrinsic *optic axis*. Calcite ($CaCO_3$), quartz ($SiO_2$), and potasssium dihydrogen phosphate ($KH_2PO_4$, KDP) are *uniaxial* or *birefringent* crystals. They have a single optic axis and two refractive indices. A *biaxial* crystal such as mica has two optic and three refractive indices. Calcite is described as "doubly refracting" because an entering light beam separates into two beams having perpendicular polarizations and different propagation velocities. Figure 2.9a depicts double refraction by a calcite crystal. The crystal plate was cut with the optic axis tilted at 45° to the faces. The incident beam splits into an "ordinary" ray (O-ray) and an "extraordinary" ray (E-ray) with perpendicular polarizations. The faster O-ray has a constant velocity and obeys Snell's law according to Equation (2.4). The velocity of the slower E-ray depends on its direction relative to the optic axis. A birefringent crystal can be fabricated as a high-quality linear polarizer. The *Nicol prism* consists of a split and cemented calcite rhomb (Figure 2.9b). The E-ray is non-deviated while the O-ray is deflected out of the crystal by internal reflection.

A *retardation plate* is a classical application of optical activity that has received recent impetus with the development of lasers. A *half-wave plate* is cut from an optically active crystal at a thickness of $\lambda/2$. This device rotates the plane of linearly polarized light by an angle $2\theta$, where $\theta$ is the angle between the incident polarization and the optic axis (Figure 2.10). A half-wave plate can be used as a fine-tuning element in a laser by rotating the plate around the Brewster angle. The thickness of the plate selects the wavelength of polarized light amplified by the laser cavity. A *birefringent filter* consists of one or more half-wave plates separated by sandwiches of polarizing films. The transmitted spectrum has a series of narrow bands, widely spaced in wavelength, which can be shifted by rotating the unit. A *quarter-wave plate* is cut at a thickness of $\lambda/4$. When linearly polarized light is incident on a quarter wave plate with the electric vector oriented at 45°to the optic axis, the horizontal and vertical projections

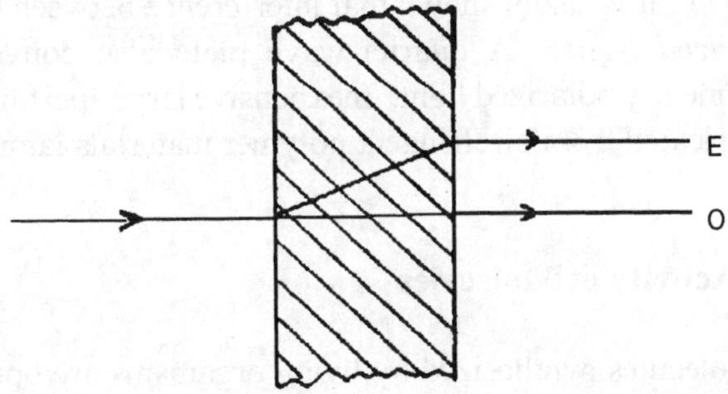


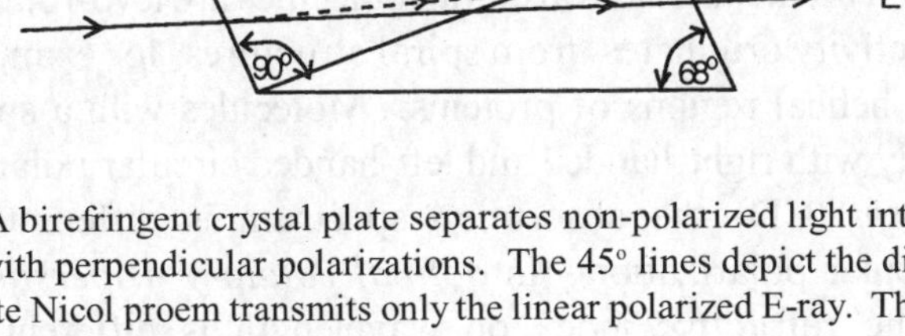


**Figure 2.9** (a) A birefringent crystal plate separates non-polarized light into two rays of linear polarized light with perpendicular polarizations. The 45° lines depict the direction of the optic axis. (b) A calcite Nicol proem transmits only the linear polarized E-ray. The O-ray is reflected out of the crystal at the cemented interface.

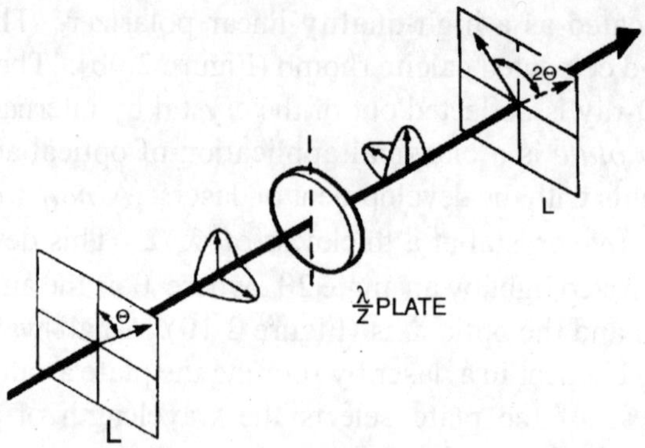


**Figure 2.10** A half-wave plate rotates the plane of monochromatic linear polarized light (L). The vertical component of the E-vector propagating parallel to the optic axis (dashed line) is unchanged, while the phase of the horizontal component is retarded by a half-wavelength. The linear polarization of the emergent wave is rotated from θ to 2θ.

Of the electric vector emerge with thee same amplitude and a relative phase shift of λ/4 (Figure 2.11)[8]. It is easily shown that interference between these rays lead to *circular polarized light*. A quarter-wave plate also converts circularly polarized light to linearly polarized light. Inexpensive large aperture retarders are now available fabricated from birefringent polymer materials laminated between glass plates.

### 2.3.2.4 Optical Activity of Molecules

Nearly all molecules synthesized by living organisms are optically active. Amino acids (except glycine) and some sugars rotate the plane of linearly polarized light. This type of optical activity termed *chirality* arises from the presence of four different substituted groups on carbon atoms. Chemical synthesis of chiral compounds leads to an equal mixture of the right-handed (dextro-rotatory)and left-handed (levo-rotatory) *stereoisomers*, each of which rotates the plane of linearly polarized light in an opposite direction. For unknown and profound reasons, virtually all natural chiral molecules are levo-rotatory. Another form of optical activity originates from spiral structures, for example, the DNA double helix and helical regions of proteins. Molecules with a spiral structure interact differently with right-handed and left-handed circular polarized light. In *circular dichroism* (CD) the absorption spectrum is different for the two directions of circular polarization. In *optical rotatory dispersion* (ORD) the dependence of the refractive index on wavelength is different for the two directions of circular polarization. CD and ORD spectra are complementary measurements. CD is measured in spectral regions of strong absorption in order

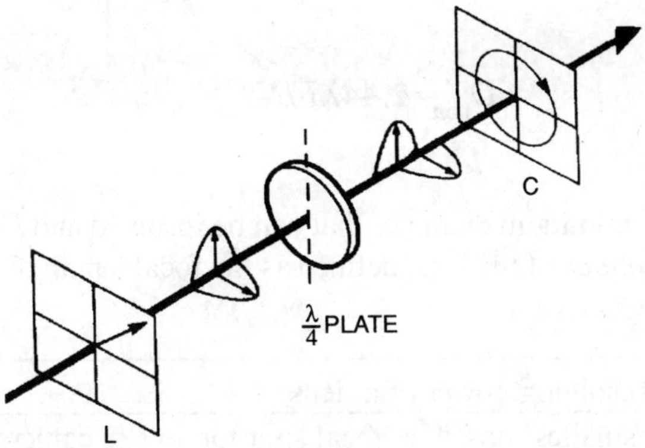


**Figure 2.11** A quarter-wave plate converts linear polarized light (L) to circular-polarized light. The incident E-vector is 45° to the optic axis (dashed line). The phase of the horizontal component is retarded by a quarter-wavelength. The transmitted transverse components combine to give circular-polarized light.

to promote the differences in absorption. ORD is measured in spectral regions of weak absorption in order to emphasize the differential refractive index. ORD spectra can be converted into CD spectra and *vice versa* by a fundamental pair of integrals called the Kronig-Kramers relations.

## 2.3.3 Diffraction of Light

Diffraction is the bending of a light ray by an object of dimensions comparable to the wavelength. The fuzzy shadow of an illuminated straight edge on a screen is an example of diffraction. Alternating light and dark diffraction fringes are easily observed by viewing a bright light through two closely spaced fingers. Diffraction occurs when an object cuts off part of an incident wave front, leading to destructive interference in certain directions depending on the shape of the object. The diffraction pattern of a thin slit illuminated with monochromatic light as viewed on a screen consists of a central bright fringe located between increasingly weaker and narrower bright fringes on each side separated by dark regions. The diffraction pattern of a circular aperture illuminated by a point source of monochromatic light consists of a central bright disc surrounded by concentric fainter rings. The spacing of diffraction fringes is a classical problem in physical optics. Diffraction limits the ability of a lens to produce a point image of a point source and to focus a collimated laser beam to a point. The resolving power of a uniformly illuminated simple lens is given by:

$$D_{\min} = 2.44\lambda F/N \tag{2.9}$$

where $D_{\min}$ is the minimum diameter that can be resolved and $F/N$ is the relative aperture or *F-number* of the lens, defined as the focal length divided by the lens diameter.

---

**Example 2.6** Resolving power of a lens

Calculate the smallest possible focal spot for an F/2 camera lens at 500 nm.

According to Equation (2.9): $D_{min}$ = 2.44 x 0.0005 mm x 2 =0.0024 mm. This result applies for an ideal lens and a perfectly collimated, uniform circular beam.

---

The complete theory of diffraction is described in standard optics texts. The angular spread of the dark diffraction fringes for a single slit illuminated by a collimated, monochromatic light beam is given by

$$\sin\theta = p\lambda/a \tag{2.10}$$

where $a$ is the slit width, $\theta$ is the angular deviation of the diffracted ray, and $p$ is an integer specifying the "order" of the diffraction. The actual spacing of the fringes on a viewing screen increases with increasing distance of the screen from the slit. The intensity of each bright fringe is given by

$$I = I_m\left[\frac{\sin\alpha}{\alpha}\right]^2 \tag{2.11}$$

where $\alpha \equiv (\pi a/\lambda)\sin\theta$ and $I_m$ is the maximum intensity of the central bright fringe. The bright fringes do not lie exactly halfway between the adjacent dark fringes. A numerical calculation leads to the values of $a$ (in radians) for the first few maxima: 0, 1.4303 $p$, 2.4590 $p$, 3.4707 $p$, etc.

**Example 2.7** Single-slit diffraction

Consider diffraction of 500 nm green light by a 2 mm width slit. Calculate: (a) the angular deflection of the first three dark fringes; (b) the relative intensity of the first three bright fringes.

(a) From Equation (2.10): $\sin\theta = (500 \times 10^{-9} / 2 \times 10^{-6})p = 0.25\, p$. Substituting $p = 1, 2$, and 3 gives $\theta_1 = 14.48°$, $\theta_2 = 30.00°$, and $\theta_3 = 48.59°$. The angular spread between the successive orders is quite large

(b) From Equation (2.11) and the values of a for bright fringes: $I_1 = 1$, $I_2 = 0.0472$, and $I_3 = 0.0165$. More than 93% of the optical power is concentrated in the broad central band.

A diffraction pattern is readily visualized by the famous two-slit experiment of Thomas Young depicted in Figure 2.12a . Two closely spaced slits $S_1$ and $S_2$ are illuminated by monochromatic light from a single slit S. The largest separation between the single slit and double slits that leads to interference fringes on the screen provides an experimental determination of the coherence length.

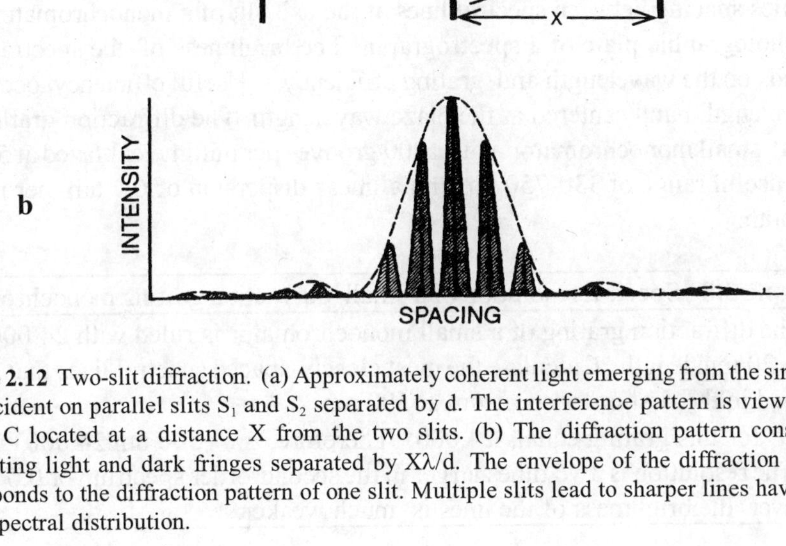


**Figure 2.12** Two-slit diffraction. (a) Approximately coherent light emerging from the single slit S is incident on parallel slits $S_1$ and $S_2$ separated by d. The interference pattern is viewed on a screen C located at a distance X from the two slits. (b) The diffraction pattern consists of alternating light and dark fringes separated by $X\lambda/d$. The envelope of the diffraction pattern corresponds to the diffraction pattern of one slit. Multiple slits lead to sharper lines having the same spectral distribution.

The two-slit diffraction pattern consists of alternating bright and dark fringes within an intensity envelope corresponding to the single slit diffraction pattern (Figure 2.12b). When the distance of the slits to the screen (*X*) is much larger than the slit spacing (*d*), the alternating bright and dark fringes are equally separated by $\lambda X/d$. The diffraction pattern for white light consists of overlapping sets of fringes belonging to the different orders. The spacing of fringes within each order is proportional to the wavelength. A double-slit can be used for wavelength separation. The fringe pattern is sharper and more intense when the two slits are replaced by a large number of closely-spaced parallel slits. This array forms a transmission *diffraction grating*. For practical reasons reflection-type diffraction gratings are used in monochromators. In a reflection grating the slits are replaced by parallel grooves cut into a metallic surface. The groove faces are tilted or "blazed" relative to the base in order to shift the brightest fringe to the first order. The first precision diffraction gratings were cut on a metal base by a ruling engine. However, ruling engines always have minute imperfections and generate errors in the diffraction pattern call "ghosts". (Note that resealed red blood cell membranes are also called "ghosts".) The best available gratings at present are fabricated from images of optical interference patterns. The bandwidth of the individual spectral lines is determined by the resolving power (*RP*) of the grating defined as $\Delta\lambda/\lambda$, where $\Delta\lambda$ is the minimum spacing of spectral lines that can be distinguished. The value of *RP* increases with the total number of grooves. The angular dispersion ($D_\theta$) of a diffraction grating specifies the angular spread between different wavelengths. $D_\theta$ increases with the total number of grooves per unit ruled length. The linear dispersion (*L*) is an equivalent parameter that specifies spacing between spectral lines at the exit slit of a monochromator or on the photographic plate of a spectrograph. The brightness of the spectral lines depends on the wavelength and grating efficiency. Useful efficiency occurs for a wavelength band centered at the blaze wavelength. The diffraction grating in a typical small monochromator with 1200 grooves per millimeter blazed at 500 nm has a useful range of 330-750 nm and a linear dispersion of 6.4 nm per mm. of slit width.

---

**Example 2.8** Spectral resolution of a small diffraction grating monochromator.

The diffraction grating of a small monochromator is ruled with 24,000 lines over a 20 mm length. Calculate the most closely-spaced spectral lines that can be resolved in the first order spectrum at 500 nm.

*RP* for this grating equals 24,000. Therefore, $\Delta\lambda$ =500 nm/24,000 = 0.021 nm. The resolution is two times larger in the second-order spectrum or 0.042 nm. However, the brightness of the lines is much weaker.

---

## 2.4 OPTICAL RADIOMETRY

*Radiometry* is the science of optical power measurements. This subject is complicated by several systems of units. *Photometric units* are based on the spectral response of an idealized human eye. Consequently, photometric units are inapplicable for IR and UV radiations. However, photometric units are still widely used in the lighting and photographic industries. *Radiometric units* are based on radiant power using SI units established by the International Commission on Illumination (CIE). This system of units is used in photobiology and photomedicine. *Actinometric units* are based on the photon content of light. Photons are generally used in photophysics and photochemistry for events that take place at atomic and molecular levels. *Radiant flux* (*F*) is the basic unit of optical power in the SI system. Since power equals energy per unit time, radiant energy (*Q*) equals the product of the flux in watts and the time in seconds. *Flux density* is a general term for flux per unit area. However, *energy density* usually refers to radiant energy per unit volume. *Luminous flux* in units of the lumen (lm) is the SI photometric equivalent of radiant flux. One lumen equals 683 W at 555 nm, which is the peak of the *visual photopic luminous efficiency function* $V(\lambda)$. The relative value of $V(\lambda)$ equals unity at 555 nm and falls to about 0.001 at 410 nm and 720 nm. Numerical values of $V(\lambda)$ are provided in standard references. All radiometric units have photometric equivalents. The two sets of units are compared in Table 2.1. A general relationship for converting radiometric quantities to the corresponding photometric quantity is

$$\textit{Photometric Quantity} = \int (\textit{Spectral Radiometric Quantity})V(\lambda)d\lambda \quad (2.12)$$

where the limits of the integration are 380 nm to 770 nm.

Radiometric properties are employed to specify the power and directionality of light emitted by a source, incident on a receiver, and reflected by a surface. In order to do this, it is necessary to describe directions in space. *Solid angle* ($\Omega$) is the three-dimensional equivalent of *polar angle* in a plane. Polar angle in units of radians (rad) is defined as the length of a circular arc centered around a point in a plane divided by the radius (Figure 2.13a). The total polar angle subtended by a circle of radius *r* about its center is $2\pi\, r/r = 2\pi$ rad or 360°. Any simple closed loop in a plane subtends $2\pi$ rad. The solid angle subtended by a spherical surface around a point in space is defined as the area of the spherical surface divided by

**Table 2.1 Radiometric and Photometric Quantities**

| Radiometric Quantity | Definition | Equation | S. I. Units | Photometric Equivalent |
|---|---|---|---|---|
| Radiant energy | | Q | J | Luminous energy (lm-s)[a] |
| Radiant flux | Radiant power | F = dQ/dt | W | Luminous flux (lm)[b] |
| Radiance | Radiant flux in a given direction per unit solid angle and per unit area perpendicular to the direction of propogation | L= dI/dA cos θ | $W/m^2$-sr | Luminance ($cd/m^2$)[c] |
| Intensity | Radiant flux emitted by a source in a given direction per unit solid angle | I = dF/dΩ | W/sr | Candlepower (cd; lm/sr) |
| Irradiance | Radiant flux arriving per unit area | E = dF/dA | $W/m^2$ | Illuminance (lux; $lm/m^2$)[d] |
| Exitance | Radiant flux emmitted or reflected per unit area | M = dF/dA | $W/m^2$ | Luminous exitance ($lm/m^2$) |
| Radiant exposure | Radiant energy arriving per unit area | H = Et | $J/m^2$ | Luminous exposure (lm-$s/m^2$) |

[a]The *talbot* is a unit of luminous energy equal to 1 lm-s.

[b]One *lumen* is equivalent to approximately 683 W at 555 nm and to fewer watts at other wavelengths (see text).

[c]Other luminance units are the *nit* (1 $cd/m^2$); the *stilb (1 cd/cm2);* the *lambert* [(1/π) $cd/cm^2$].

[d]Another illuminance unit is the *foot-candle* (1 $lm/ft^2$).

the square of the radius (Figure 2.13b). The units of solid angle are steradians (sr). The total solid angle subtended by a spherical shell about its center is $4\pi r^2 / r^2 = 4\pi$ sr. Any simple closed surface on space subtends $4\pi$ sr.

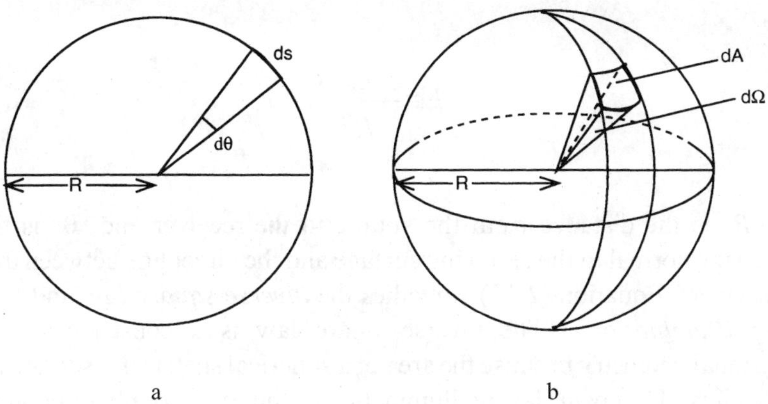


**Figure 2.13** (a) The differential polar angle dθ at a point in a plane subtended by a circular arc d*s* is given by: dθ = d*s*/*R* radians where *R* is the radius of the circle. (b) The differential solid angle dΩ at a point in space subtrended by spherical surface dA is given by: dΩ at a point in space subtended by spherical surface dA is given by: $d\Omega = dA/R^2$ steradians where *R* is the radius of the spherical shell.

---

**Example 2.9** Solid angle

Show that the integration of the differential solid angle dΩ over the total solid angle is 4π sr for a spherical shell of constant radius *r*.

It is convenient to use spherical coordinates with θ as the polar angle, ϕ as the azimuthal angle, and *r* as the radius (Figure 2.14b). The differential surface element is: $da = (r\,d\theta)(r \sin\theta\, d\phi)$ and $\cos\theta = 1$ for this geometry. Performing the integration:

$$\Omega = r^2 \int_0^{\pi} \sin\theta d\theta \int_0^{2\pi} d\phi / r^2$$

---

*Intensity* (*I*) is the radiometric property describing the directionality of radiant power emitted by a light source. For a small source, or a surface element of a large source, intensity is defined as the flux emitted in a given direction per unit solid angle in that direction (Figure 2.14a). If a source of intensity *I* emits uniformly in all directions, the total emitted flux is $4\pi I$. The corresponding photometric quantity of luminous intensity is *candlepower* with the SI unit of *candela* (cd). *Irradiance* (*E*) is the fundamental unit of illumination, defined as the incident flux on a unit area of receiving surface (Figure 2.14b). The corresponding quantity for an emitting or reflecting surface is the *exitance*. The irradiance at a small surface illuminated by a point source of uniform intensity I is given by

$$E=\frac{I\cos\theta}{R^2} \tag{2.13}$$

where $R$ is the distance from the source to the receiver and $\theta$ is the angle between the normal to the receiving surface and the direct line between the source and receiver. Equation (2.13) embodies the *inverse-square law* and the *cosine law of illumination.* The inverse-square law is a consequence of three-dimensional geometry because the area of a spherical shell increases as the square of the radius. The cosine law of illumination is important for phototherapy. Light delivery systems are designed to maximize the perpendicular incidence of the therapeutic radiation. The light intensity emitted by a practical source depends on the angle between the emitting surface and the receiver. The *radiance* (*L*) specifies the intensity emitted in a given direction per unit projected area in that direction (Figure 2.14c). The equivalent definition of radiance applies for reflected light. The photometric equivalent of radiance is *luminance* (lm/-sr-$m^2$). An object is "bright"when its image on the retina or other receiver has a high flux density in a narrow angular spread. A low-power light source can be very bright if the emitted light is confined to a narrow solid angle. For example, a 1 mW helium-neon laser viewed along the axis is much brighter than a 300-W projection lamp. The radiance of a practical light source depends on the wavelength. *Spectral radiance* is defined as the radiance per unit bandwidth. The spectral radiance of the extraterrestrial sun at 633 nm is 0.0002 (W/$m^2$ -sr-nm) compared to 10,000 (W/$m^2$ -sr-nm) for a small helium-neon laser.

In 1760 Lambert observed that sunlight reflected by a white painted wall appeared equally bright at all viewing directions. He correctly deduced that the scattered intensity distribution does not depend on the direction of the incident light and decreases with the cosine of the viewing angle from the vertical. *Lambert's cosine law* has been verified for many types of paper and powder layers. Such materials consist of many small reflectors oriented in random directions. The reflected light is referred to as *diffuse light* to distinguish it from directed light. Light sources that obey Lambert's cosine law are referred to as *lambertian*. The Sun is a lambertian source.

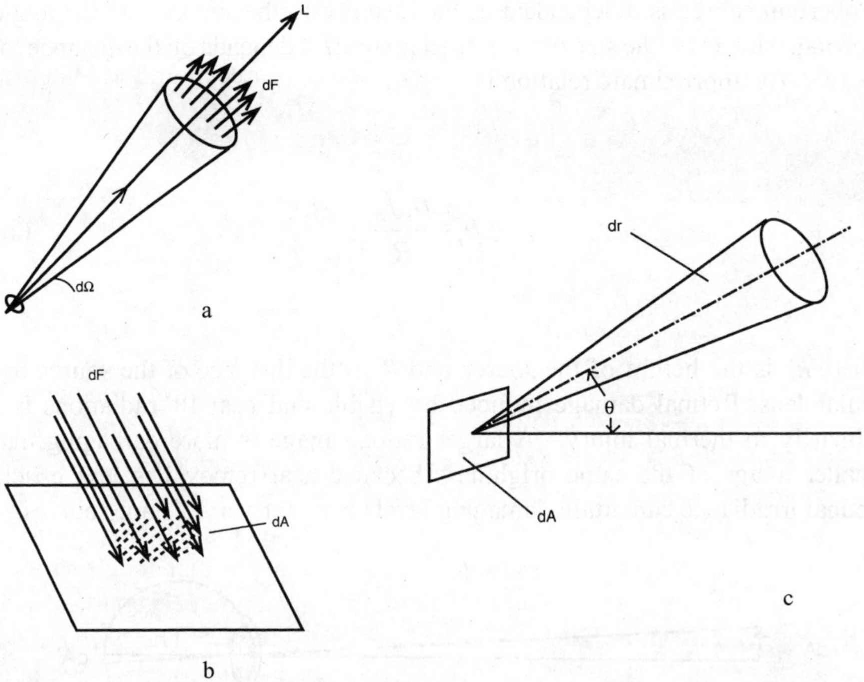


**Figure 2.14** (a) Intensity (*I*) is a measure of the radiant flux per unit solid angle emitted in a given direction. (b) Irradiance (*E*) is a measure of the radiant flux per unit area incident on a planar surface. (c) Radiance (*L*) is a measure of the intensity per unit projected area or "brightness".

The radiometry associated with the human retina is important in connection with optical hazards. Figure 2.15 depicts the imaging of a distant lambertian source on the retina, where $dA$ is a surface element of the source, $dA'$ is a surface element of the retinal image, and $A_p$ is the area of the pupil. A straightforward calculation of the irradiance at the retina ($E_{ret}$) leads to

$$E_{ret} = \frac{T_c L_s A_p}{f_e^2} \tag{2.14}$$

where $T_e$ is the overall transmission coefficient of the eye, $L_s$ is the source radiance, and $f_e$ is the focal length of the eye. The subjective brightness of a lambertian source is independent of the distance to the source and the angle of viewing. However, the size of the retinal image ($h_r$) depends on the distance to the source. An approximate relation is

$$h_r = \frac{h_s f_e}{R} \qquad (2.15)$$

where $h_s$ is the height of the source and $R$ is the distance of the source to the ocular lens. Retinal damage induced by visible and near-IR radiations is due primarily to thermal injury. A larger retinal image is more damaging than a smaller image of the same brightness because heat removal is less efficient. Retinal irradiance can attain damaging levels even for very distant sources.

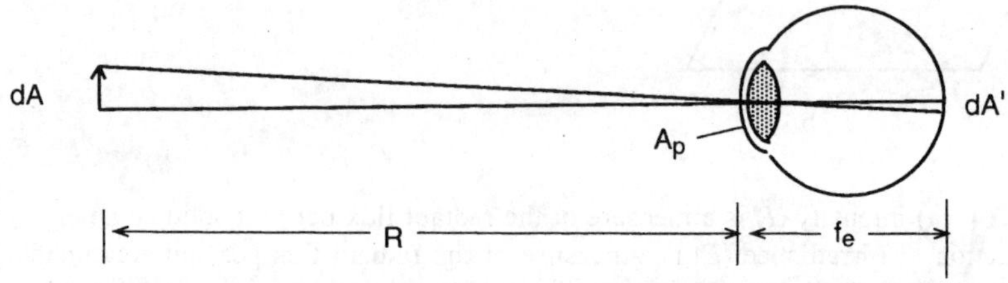


**Figure 2.15** Retinal imaging. A lambertian source of area d$A$ located a distance $R$ from the ocular lens forms an image on the retinal of area d$A'$.

---

**Example 2.10** Direct viewing of the sun

The sun is 1.5 x $10^{11}$ m from the earth with a diameter of 1.4 x $10^9$ m. The approximate radiance of the noon sun from 400-1400 nm is 1000 W/cm $^2$-sr.

(a) Calculate the size of the sun's image on the retina. (b) Calculate the irradiance of the sun on the retina for a 2 mm pupil and $T_e$ = 1. (c) For a thermal injury damage threshold of 100 J/cm$^2$, calculate the maximum time of direct viewing of the sun. The focal length of the normal human eye is 1.7 cm.

(a) The size of the sun's image on the retina from Equation (2.15) is: $h_r$ = 1.4 x $10^9$ m x 1.7 cm/1.5 x 10 $^{11}$ m = 0.015 cm.

(b) The retinal irradiance from Equation (2.14) is: $E_r$ = [($\pi$ /4) x (0.2 cm) 2] x 1000 W/cm$^2$ -sr/(1.7 cm)$^2$ = 11 W/cm$^2$

(c) The maximum safe time of direct viewing is: 100 J/cm$^2$ /11 W/cm$^2$ = 9s. The safe viewing time is much briefer with a binocular or telescope owing to the magnified retinal image. A 10 times magnification reduces the safe viewing time to less than 1 sec!

## 2.5 LIGHT SCATTERING

Light scattering is one of the most fascinating of all optical phenomena. The white appearance of cumulus clouds against the blue sky and the red sunset are manifestations of light scattering. Light scattering also is one of the most important factors involved in light dosimetry because biological tissues and fluids are highly inhomogeneous substances. Microscopic entities in tissues induce multiple scattering of incident light leading to the appearance of turbidity or "cloudiness". In many cases the poor penetration of visible light into a tissue or fluid results from its light scattering and not its absorptive properties.

The correct explanation for atmospheric light scattering was provided by Lord Rayleigh in 1871, three decades before the first elucidation of atomic structure. Rayleigh proposed that the oscillating E-field of a light wave initiates synchronous vibrations of electrons in the scattering particles. The vibrating electrons radiate scattered light at the same wavelength as the incident light. Owing to the transverse character of EM waves, the intensity and polarization of scattered light in a given direction depend on the incident polarization and the angle between the incident and scattered directions. Using only dimensional arguments, Rayleigh deduced that the scattered intensity in a given direction decreases with the inverse-fourth power of the wavelength. The strong dependence of *Rayleigh scattering* on wavelength explains the blue daytime sky, because molecules and fine dust in the atmosphere preferentially scatter blue light to the earth's surface. However, when the sun is close to the horizon, the more weakly scattered red light is viewed preferentially by line-of-sight.

**Example 2.11** Rayleigh scattering

Compare the total scattered intensity of blue light (400 nm) and red light (700 nm) for Rayleigh scattering by small particles.

The scattered intensity for blue light is: $(700/400)^4$ = 9.4 times stronger than red light.

Light scattering by very small particles is adequately explained by Rayleigh's theory. More generally, light is scattered by local variations of the refractive index. For example, the white appearance of milk is due to light scattered by butterfat particles. Gans and Debye explained light scattering for larger particles as arising from interference between rays originating from different scattering sites. This theory was applied to ideal spherical and cylindrical macromolecules and the random coil structures of globular proteins. The rigorous theory of light scattering is based on Maxwell's electromagnetic equations. The exact solution for an isolated sphere is referred to as Mie scattering, after Gustave Mie who was one the first to solve the problem in 1908. A key result of Mie theory is that particles much larger than the light wavelength scatter predominantly in the forward direction. Figure 2.16a depicts the angular scattering pattern for Rayleigh scattering of ordinary light by a small particle, which has equal lobes in the forward and backward directions. Mie scattering by an insulating sphere is highly forward-directed. The angular pattern in Figure 2.16b was calculated for scattering of green light by a 0.5 μm diameter sphere. The scattered intensity in the forward direction is 200 times stronger than the backward direction.

Light scattering within a turbid material leads to spreading of an incident light beam and loss of directionality. The directionality of the individual scattering events is lost in multiple scattering after several steps and the rays acquire approximately random directions. Light within the material is diffuse. The *energy fluence rate* ($\Phi$) is a useful radiometric quantity for diffuse light within a turbid medium. (Figure 2.17). This definition of flux density is defined as the total flux incident on a small spherical surface centered around a point in the medium divided by the cross-sectional area of that sphere. The energy fluence rate is equivalent to the average intensity, calculated by integrating the radiance over the entire solid angle.

$$\Phi = \int_{4\pi\ sr} L d\Omega \qquad (2.16)$$

The units of energy fluence rate are the same as irradiance. However, $E$ and $\Phi$ measure flux density in different geometries. A small planar detector measures $E$, while an isotropic probe measures $\Phi$. For perfectly isotropic light $\Phi = 4E$; for perfectly collimated light $\Phi = E$. The energy fluence rate is also referred to as *spherical irradiance* and *space irradiance*. The term *photon fluence rate* is used when the optical power is expressed as photons per unit area per unit time.

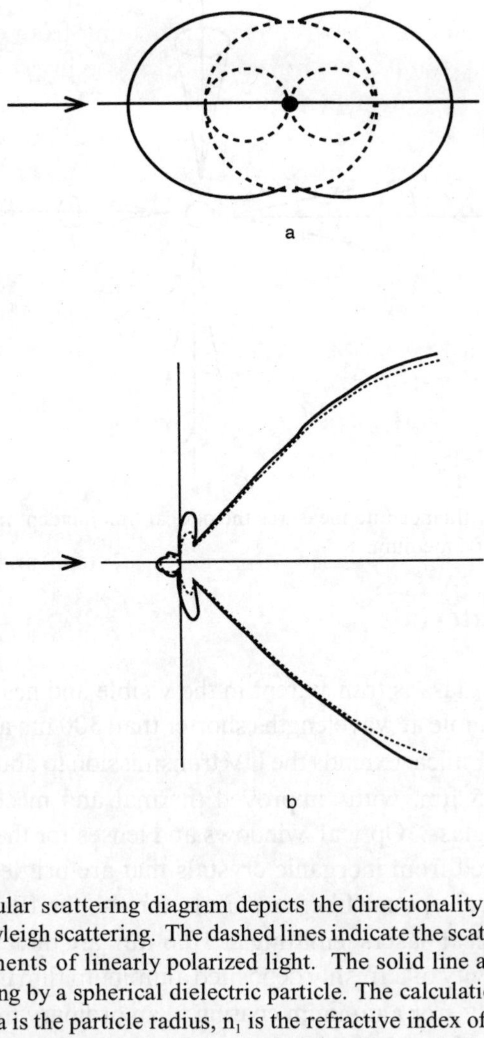


**Figure 2.16** An angular scattering diagram depicts the directionality of light scattering by a single particle. (a) Rayleigh scattering. The dashed lines indicate the scattering directions for two perpendicular components of linearly polarized light. The solid line applies for nonpolarized light. (b) Mie scattering by a spherical dielectric particle. The calculation is for $2\pi a n_r/\lambda=4$ and $n_r=n_1/n_2=1.25$, where a is the particle radius, $n_1$ is the refractive index of the sphere and $n_2$ is the refractive index of the medium. The solid line and dashed line apply for vertical and horizontal polarization components, respectively.

## 2.6 PRACTICAL OPTICAL DEVICES

Virtually all of the passive components used in optical systems were invented in the 19th century or earlier. However, the current demands of precise control and high-power applications have led to new implementations of these venerable devices.

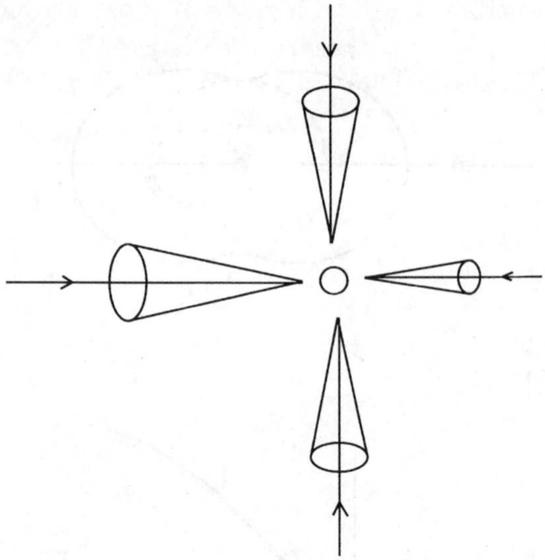

**Figure 2.17** The energy fluence rate measures the optical flux incident from all directions at a small spherical region of a medium.

## 2.6.1 Optical Materials

Ordinary optical glass is transparent in the visible and near-IR regions. The transmission is negligible at wavelengths shorter than 300 nm and longer than 2.8 μm. Synthetic fused silica extends the UV transmission to about 200 nm and the IR transmission to 5 μm, with improved thermal and mechanical properties compared to optical glass. Optical windows and lenses for the far-IR region are of necessity fabricated from inorganic crystals that are brittle and hygroscopic. There are no practical optical fiber materials for the far-IR region at present. Carbon dioxide surgical lasers emitting at 10.6 μm are directed by systems of mirrors. The reflectivity of a freshly deposited aluminum film is 85-95% from 0.2 μm to 30 μm. The silver or aluminum coating in an ordinary mirror is located on the rear surface of a glass plate to reduce oxidation, in which case the spectral range of reflection is limited by the transmission of the glass. The full spectral range of the mirror coating can be utilized with a front-surface mirror coated with a transparent protective layer. Gold mirrors are good reflectors from 0.8 μm to 30 μm and do not require a protective coating. Laser mirrors must be highly reflecting only over a narrow wavelength range, but they must withstand very high optical power levels. They are usually fabricated from dielectric coatings on a hard glass or fused-silica substrate. The basic construction of a laser mirror consists of alternating quarter-wave stacks of high refractive index and low refractive index materials. Constructive interference between reflected rays differing by successive $\lambda/2$ path differences leads to high reflectivity in a narrow bandwidth, typically exceeding 99.5% for normal incidence.

### 2.6.2 Imaging with Lenses and Mirrors

Lens design for cameras and optical instruments is a highly specialized subject which the "end-user" is seldom required to master. However, the practitioner of phototherapy may be required to design or modify a simple optical system for light delivery or measurement. This section provides a brief review of optical imaging that may assist in this task.

Simple lenses used for ordinary imaging have spherical surfaces. They are described as "convex", "concave", and "plano". For example, a plano-convex lens has one outward facing spherical surface and one flat surface. The elementary theory of imaging is applicable for *paraxial rays* which are close to the optical axis. The *focal length* of a lens or mirror ($F$) is the distance from a plane through the optical center or *principal plane* at which parallel rays either converge or diverge, depending on the curvature of the surfaces. In physiological optics the focal length is expressed in *diopters* defined as the reciprocal of the focal length in meters. A "thin" lens can be treated as having a single principal plane with equidistant foci each side. The distances of the object ($p$) and image ($q$) from the principal plane are related by the expression

$$\frac{1}{p}+\frac{1}{q}=\frac{1}{F} \tag{2.17}$$

The effective focal length of two thin lenses separated by an air distance $d$ is given by:

$$\frac{1}{F}=\frac{1}{F_1}+\frac{1}{F_2}-\frac{d}{F_1F_2} \tag{2.18}$$

Equation (2.17) applies also for spherical mirrors. The numerical applications of these formulas requires sign conventions: $F$ is taken as positive for a lens that forms a real image and negative for a lens that forms a virtual image. A *real image* can be projected on a screen. A *virtual image* is perceived at those points in space from which divergent rays appear to originate. For example, a camera lens forms a real image on the film, while a plane mirror forms a virtual image. Assuming that rays travel from left to right, the object distance $p$ is taken as positive if the object is to the left of the lens or mirror surface and negative if the object is to the right of the lens or mirror surface and vice-versa for the image distance $q$. The ratio of image size to object size or magnification ($M$) is given by

$$M = -\frac{q}{p} \tag{2.19}$$

The sign of $M$ is positive for an upright image and negative for an inverted image.

Ray tracing is a graphical approach to image analysis for simple optical systems that avoids the requirement for sign conventions. For paraxial rays the basic principles of ray tracing are: (1) A ray parallel to the optic axis is refracted through a focal point. (2) A ray through a focal point is refracted parallel to the optic axis. (3) A ray through the lens center is non-deviated. In Figure 2.18a, a double-convex lens forms an inverted real image of an object located outside the focal plane, for example, a simple camera lens. Figure 2.18b shows that an upright virtual image is formed by an object located inside the focal plane by a double-convex lens. This arrangement is used for a hand magnifier. Figure 2.18c shows an upright, virtual image formed by a double-concave lens for an object located outside the focal plane. Light from a small source can be collimated by collecting the light with a convex lens followed by convex lens to form a collimated beam.[9] The light-gathering efficiency of a lens system is specified by the F-number ($F/N$). A "fast" lens has a low $F/N$. Numerical aperture ($N.A.$) is an equivalent parameter defined by

$$N.A. = n_0 \sin\alpha = \frac{1}{(2F/N)} \tag{2.20}$$

where $\alpha$ is the half-acceptance angle of the lens and $n_0$ is the refractive index of the external medium.

The high-quality lenses used in microscopes and other precision optical instruments are fabricated of high-refractive index optical glass. By judicious use of multiple elements and different types of glasses it is possible to minimize imaging errors, including chromatic aberration (change of focal length with wavelength), spherical aberration (failure to form a perfect image of a point source on the axis), and astigmatism (off-axis rays are focused as an ellipse instead of a circle). A variety of special lenses are employed in modern optical systems. A *Fresnel lens* consists of concentric stepped rings, each of which bends light as if it were a small portion of a conventional spherical lens. With this concentric groove structure there is no inherent physical limit to the *F/N*. Fresnel lenses have been fabricated from acrylic plastic with very large diameters and high optical speed, typically *F/1*. They are used primarily for light collection in the visible and near-IR regions. A *gradient index lens* (GRIN lens) uses continuously variable refractive index for focusing. GRIN lenses are available in the form of cylindrical rods in which the refractive index increases with the radial distance from the axis. The focusing properties depend on the length of the rod. A GRIN lens can form a real image directly on or very close to the lens surface with very low spherical aberration. GRIN lenses are used to couple a narrow laser beam into an optical fiber or a small photodetector. A *ball lens* is a sphere with a uniform refractive index. Optical fibers are often coupled with a ball lens because the alignment is easier than with a spherical thin lens and they have high coupling efficiency. A *cylindrical lens* has one or two cylindrical surfaces and, therefore, it refracts in only one plane. They can be used to image a point source on a line and to change the height of an image without changing the width. An *aspheric lens* uses one or more non-spherical faces to provide a low F/*N* typically below *F*/1. In many applications it is desirable to use mirrors instead of lenses because mirrors have no chromatic aberration. Special mirror shapes are available in addition to standard plane and spherical mirrors. A parabolic mirror brings a parallel incident beam to a focus and delivers collimated light from a point source. A line source is focused on a parallel line with an ellipsoidal mirror. Ellipsoidal reflectors are employed with flash-lamp pumped lasers to achieve very efficient light delivery.

### 2.6.3 Optical Fibers

Long optical fibers are used in communications to transport light signals over long distances. The short optical fibers designed for phototherapy applications provide for efficient coupling to lasers and other small, high-power light sources. The basic operating principles are the same for the two types of optical fibers. Light is transmitted down the fiber by a succession of internal reflections until it exits at the far end. The basic design of a *step-index fiber* consists of an inner core and a concentric outer cladding of slightly lower refractive index. (Figure 2.19). A *graded-index fiber* has a core in which the refractive index decreases gradually from the center. The "speed" of an optical

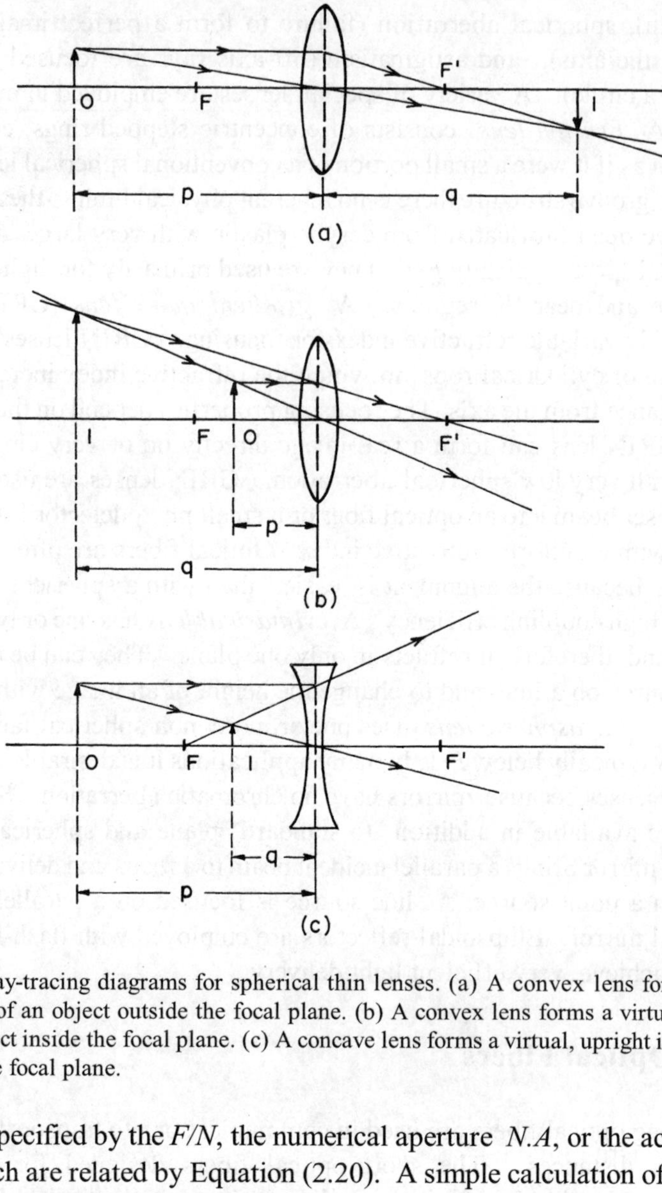


**Figure 2.18** Ray-tracing diagrams for spherical thin lenses. (a) A convex lens forms a real, inverted image of an object outside the focal plane. (b) A convex lens forms a virtual, upright image of an object inside the focal plane. (c) A concave lens forms a virtual, upright image of an object inside the focal plane.

fiber can be specified by the *F/N*, the numerical aperture *N.A*, or the acceptance angle $\alpha$ , which are related by Equation (2.20). A simple calculation of the *N.A.* of a step-index fiber based leads to

$$N.A. = \frac{\sqrt{(n_1^2 - n_2^2)}}{n_0} \tag{2.21}$$

where $n_1$ and $n_2$ are the refractive indices of the core and cladding, respectively, and $n_0$ is the refractive index of the external medium.

**Example 2.12** Properties of a step-index optical fiber

A fused-silica optical fiber has an inner core refractive index of 1.457 at 633 nm and a plastic outer core with a refractive index of 1.432. Calculate: (a) the numerical aperture; (b) the *F/N*; (c) the full acceptance angle in air.

(a) From Equation (2.15): $N.A. = \sqrt{(1.457^2 - 1.432^2)/1} = 0.27$

(b) $F/N = 1/(2 \times 0.27) = F/1.9$

(c) $\alpha = 2 \arcsin(0.27) = 31.3°$

The acceptance angle for this optical fiber embedded in tissue would be 25° for $n_0 = 1.4$.

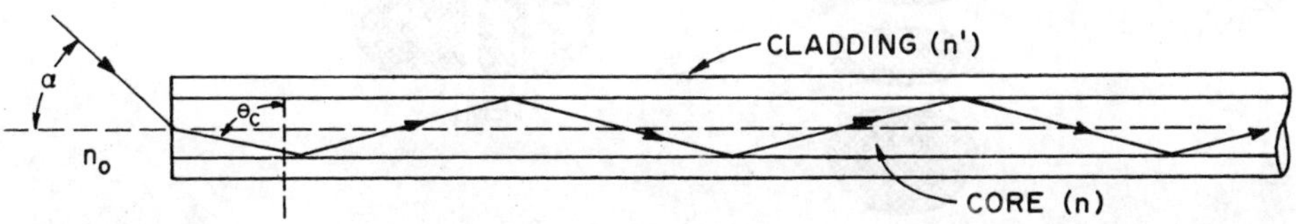


**Figure 2.19** Optical paths in a step-index optical fiber. The angle a is the maximum acceptance angle for propagation of the incident ray. The numerical aperture is given by equation 2.21. From Grossweiner, L.I., *The Science of Photobiology*, 2nd ed., Plenum Press, New York, 1989. With permission.

Electromagnetic theory shows that optical fibers are cylindrical optical cavities. This property is relevant mostly for communications fibers. However, similar considerations are involved in laser physics as described in Section 3.6. Light propagation along an optical fiber establishes a set of standing wave patterns or *optical modes* perpendicular to the fiber axis. Each mode corresponds to a specific linear polarization of the transmitted light and a unique spatial pattern of the E-field. A *single-mode fiber* transports only one linearly polarized mode in which the radial E-field intensity has a gaussian (bell-shaped) profile (Figure 2.20a). The properties of commercial optical fibers are specified by the *normalized frequency parameter* or *V- number* defined by

$$V = 2\pi \frac{a}{\lambda_0}(N.A.) \qquad (2.22)$$

where $a$ is the core radius and $\lambda_0$ is the wavelength in the external medium. The total number of modes equals $V^2/2$. A step-index fiber becomes single mode when $V < 2.405$.

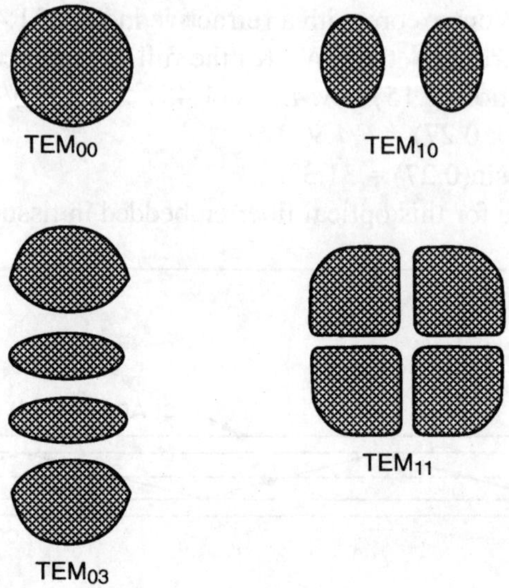


**Figure 2.20** Projection of some low-order tansverse modes of a laser cavity. The $TEM_{00}$ mode has a gaussian irradiance profile.

---

**Example 2.13** Modes in a step-index fiber

A 100 μm diameter optical fiber has a *N.A.* equal to 0.27 at 633 nm. Calculate: (a) the *V*-number; (b) the number of linearly polarized modes; (c) the maximum radius for single-mode operation.

(a) $V = 2\pi \times 0.27 \times 50 \times 10^{-6}/633 \times 10^{-9} = 134$

(b) $134^2/2 = 9000$ modes.

(c) The maximum radius for single-mode operation is given by: $2\pi \times 0.27 \times a / 633 \times 10^{-9} < 2.405$, leading to $a = 0.90$ μm.

---

A multi-mode fiber can transport a much higher optical power than a single-mode fiber, about 3000 times higher than in Example 2.13. The advantage of single-mode operation lies in lower *inter-modal dispersion*. Since each mode in a multi-mode fiber propagates with a different optical path, there is a large variation in the arrival times of the various modes. This time delay between the arrival of the slowest and fastest modes sets an upper limit on the optical bandwidth and limits the amount of information that can be transmitted. Intermodal dispersion is absent in a single-mode fiber and the bandwidth is limited by the dependence of refractive index on wavelength. Only laser sources can couple sufficient power into a small-diameter single-mode fiber and accurate alignment is required at inter-fiber junctions. Coupling a laser into an optical fiber is illustrated in Figure 2.21. The general guidelines for efficient coupling are: (1) The focused spot should be comparable to the core size; (2) The *N.A.* of the lens should not exceed the *N.A.* of the fiber.

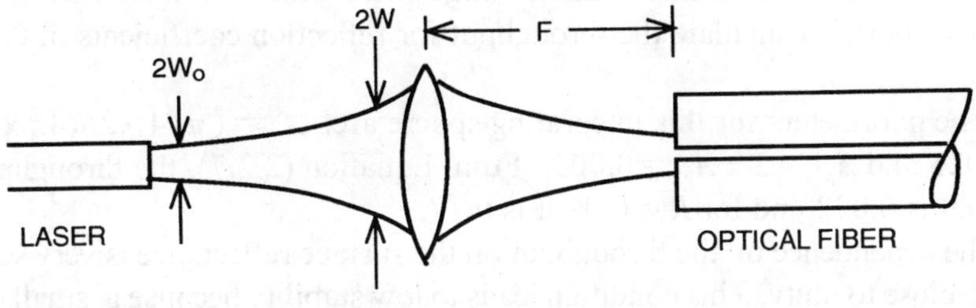


**Figure 2.21** Coupling a laser beam into an optical fiber. For optimal coupling the diameter of the focused laser beam equals the diameter of the fiber core, and the numerical aperture of the lens equals the numerical aperture of the optical fiber.

## 2.6.4 The Integrating Sphere

The integrating sphere is a venerable device with many new applications. The basic construction of an integrating sphere consists of a hollow spherical shell with a highly reflecting inner surface and small ports to admit and extract radiation (Figure 2.22). Light rays entering the sphere at different angles are "homogenized" by multiple reflections. The integrating sphere is an ideal lambertian light source. An entering ray that strikes any point of the interior surface from any angle illuminates the entire surface of the sphere at constant irradiance. Integrating spheres are used for measurements of total radiant power, to collect light from a wide angle source, as uniform light sources, laser beam attenuators, and for light beam mixing. The addition of an integrating sphere to a spectrophotometer makes it possible to measure reflection spectra of opaque materials. Reflection meaurements are especially useful for thin layers such as paper and for highly opaque materials. The throughput (H) of an integrating sphere is the ratio of the exiting flux to the input flux. A ray tracing calculation leads to

$$H = A_e \frac{R}{[1 - R(1 - A_p)]} \tag{2.23}$$

where $A_e$ is the ratio of the exit port area to the surface area of the entire sphere, $A_p$ is the ratio of all port areas to the sphere area, and $R$ is the reflection coefficient of the interior surface.

---

**Example 2.14** Throughput of an integrating sphere

Consider a 20 cm diameter integrating sphere with 20 mm diameter entrance and exit ports. Calculate the throughput for reflection coefficients of 0.90 and 0.99.

The parameters for this integrating sphere are: $A_e = (\pi/4) \times 2^2/(4\pi \times 10^2) = 0.00025$ and $A_p = 2 \times A_e = 0.005$. From Equation (2.2.7), the throughput for $R = 0.9$ is 0.022 and for $R = 0.99$ it is 0.17.

The dependence of the throughput on the surface reflectance is very sensitive for $R$ close to unity. This condition leads to low stability because a small change in the reflecting coating can have a large effect on the properties. A less efficient integrating sphere with R $\approx$ 0.90 is preferable.

---

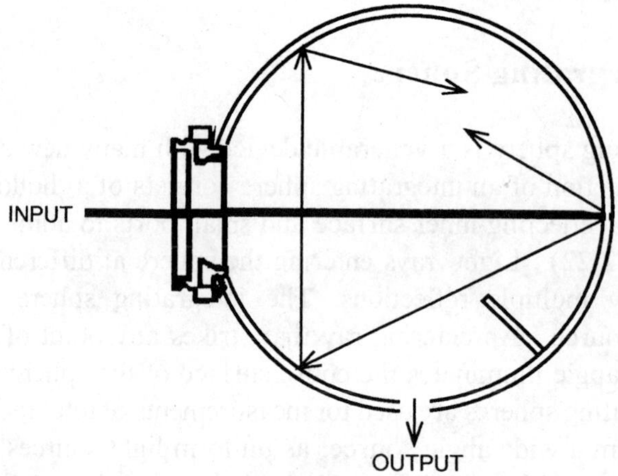


**Figure 2.22** An integrating sphere. Light entering the sphere at any angle is diffused by multiple reflections within the sphere. Adapted from Oriel Corporation *catalog Light Sources, Monochromators and Spectrographs, Detectors and Detector Systems, Fiber Optics*, Vol II, 1994, Stratford, CT. With permission.

## 2.6.5 Coupling of Optical Components

The basic principles of optical coupling are illustrated for a two-lens system in Figure 2.23, where $h_1$ and $h_2$ are the heights of a source and receiver, respectively, and $\theta_1$ and $\theta_2$ are the half-angles subtended by the collecting and the focusing lens. The theory of paraxial ray imaging leads to the relationship:

$$\theta_1 h_1 = \theta_2 h_2 \qquad (2.24)$$

If $h_1 = h_2$, Equation (2.24) gives the "rule of thumb" that the *F/N* subtended by the optical system at the receiver should be equal to or lower than the *F/N* subtended at the source.

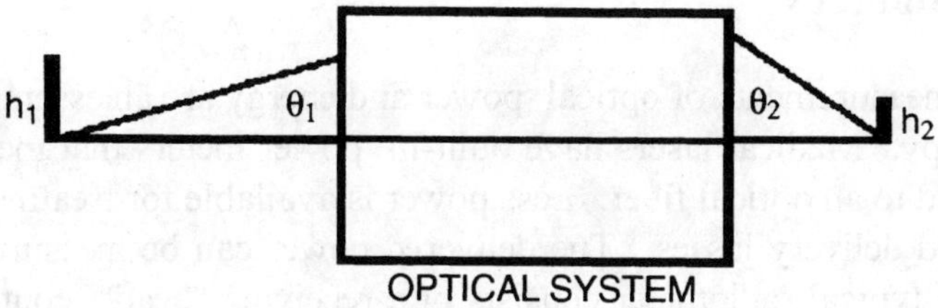


**Figure 2.23** Coupling of optical components. An object of height $h_1$ subtends an acceptance angle $\theta_1$. The image formed by the optical system has a height $h_2$ and subtends an exit angle $\theta_2$. The most efficient coupling satisfies Equation (2.24).

---

**Example 2.15** Optical coupling of a small lamp to a receiver

An arc lamp is focused on a 2 mm high receiver using an *F/1.5* collecting lens and a *F/4* focusing lens. (a) Calculate the maximum length of source that can be imaged. (b) If both lens have 50 mm apertures, calculate the focal length of each lens and the magnification.

(a) The approximate values of $\theta$ for the collecting and focusing lenses are: $\theta_1 = 1/(2 \times 1.5) = 0.33$ rad = 19°; $\theta_2 = 1/(2 \times 4) = 0.125$ rad = 7°. According to Equation (2.28): $\theta_1 h_1 < 2 \times 7 = 14$ mm-deg. The longest source imaged is: 14/19 = 0.7 mm.

(b) The focal lengths are: $F_1 = 50 \times 1.5 = 75$ mm; $F_2 = 50 \times 4 = 200$ mm. The magnification of the source at the receiver is given by: $M = F_2 / F_1 = 200/75 = 2.7$.

---

Equation (2.24) is applicable for ordinary light sources. Lasers require a different treatment that is described below in connection with launching light into an optical fiber. Equation (2.24) is a special case of a more general relationship relating the flux exiting a source to the flux at the receiver

$$A_1\Omega_1 = A_2\Omega_2 \tag{2.25}$$

where $A_1$ and $A_2$ are the cross-sectional areas of the source and receiver and $\Omega_1$ and $\Omega_2$ are the solid angles subtended by the input and exit apertures, respectively.[10] The angle-averaged radiance is defined by: $L_{ave} = \Gamma/A\Omega$, where $\Gamma$ is the radiant flux . Equating the flux leaving the source to that reaching the receiver shows that $L_{ave}$ (source) = $L_{ave}$ (receiver). This calculation proves that *an image cannot be brighter than the source.*

### 2.6.6 Radiometers

Accurate measurements of optical power and energy are an essential aspect of phototherapy. Medical lasers have built-in power meters that measure the power delivered to an optical fiber. Less power is available for treatment owing to coupling and delivery losses. The delivered power can be measured with a radiometer. A typical radiometer consists of a receiving "head" containing an optical sensor connected to a calibrated readout device. Radiometers are deceptively simple instruments because the validity of the reading depends on the measurement geometry and spectral match between the light source and the photodetector. Two basic types of photodetectors are available. A *thermal detector* responds to the heat generated by absorption of radiant energy. A *photoelectric detector* converts light to an electrical signal. The operating characteristics of radiometers include the useful wavelength range, the lowest and highest power levels that can be measured, and the ability to measure pulsed and CW light. Commercial photodetectors are characterized by standard parameters: (1) responsivity - detector output per unit optical power input; (2) response time constant - time required for the photodetector output to reach 63% of the maximum value; (3) linearity - range of input irradiance over which the output signal is a linear function of the input; (4) detectivity , a figure of merit indicating the minimum measurable optical power exceeding electrical noise.

The response of a thermal detector is independent of wavelength over a very wide spectral range. The thermopile is an early type of thermal detector. This device consists of an array of closely-spaced junctions of two dissimilar metals

(Figure 2.24a ). The "hot" junctions are illuminated by the optical radiation and the "cold" junctions are maintained at a constant temperature. The temperature difference between the hot and cold junctions generates a thermoelectric voltage proportional to the absorbed radiant power. The junctions in a modern thermopile are fabricated from a thin metal film coated with an absorbing layer of black paint or metal oxide. Thermopiles have a slow time response and are most useful for CW measurements. The low responsivity limits the applications to relatively strong light sources. Another type of thermal detector is fabricated from a *pyroelectric crystal* such as lead zirconate and lithium tantalate. In these crystals a permanent electric field exists across opposite faces. Thermal expansion induced by absorption of radiation induces a voltage change (Figure 2.24b). A typical pyroelectric sensor is about 100 times more sensitive than a thermopile and has a faster time response. However, only pulsed and "chopped" CW signals can be measured. The "flat" spectral response of thermal detectors makes them useful for measurements of total power delivered by a very broad wavelength light source such as sunlight.

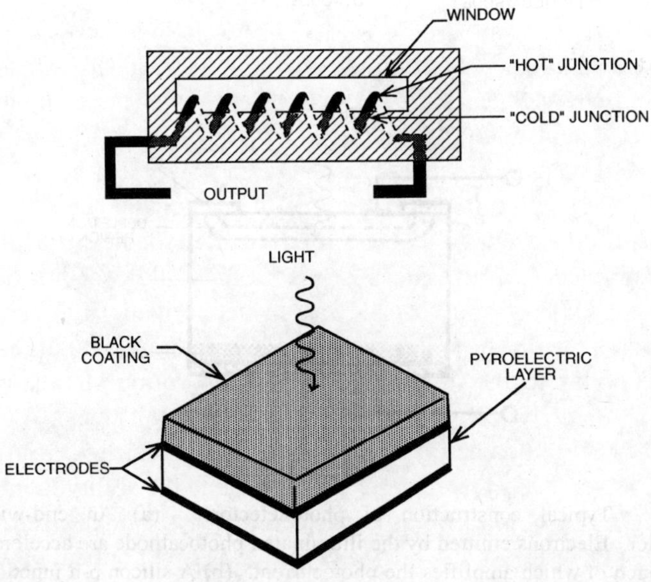


**Figure 2.24** Typical construction of thermal detectors. (a) A thermopile is an alternating array of two dissimlar metals. (b) A pyroelectric detector is a ferroelectric crystal in which heating alters the "built-in" electric field across the external faces.

A photodetector converts incident photons to an electric current. The useful wavelength range and spectral sensitivity vary widely with the light-sensitive element. A vacuum phototube is an early type of photoelectric detector in which electrons are emitted by an illuminated photocathode and a collected by an anode. The primary photocurrent is amplified many times in a *photomultiplier* (Figure 2.25a). Electrons emitted by the photocathode are accelerated by series of "dynodes", each of which emits a higher electrical current than it receives. Photomultipliers are fabricated with many different photocathodes, covering the spectral range from 80 nm in the far-VU to 1100 nm in the near-IR. They are rugged, highly sensitive devices, and respond to both CW and light pulses in the nanosecond range. A faster time response is achieved in a *microchannel plate* that uses an array of approximately one million electron-multiplying elements. Microchannel plates are used in scanning electron multipliers and image

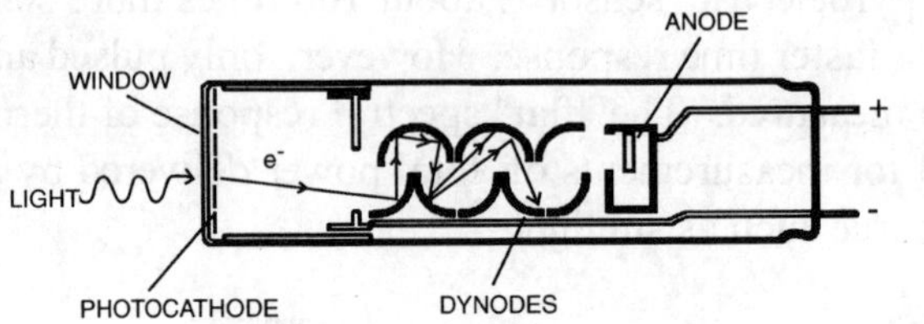


a

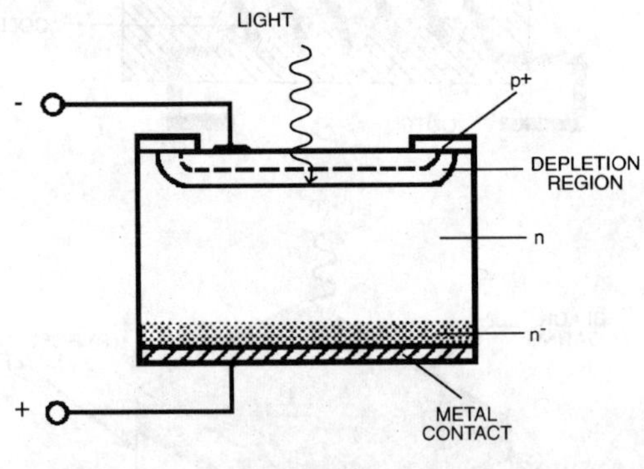


b

**Figure 2.25** Typical construction of photodetectors. (a) An end-window vacuum photomultiplier. Electrons emitted by the illuminated photocathode are accelerated by a series of dynodes, each of which amplifies the photocurrent. (b) A silicon p-n junction photodiolde. Light absorption in the $p^+$, depletion, and n regions generates electrons and holes that are accelerated by the built-in voltage into the $n^-$ and $p^+$ regions, respectively. In unbiased operation, the charge separation is measured with a sensitive voltmeter. Application of reverse bias generates a photocurrent. Adapted from Oriel Corporation *catalog Light Sources, Monochromators and Spectrographs, Detectors and Detector Systems, Fiber Optics*, Vol II, 1994, Stratford, CT. With permission.

intensifiers. Except for measurements of ultrashort pulses, photomultipliers have been largely superceded by solid state devices. A review of semiconductor physics in Section 3.6.3 is applicable for both solid state photodetectors and semiconductor lasers.

## 2.7 WAVELENGTH SELECTION

A wavelength selection device is used to modify the spectral output of a light source. Virtually any optical phenomenon can be adapted for wavelength selection, including absorption, dispersion, interference, diffraction, birefringence, and light scattering. Color glasses incorporating inorganic salts with selective absorptions are the least expensive type of color filter, although the characteristics are limited by the available materials. A "long-pass" filter is strongly absorbing below a cutoff wavelength and transmitting at longer wavelengths and *vice versa* for a "short-pass" filter. A "band-pass" filter is highly transmitting over one or several wavelength bands. More expensive interference filters may be fabricated as required, adding an integral blocking layer. Narrow bandwidth is achieved at the expense of throughput in practical interference filters by using highly reflective coatings. An optical cavity consisting of a dielectric (or air) layer located between parallel reflectors is referred to as a *Fabry-Perot*. The Fabry-Perot units can be stacked for a further reduction in bandwidth. Metallic mirrors used in earlier designs have high absorption losses. Modern interference filters utilize multilayer dielectric reflectors. A stack of alternate high-refractive index and low-refractive index layers, each of $\lambda/4$ thickness, has a very high reflectance at wavelength $\lambda$. A similar design is used for a *hot mirror* that transmits UV and visible light and reflects IR, and for a *cold mirror* that reflects UV and visible light and transmits IR. Hot mirrors are used to remove IR heat rays from arc lamps. A variable wavelength selection device is the key component in all instruments used for spectral measurements. A *monochromator* selects a narrow spectral bandwidth out of incident light. The early prism instruments have been superceded by diffraction gratings that are used on virtually all modern monochromators.. The addition of visual detection system to a monochromator gives a *spectrometer*. A *spectrograph* is a spectrometer in which the output spectrum is recorded on a photographic plate or a photodiode array. Fast temporal changes in output spectra can be measured with a *streak camera* in which a photographic film is displaced at a high speed. A *spectroradiometer* is a spectrometer plus a photodetector for measurements of radiant power or energy. A *spectrophotometer* consists of a monochromator with a light source at the input and a photodector at the output. This device is used to record absorption or emission spectra of samples placed in the optical path.

NOTES

1. A Fourier series expresses an arbitrary function as an infinite sum of discrete sine and cosine functions. This representation is very convenient for periodic wave motion. Each term in a Fourier series corresponds to a monochromatic sinusoidal wave. A Fourier integral extends the Fourier series to the expansion of non-periodic functions. In a Fourier integral the discrete terms in the Fourier series are replaced by an integral over pairs of conjugate variables such as frequency and time.

2. The International System of units (SI) are standard in physics. The SI base units are: length (meter, m), mass (kilogram, kg), time (second, s), and amount of substance (mole, mol). Standard prefixes are used when dealing with very large or very small numbers, Some of the most widely used prefixes are kilo ($10^3$), mega ($10^6$), giga ($10^9$), and tera ($10^{12}$) for large quantities and deci ($10^{-1}$), centi ($10^{-2}$), milli ($10^{-3}$), micro ($10^{-6}$), nano ($10^{-9}$) pico ($10^{-12}$), and femto ($10^{-15}$) for small quantities. Other SI units are defined for specific properties including energy (joule, J), power (watt, W), electric potential difference (volt, V), and magnetic flux (weber, Wb).

3. Many molecules including water have a built-in separation of electric charge which is equivalent to an electric dipole. An alternating electric field induces a back and forth motion of the dipoles referred to as a dielectric interaction. The friction associated with this motion heats the material. Microwave ovens utilize this effect.

4. The electron volt (eV) is the amount of kinetic energy acquired by an electron when it accelerated by a potential difference of one volt. The utility of this unit is based on the fact that the same energy increment applies for any particle having the same charge as the electron, for example, a proton. However, the 1 eV proton is moving about 40 times slower than a 1 eV electron owing to the heavier mass.

5. This calculation is slightly ambiguous because according to Equation (2.1), changes in $V$ can be related to changes in $f$, $\lambda$, or both. Example 2.2 assumes that $f$ is the same in air and glass while $\lambda$ differs in each medium. This assumption is justified by classical electromagnetic theory. For the simpler case of elastic waves in adjoining media the particles at an interface between different media must oscillate at the same frequency, for example, vibrations of a thin string joined to a heavier cord.

6. The reader familiar with ultrasound may observe that Equation (2.4) resembles the expression for reflection ultrasound at the interface between two regions of different specific acoustic impedance $Z$. Equation (2.6) for normal incidence gives the ultrasound reflection coefficient with $n_r$ replaced by $Z_2/Z_1$.

7. Soliton motion describes a unique type of wave in which spreading does not take place. Soliton waves were first reported in 1834 by Russell in the form of a high-speed "solitary" wave that continued without change of form or speed in the Glasgow-

Edinburgh canal. Very short optical pulses can propagate in optical fibers as solitons.

8. In circularly polarized light, the mutually perpendicular vectors representing the E-field and B-field rotate together at the wave frequency and constant amplitude. It is easily shown that linearly polarized light can be resolved into two equal-amplitude circular polarizations rotating in opposite directions.

9. More complicated optical systems can be analyzed with ray tracing computer programs. A general analytical technique applicable for paraxial rays uses ray matrices. The action of each optical element is represented by a "2 x 2" array or "ABCD" matrix. The effect of a complex optical system on an incident ray is calculated by the rules of matrix algebra. This method is suitable for cascaded groups of lenses, mirrors, and spatial regions of variable refractive index or "ducts".

10. A geometrical calculation for paraxial rays shows that W is related to q by: $W = 2p\,(1 - \cos q\,)^2 \approx pq^2$. Substituting for W and taking the square root of both sides leads to Equation (2.24).

# CHAPTER 3

# PHOTOTHERAPY LIGHT SOURCES

Natural sunlight has been used to treat skin disorders since antiquity. The widespread use of artificial light sources for phototherapy started in the 19th century. The most important characteristics of a phototherapy light source are: (1) physical dimensions; (2) total radiant power; (3) directionality and uniformity of the emitted light; and (4) spectral intensity distribution. Additional factors for practical applications are the cooling requirements, useful lifetime, replacement cost, and optical, thermal, and electrical hazards. Phototherapy light sources include sunlight, incandescent lamps, fluorescent lamps, electric arcs, light emitting diodes (LEDs), and lasers. Conventional lamps are usually employed for phototherapy applications where a wide treatment field is illuminated. Lasers are most useful for applications where therapeutic light is delivered *via* an optical fiber to a small treatment field. Until recently it was only by fortuitous coincidence that a commercial light source was available for a given application. The development of light sources for specific phototherapy applications is currently a major part of the technology.

## 3.1 SUNLIGHT

Sunlight may be the only practical light source for medical applications in underdeveloped regions lacking modern resources. The sun is a sphere of luminous gas about 1,4000,000 km in diameter located 150,000,000,000 m from the earth. The temperature of the sun ranges from about 15,000,000 K in the core to 4,000 K about 500 km above the outer surface. The normal solar spectrum emitted by the outer region of the sun or *photosphere* has an average temperature about 6000 K. By comparison, a high-pressure xenon arc lamp operates at 6,500 K and the filament temperature in an incandescent lamp is about 3000 K. About 50% of the energy in sunlight is visible light, about 40 % is IR, and about 10% is UV. Sunlight is filtered by gases in the solar atmosphere. The alternating regions of light and dark in the solar spectrum are *Fraunhofer lines*. There are about 25,000 Fraunhofer lines between 0.3 μm and 1 μm. The solar radiation at the surface of the earth is modified by the earth's atmosphere. *Air mass* is the relative amount of atmosphere traversed by sunlight. The value of air mass depends on the latitude and the time of day. Sunlight reaching the surface of the earth consists of direct rays from the sun and *skylight* resulting from scattering by the earth's atmosphere. The solar spectrum at sea level with the sun directly overhead increases from 290 nm to its maximum at 550 nm and then decreases gradually to about 3 μm. The ozone layer in the stratosphere is responsible for the UV cutoff which varies with location and the time of the year. Sunburn is caused mostly by UV-B and is strongly dependent on the location, time of year, and time of day. The UV-A irradiance is about 5 $mW/cm^2$ which is the order of 100 times more intense than UV-B.

## 3.2 THERMAL RADIATION

All material objects emit thermal radiation. The spectral distribution of an idealized *black body* (BB) should obey the Planck's radiation law (Section 2.2). A BB is a hypothetical perfect emitter and a perfect absorber of thermal radiation.[1] The Planck law predicts important properties of thermal radiation that were first deduced experimentally. The radiant power emitted by a BB is related to absolute temperature by the Stefan-Boltzmann law

$$E = BT^4 \tag{3.1}$$

where $B$ is the Stefan-Boltzmann constant ($5.67 \times 10^{-8}$ $W/m^2 - K^4$) and $T$ is in K. The BB spectrum shifts towards the blue with increasing temperature (Figure 3.1). The maximum emission wavelength obeys Wien's displacement law

$$\lambda_{max}(\mu m) = \frac{2898}{T} \tag{3.2}$$

$\lambda_{max}$ is about 10 μm at room temperature and 1.2 μm for a hot tungsten filament. The emission spectrum of the sun is close to that of a *black body* at 6000 K with the maximum emission at 550 nm green light.[1] The spectral irradiance of thermal radiation emitted by a real object depends on the material and surface treatment. The ratio of the actual emitted power at a given wavelength to that of a BB is the *emissivity*. The typical emissivity of a tungsten filament is about 0.4 while that of human skin is approximately 0.98 at 3 μm - 4 μm. The concept of *color temperature* is used in the photography and lighting

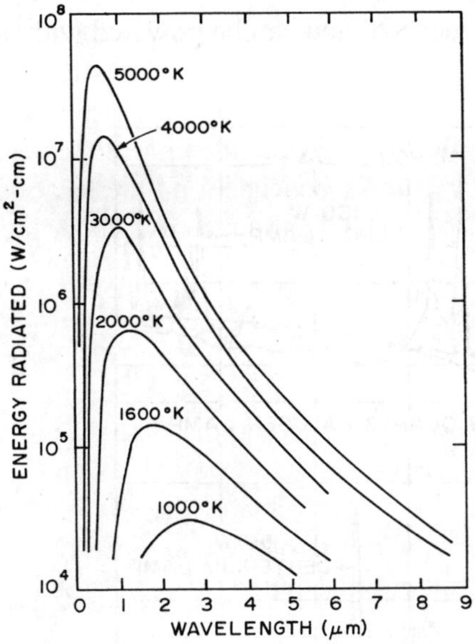


**Figure 3.1** Spectral irradiance of blackbody radiation. The total emitted radiation increases with the fourth power of the absolute temperature. According to Wien's law, the product of the absolute temperature and wavelength of maximum emission is a constant. From Grossweiner,L.I., *The Science of Photobiology*, 2nd ed., Plenum Press, New York, 1989. With permission.

industries to relate the color balance (chromaticity) of a lamp in the visible region to a BB radiator. The color temperature may have little relationship to the physical temperature. Some values of color temperature are: extraterrestrial sun, 5900 K; clear blue sky, 12,000-25,000 K; photographic flash lamp, 3800 K; "cool white" fluorescent lamp, 4300 K; quartz-tungsten-halogen (QTH) lamp, 3300 K. The high color temperature of the daylight sky results from the strong contribution of scattered blue light which corresponds to an extremely hot BB.

## 3.3 INCANDESCENT LAMPS

An incandescent lamp consists of a tungsten filament in a glass envelope filled with an inert gas and heated by an electric current to about 2500 K. The emitted light is the thermal radiation of the hot filament. The useful lifetime of an incandescent lamp is limited by the deposition of black metal on the envelope and oxidation of the filament. A quartz-tungsten-halogen (QTH) lamp achieves higher operating temperatures by employing a thermally-resistant fused-silica envelope containing a small quantity of a halogen mixture. The halogen vapor extends the life of the lamp by reacting with the evaporated metal film on the hot envelope and returning it to the heated filament. The color temperature of an incandescent lamp is higher than the physical temperature because the wavelength variation of tungsten emissivity shifts the spectrum to the blue. The spectral output of a 100-W QTH lamp operating at 3300 K color temperature is shown in

Figure 3.2. A 250-W QTH lamp costs under $20 and can be powered with an inexpensive step-down transformer.

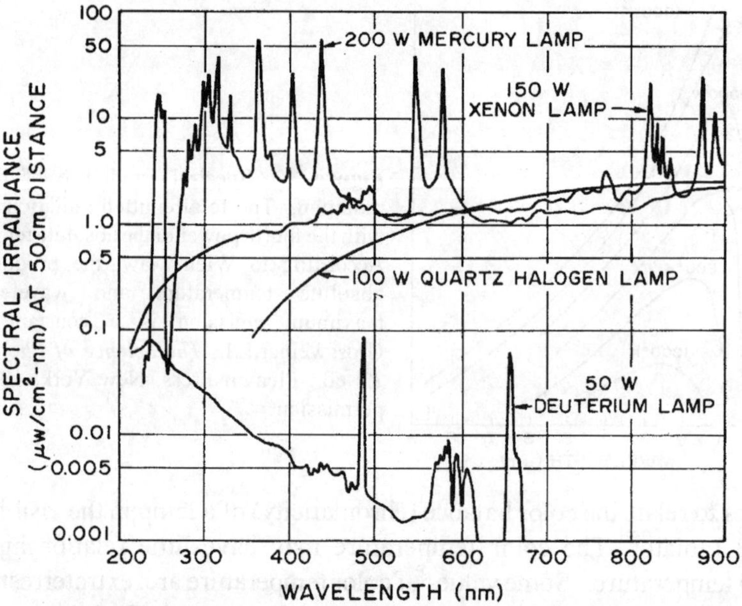


**Figure 3.2** Spectral irradiance of some incandescent lamps. From Oriel Corporation catalog *Light Soursecs, Monochromators & Spectrographs, Detectors & Detector Systems, Fiber Optics,* Vol. II, 1994, Stratford, CT. With permission.

## 3.4 ARC LAMPS

A gaseous discharge is established by passing an electric current *via* metallic electrodes through a gas or vapor. There are many different types of gaseous discharges depending on the composition of the filling gas, the electrode configuration, and the operating conditions. Commercial lamps are categorized as brush discharges, corona discharges, sparks, glow discharges, and arcs. Each type of discharge has fairly well-defined characteristics. In the Geissler tube, dating to 1855, the application of a high voltage to a low-pressure gas in a glass vessel generates a line spectrum characteristic of the filling gas. Practical examples of this device are the neon tube and sodium vapor lamp. The intimate mixture of electrons and ions in the glowing region of a discharge constitutes a plasma.[2] A corona discharge is established between a positive high voltage point and a plane cathode at a higher gas pressure and higher voltage. This term originates from the appearance of the outermost region of the Sun's atmosphere, which consists of an extremely hot, low-density plasma. A corona converts to an arc at sufficiently high current densities. A discharge can be established by injecting high-voltage electrons directly into a gas through a thin foil. In an "electrodeless" discharge the gas is excited by an external source of radio waves or microwaves. An arc is essentially a high-density plasma with "negative" electrical resistance, *i.e.*, the current flow decreases with increasing applied voltage. The arc properties depend on the type of electrodes and the operating pressure of the filling gas or vapor. A low-pressure mercury arc emits approximately 90% of the UV transmitted by the quartz envelope at 254 nm. They are often used as "germicidal" lamps, especially in Europe where they are the primary treatment for water systems to kill bacteria and viruses. The output of medium and high-pressure arcs consist of broadened spectral lines characteristic of the filling gas superimposed on a bright continuous background (Figure 3.2). An arc lamp requires a starting device because a higher current is required to initiate the arc than to maintain it. The full intensity of a "cold cathode" arc is not attained for some minutes. "Hot cathode" arcs incorporate a heated tungsten filament in the cathode for faster startup. High-pressure mercury and xenon arcs are the brightest of all non-laser light sources. They are long-lived (typically 1000 hours), relatively expensive, and require a regulated high-current power supply, an ignition device, and a cooled lamp housing. Medium pressure mercury arcs are used for UV-B phototherapy of skin disorders (see Chapter 12). A "solar simulator" is essentially a high- pressure xenon arc with a UV-reflecting filter or "cold mirror" to reduce the UV output. An additional filter is used to remove the xenon spectral lines above 800 nm (Figure 3.2). Metal halide lamps are a recent development in which the addition of a metal halide salt to a medium or high-pressure mercury arc leads to a relatively uniform spectral output from 295-450 nm.

## 3.5 FLUORESCENT LAMPS

A fluorescent lamp is a low-pressure mercury arc with a luminescent coating or *phosphor* on the inner surface of the glass envelope. The relatively weak emission lines of the mercury arc are superimposed on the broad luminescence bands of the phosphor. The output may vary considerably for small changes of the phosphor composition including lamps from different manufacturers having the same designation. The spectra of some common fluorescent lamps are shown in Figure 3.3. Table 3.1summarizes the typical characteristics of commercial fluorescent lamps, where $\lambda_{max}$ is the maximum emission wavelength of the phosphor and $\Delta\lambda$ is the full bandwidth at half-maximum intensity. "Daylight" (D), "blue" (B), and "special blue" (BB) lamps are used for phototherapy of neonatal jaundice. Special fluorescent lamps used in phototherapy of skin diseases include the "sunlamp" (FS) emitting 60% UV-B (270-390 nm), the earlier type of "PUVA lamp" emitting 20% UV-A (320-340) with less than 3% UV-B, and the "UV-A/UV-B" lamp emitting about 30% UV-A (320-340 nm) and 8% UV- B. The useful features of fluorescent lamps include uniform irradiance, low cost, cool operation, and lifetimes about 200 hours.

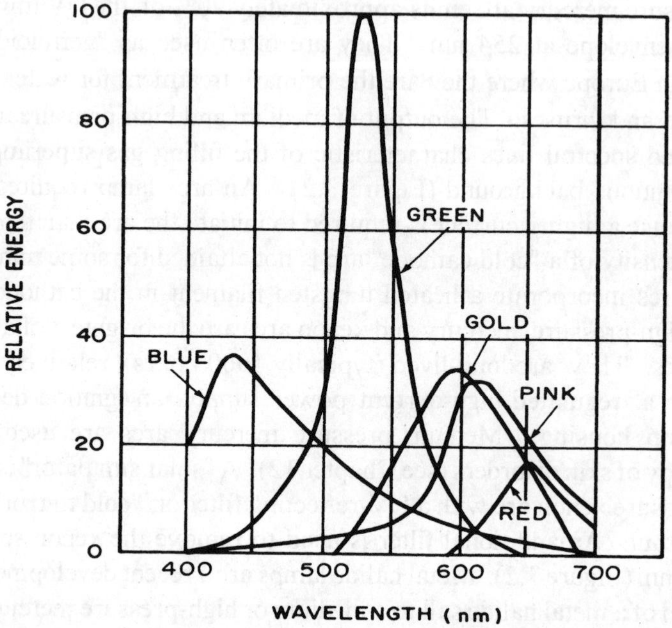


**Figure 3.3** Emission spectra of some fluorescent lamps of equal wattage. The superimposed spectral lines of the low-pressure mercury arc are not shown. From Grossweiner, L.I., *The Science of Photobiology*, 2nd ed., Plenum Press, New York, 1989. With permission.

**Table 3.1 Characteristics of commercial fluorescent lamps***

| Lamp Type | $\lambda_{max}$ (nm) | $\Delta\lambda$ (nm) |
|---|---|---|
| Black Light (BL) | 350 | 40 |
| Blue Light Black | 365 | 70 |
| Blue (B) | 420 | 35 |
| Special Blue (BB) | 450 | 35 |
| Green (G) | 525 | 40 |
| Gold (GO) | 580 | 75 |
| Daylight (DA) | 505 | 135 |

*The exact values differ for various manufacturers

## 3.6 LASERS

A lamp is a device that converts chemical, thermal, or electrical energy to light. The microscopic mechanisms usually involve a sequence of random events leading to unsynchronized emission of photons. For example, thermal radiation is emitted by vibrating atoms and arc lamps utilize electronic transitions between energy states of the filling gas. The unique feature of a laser is the synchronization of the microscopic light-emitting events. Laser light originates from the quantum transitions between energy levels of an *active medium*, which may be a gas, liquid or solid. The only limiting factors are the properties of the atomic and molecular energy levels of the active medium and the feasibility of achieving a practical design. Laser action has even been accomplished in vegetables. Laser light is highly collimated, highly coherent, and approximately monochromatic. The required synchronization of light emission in a laser is achieved by the process of *stimulated emission*. Accordingly, the term "laser" is an acronym for "light amplification by stimulated emission of radiation". The laser concept evolved from the *maser*, an earlier type of device that amplifies microwave radiations. The first operating maser was developed by Townes and coworkers at Columbia University in 1953. The ammonia maser operating at a wavelength of 1.256 cm is a very low power device with application as a frequency standard or clock. A tunable solid state maser utilizing the splitting of "impurity" $Cr_3^+$ energy levels in crystalline ruby ($Al_2O_3$) by a magnetic field is used as an amplifier in radio telescopes. In 1960, Maiman at Hughes Research Laboratories constructed a ruby laser emitting at 694 nm, and in the same year Javan at Bell Telephone Laboratories developed the helium-neon gaseous discharge laser emitting at 633 nm. Other types of lasers followed very rapidly. Table 3.2 summarizes the properties of some lasers used for biomedical applications.

### 3.6.1 General Principles of Laser Action

Laser light originates from the fluorescence emitted by atoms or molecules of the active medium. Fluorescence photons are normally emitted in many directions at random times. Owing to the short lifetime of fluorescent states, typically nanoseconds, only a small fraction of the atoms or molecules of the active medium are excited at any time. The random nature of fluorescence is replaced by laser emission when the population of an upper energy state exceeds that of the lower energy state. This condition is *population inversion*. Population

**Table 3.2 Characteristics of some biomedical lasers**

| Active Medium | Principle Wavelengths (μm) | Mode |
|---|---|---|
| | **Solid State** | |
| Doubled Nd-YAG/KTP | 0.532 | pulsed |
| Ruby | 0.694 | pulsed |
| Alexandrite | 0.72-0.78 | |
| Neodymium-YAG | 1.064, 1.330 | pulsed, CW |
| Holmium-YAG | 2.10 | pulsed |
| Erbium-YAG | 1.54, 2.94 | pulsed |
| | **Gas** | |
| Helium -Cadmium | 0.325 | CW |
| Nitrogen | 0.337 | pulsed, CW |
| Argon ion | 0.488-0.514 | CW |
| Helium-Neon | 0.554, 0.594 0.663, 1.152 | CW |
| Krypton ion | 0.46-0.68 | CW |
| Gold vapor | 0.528 | pulsed |
| Hydrogen fluoride | 2.6-3.3 | pulsed, |
| Carbon dioxide | 10.6 | pulsed, CW |
| | **Liquid** | |
| Organic dye | 0.3-0.9 | pulsed, CW |
| | **Excimer** | |
| Argon-Fluorine | 0.193 | pulsed |
| Krypton-Chlorine | 0.222 | pulsed, CW |
| Krypton-Fluorine | 0.248 | pulsed |
| Xenon-Chlorine | 0.308 | pulsed |
| Xenon-Fluorine | 0.351 | pulsed |
| | **Semiconductor** | |
| Gallium arsenide | 0.78-0.95 | pulsed |
| Gallium-Aluminum arsenide | 0.75-0.90 | pulsed, CW |

inversion is achieved by *pumping* the active medium. (Laser science is replete with jargon.) The pumping process may be continuous (CW) or pulsed, depending on the active medium and design of the laser. The active medium is enclosed in an optical cavity, typically a Fabry-Perot, in order to confine the emitted light and promote synchronous fluorescence of many atoms. The dimensions of an optical cavity support a limited number of electromagnetic standing waves or "modes", analogous to the mechanical standing waves in a musical instrument. Figure 3.4 illustrates the basic arrangement of a ruby laser. Optical pumping is provided by a gas-filled flash lamp (L) adjacent to the ruby rod (R) within a reflector (RF). The optical cavity is formed by a 100% reflecting mirror ($M_1$) and a parallel mirror ($M_2$) with a low transmission for extraction of the laser light. The ruby laser emits a pulse of red light, shorter in duration than the pumping flash, with a substructure consisting of very short

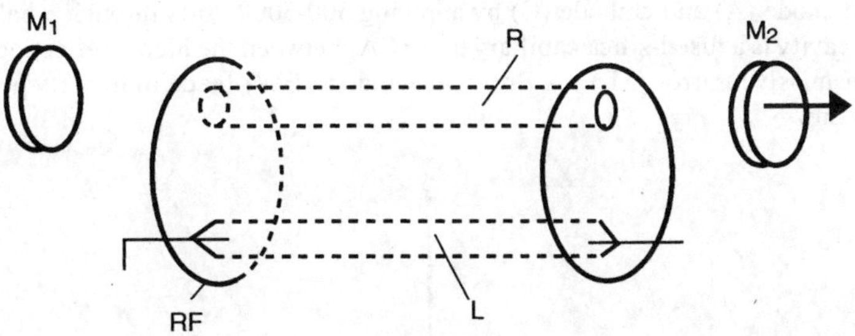


**Figure 3.4** Construction of a basic pulsed ruby laser. The ruby rod (R) and gas-filled flash lamp (L) are located in a reflector (RF). The optical cavity is established by the highly reflecting mirror $M_1$ and the transmissive mirror $M_2$. The laser pulse at 694 nm has a spiked structure; see Figure 3.19a.

"spikes". Gas and vapor lasers cannot be pumped optically because gases are not sufficiently absorbing to induce population inversion. However,.population inversion is achieved by collisions between atoms or ions of the gas and energetic electrons generated in a discharge. The design of a He-Ne laser is shown in Figure 3.5. The arc is established between the anode (A) and cathode (C) by passing current through the gaseous mixture (G) in a quartz capillary tube (CA) with the end window (BW) tilted at the Brewster angle. The spectral output consists of a very sharp 633 nm line with several satellite lines on each side. New types of He-Ne lasers can emit yellow and green light. Solid state diode lasers are fabricated from semiconductor crystals. Emission of laser light is activated by passing a high electrical current. A rudimentary p-n junction semiconductor laser is depicted in Figure 3.6. The mechanism requires an elementary knowledge of semiconductor physics which is described in Section 3.6.3.

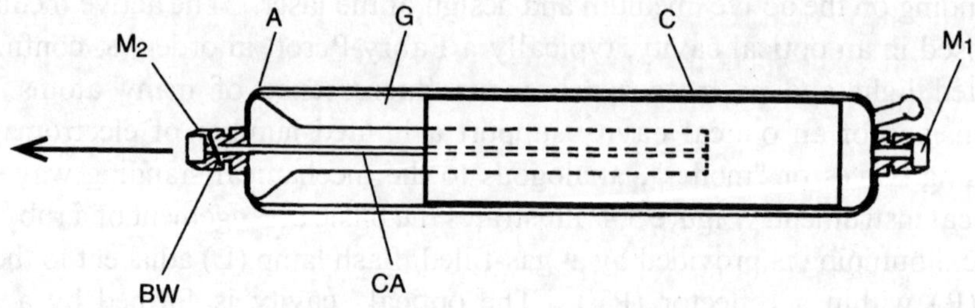


**Figure 3.5** Construction of a modern helium-neon laser. The filling gas (G) is a low pressure mixture of helium (about 90%) plus neon (about 10%). A gaseous discharge is established between the anode (A) and cathode (C) by applying 500-5000 volts through a ballast resistor. The optical cavity is a fused-silica capillary tube (CA) between the highly reflecting mirror ($M_1$) and the transmissive mirror ($M_2$). A Brewtser window (BW) leads to linearly polarized laser light at 633 nm.

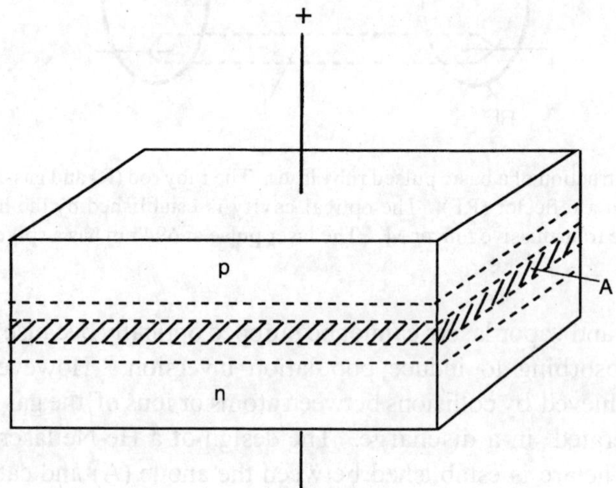


**Figure 3.6** Basic construction of a semiconductor laser. The laser consists of a thin crystal of a heavily doped p-n junction operated in forward bias. Laser light is emitted from the active region as a widely divergent beam. The polished faces of the semiconductor crystal act as the windows of the optical cavity.

### 3.6.2 Laser Action at the Atomic Level

The theoretical basis of laser action is implicit in the 1917 radiation theory of Einstein. The quantization of atomic energy levels was shown by the Bohr model of the hydrogen atom and subsequent developments. Einstein considered an assembly of identical atoms bathed in light and in thermal equilibrium with the environment. Two atomic levels are depicted in Figure 3.7 in which $N_1$ atoms occupy the lower energy state $E_1$ and $N_2$ atoms occupy the upper energy state $E_2$. In the absence of light the relative number of atoms in each state is determined by the Boltzmann statistical distribution according to

$$\frac{N_2}{N_1} = \exp\left[-\frac{(E_2 - E_1)}{kT}\right] \tag{3.3}$$

where T is the absolute temperature.[3] The population of the upper state cannot exceed that of the lower state in thermal equilibrium because $N_2 = N_1$ at hypothetical infinite temperature.[4] The most efficient optical pumping takes place when the energy of the incident photons equals the energy difference between the two states. This is the *Bohr frequency condition*

$$hf = E_2 - E_1 \tag{3.4}$$

In wavelength units Equation (3.4) can be expressed as

$$\lambda(nm) = 1239/\Delta E \tag{3.5}$$

where $\Delta E$ is the energy difference between the states in eV.

**Example 3.1** Occupation of energy levels in thermal equilibrium

The energy separation of the $n = 2$ level and $n = 3$ level of atomic hydrogen is 1.88 eV. Calculate: (a) the emission wavelength corresponding to this optical transition; (b) The fraction of atoms in the upper $n = 3$ state at 300 K.

(a) According to Equation (3.5) the wavelength corresponding to the energy separation is: 1239/1.88 eV = 659 nm.

(b) From Equation (3.3): $N(n{=}3)/N(n = 2) = \exp-[1.88 \text{ eV}/(1.38 \times 10^{-23} \text{ J/K} \times 300 \text{ K})] = \exp(-72.7) = 10^{-32}$.

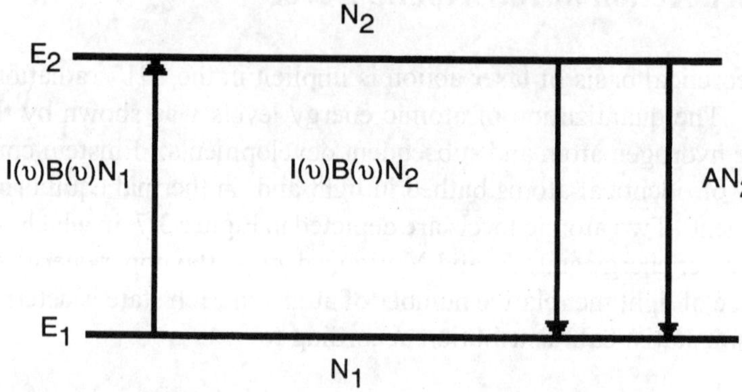


**Figure 3.7** In Einsteinian radiation theory, identical atoms are in thermal equilibrium with a heat reservoir. $N_1$ atoms occupy the lower energy state at $E_1$, and $N_2$ atoms occupy the upper energy state at $E_2$. The imposition of light of spectral intensity distribution I(ν) induces absorption by atoms in state 1, and stimulated emission by atoms in state 2, at rates proportional to coefficients B and B', respectively. Atoms in state 2 also undergo spontaneous emission at a rate proportional to coefficient A. Assuming the Bohr frequency condition, $\nu = (E_2 - E_1)/h$, B'=B, and that thermal equilibrium is maintained, this leads to the Planck spectral distribution of thermal radiation.

Equation (3.3) shows that heating alone cannot excite a significant fraction of atoms from the lower state to the upper state. However, population inversion can be achieved by pumping, in which case the energy levels are not in thermal equilibrium. The fundamental relation between light emission and the atomic parameters is easily derived. When an atomic system is pumped with radiation satisfying Eq. (3.4), the number of atoms excited to the upper state per second can be expressed as: $B(f)\,\mathbf{I}(f)\,N_1$, where $\mathbf{I}(f)$ is the light intensity at the pumping frequency *f*, $N_1$ is the the number of atoms in the lower state and $B(f)$ is an efficiency parameter that depends on the absorption coefficient of the atoms at frequency *f*. Similarly, interactions of photons with atoms in the upper state induce downward transitions at a rate: $B'(f)\mathbf{I}(f)N_2$. where $B'$ is the *Einstein stimulated emission coefficient*. The atoms in the upper state also emit photons spontaneously at a rate $AN_2$ , where $A$ is the *Einstein spontaneous emission coefficient*. Therefore, the total rate of downward transitions is $[A + B'(f)\mathbf{I}(f)]N_2$,. Einstein made the key assumption that $B' = B$, arguing that electronic transitions at the level of single atoms must be reversible. For steady conditions the rate of upward transitions equals the rate of downward transitions. However, the Boltzmann relation in Equation (3.3) must apply if the system is in thermal equilibrium at temperature *T*. Solving for $N_2/N_1$ and equating leads to the expression for I(*f*)

$$I(f) = \frac{A/B}{[\exp(-hf)/kT - 1]} \tag{3.6}$$

Equation (3.6) has the same form as the thermal radiation expression derived by Planck sixteen years earlier by quantization of thermal radiation. Exact agreement with Planck's result obtains by introducing $A/B = hf^3/c^2$ as calculated from classical electromagnetic theory. Explicit expressions for the Einstein *A*-coefficient and *B*-coefficient have been derived and are verified experimentally from spectral measurements.[5]

Stimulated emission is the fundamental interaction responsible for laser action. Optical pumping cannot induce overpopulation in the two-level system of Figure 3.7 because the most efficient process leads to $N_1 = N_2$. This limitation is circumvented in a *three-level laser* in which the $E_2$ state is populated indirectly by fast, non-radiative transitions from a higher-energy state $E_3$. The higher energy state at $E_3$ may be pumped optically, by collisions, or any other process that transfers energy to atoms. The ruby laser is a three-level system (Figure 3.8). The upper state in a ruby crystal is a continuum of closely spaced energy levels (C) populated by optical pumping (P) from the ground level (G). Levels C populate intermediate levels (F) via non-radiative transitions (N). Laser emission (L) takes place by radiative transitions F ⇒ G.

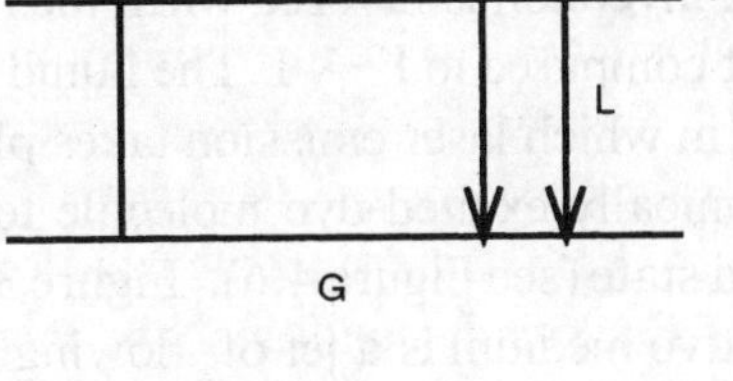


**Figure 3.8** A simplified energy level diagram for the ruby laser. The higher energy states of the $Cr^{3+}$ ions are broadened by the crystal field into a continuum of levels (C) that are populated by optical pumping (P) from the ground level (G). Levels C populate the narrow intermediate levels F by fast non-radiative transitions in competition with stimulated emission to G. Spontaneous emission from levels F to G is about $10^5$ less likely than from levels C to G; this favors overpopulation of levels C. Laser emission corresponds to the radiative transitions (L) from levels F to G. This three-level system has low efficiency owing to the high population of G.

A three-level laser requires fast, non-radiative decay of level $E_3$ compared to the decay lifetime of state $E_2$ and a sufficiently high optical-pumping rate to maintain the overpopulation of level $E_2$.[6] This situation is difficult to achieve by CW pumping and the ruby laser is normally pumped by a strong light flash. CW operation is facilitated in a four-level laser in which the lower energy state involved in laser action $E_2$ lies above the highly-populated ground state (Figure 3.9). The active medium in a neodymium-yttrium aluminum garnet (Nd-YAG) laser is a garnet rod ($Y_3Al_5O_{12}$) containing $Nd^{3+}$ ions. Laser emission takes place

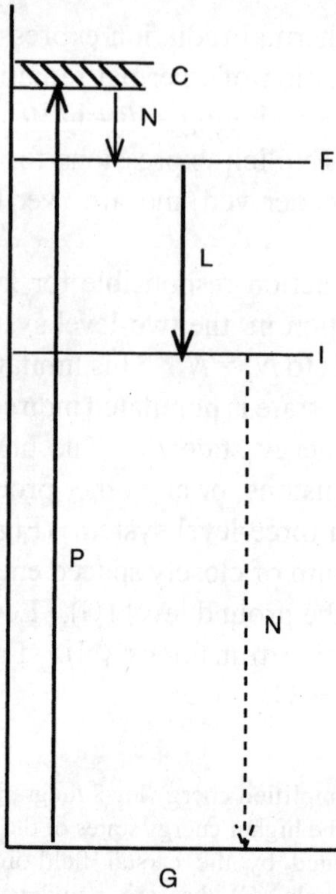


**Figure 3.9** Simplified energy level diagram for a four-level laser. Levels F are populated by fast non-radiative transitions from the pumped continuum of levels C. Laser emission (L) corresponds to radiative transitions from level F to level I. Level I is depopulated by fast non-radiative transition (N) to the ground state (G). In the Nd-YAG laser, levels C are groups of broadened $Nd^{3+}$ lines, level F consists of two closely-spaced narrow energy levels, and level I consists of three levels lying about the ground state, each of which is split into a series of closely spaces levels. The strongest laser emission is at 1.064 μm.

between $Nd^{3+}$ energy levels separated by 1.06 μm. The fluorescent level (F) is pumped *via* the continuum of upper energy levels (C) as in the ruby laser. Laser emission takes place from F ⇒ I. Population inversion is favored when the non-radiative transitions C ⇒ F and I ⇒ G are fast compared to F ⇒ I. The liquid dye laser is an optically-pumped, four-level laser in which laser emission takes place from the $S_1$ fluorescent state of the electronically-excited dye molecule to an upper higher vibrational level of the $S_0$ ground state (see Figure 4.6). Figure 3.10 illustrates one type of CW dye laser. The active medium is a jet of flowing dye

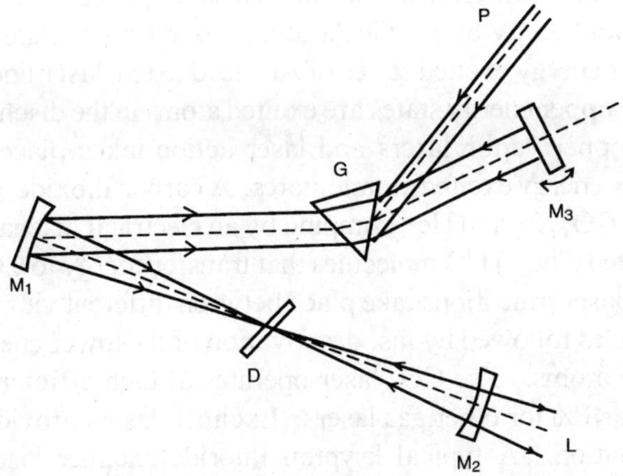


**Figure 3.10** A CW dye laser utilizing a folded three-mirror optical cavity. The active medium is a flowing dye jet (D) tilted at the Brewster angle to the axis of the optical caavity formed by mirrors ($M_2$ and $M_3$). The glass prism, having Brewster window faces (G), serves as a coarse tuning element for the pumping light (P) provided by a CS laser. Fine tuning over about 100 nm is accomplished by a totaing mirror ($M_3$). The spectral linewidth of a dye laser using DCM dye is 0.05-0.3 nm, depending on the tuning element. Adapted from Morstyne, G. and Kaye, A.H., *Phototherapy of Cancer*, Harwood Academic Publishers, Yverdon-Les-Baines, Switzerland, 1990. With permission.

(D) tilted at the Brewster angle to the axis of the optical cavity formed by a highly reflecting mirror $M_1$ and a transmitting mirror $M_2$ or the laser beam (L). The external pumping beam (P) enters the optical cavity through a glass prism (G) with the faces inclined at the Brewster angle. Rotating mirror $M_3$ provides for wavelength fine tuning. The available laser dyes cover a broad wavelength range from about 375-950 nm. Red-light emitting dyes suitable for photodyamic therapy include rhodamine 6G (570-665 nm) and DCM dye (620-740 nm).

Gas lasers are four-level systems that are pumped by an electric discharge. The filling gas in the He-Ne laser is a mixture of low-pressure He plus about 10% Ne. Collisions between He atoms and high-energy electrons generated by the discharge excite metastable He atoms (He*) lying about 20 eV above the ground state. The excitation energy of the He* atoms is transferred to Ne atoms by collisions. Allowed optical transitions from the resultant Ne* atoms lead to laser emissions at 544 nm, 633 nm, 1.15 μm, and 3.39 μm. The other wavelengths in the 633 nm He-Ne laser are eliminated by a multilayer dielectric film which acts as a reflection interference filter. The He-Ne laser is an inefficient device but it has excellent beam quality. Argon atoms are pumped directly by an electric discharge in the more powerful argon-ion laser. Multiple collisions of Ar atoms with electrons in the discharge generate energetic argon ions ($Ar^+$) lying about 35 eV above the Ar atom ground state. Allowed radiative transitions to lower energy excited states of $Ar^+$ lead to ten laser lines in the blue-green region. The upper energy states are excited atoms in the discharge-pumped gold vapor and copper vapor lasers and laser action takes place by radiative transitions to lower energy excited atomic states. A carbon dioxide laser contains a mixture of $CO_2$, CO, Xe, and He. Pumping by an electrical discharge generates vibrationally-excited ("hot") CO molecules that transfer energy to $CO_2$ molecules by collisions. The laser transitions take place between different vibrational modes of the $CO_2$ molecules followed by fast deactivation of the lower energy states *via* collisions with He atoms. The $CO_2$ laser operates at high efficiency, typically 15%, compared to <1% for other gas lasers. Excimer lasers provide high power levels of UV radiation. A typical krypton fluoride excimer laser contains a mixture of Kr, $F_2$, and a buffer gas such as He or Ne. An electrical discharge generates KrF excimers that decays to Kr and F atoms by emitting UV. Excimer surgical lasers are "cold lasers" that vaporize tissue without heat generation.

## 3.6.3 General Principles of Semiconductors

A semiconductor crystal is the active medium in a semiconductor laser. The operation of this type of device differs from gas-phase lasers such as He-Ne and solid state crystal lasers such Nd-YAG because laser action takes place at junctions between semiconductor layers. An elementary understanding of semiconductor physics is required to explain the action mechanism. A reader who is familiar with this subject may wish to go directly to Section 3.6.4.

### 3.6.3.1 Elements of Semiconductor Physics

The electrical properties of solids are explained by the theory of energy bands. When a large number of atoms are assembled to form a crystalline lattice, the discrete energy levels of the atomic electrons combine to form *energy bands* consisting of a continuum of closely-spaced electronic states. The electrical properties of the solid are determined by the distribution of electrons in the energy bands. According to the Pauli exclusion principle a maximum of two electrons can occupy an energy state. Except for the transition metals, the narrow energy bands originating from weak interactions between inner-shell electrons are not involved in the electrical and magnetic properties of solids. The highest energy band that is fully occupied by electrons is termed a *valence band* (VB). The next higher energy band is a *conduction band* (CB). The CB in a metal is partially filled by electrons up to a maximum energy referred to as the Fermi energy ($E_F$) (Figure 3.11). The probability is ½ that an energy level at $E_F$ is occupied by electrons. Metals are good conductors of electricity because an applied electric field can excite mobile *conduction electrons* located near $E_F$ to vacant higher-energy levels. When two or more conductors are in thermal equilibrium, the flow of electrons establishes the same value of $E_F$ in each constituent. This effect leads to a voltage between two dissimilar metals in contact. In an insulator, a fully-occupied VB and an empty CB are separated by a band of inaccessible energies of width $E_g$ referred to as a *forbidden gap* (Figure 3.11b). $E_F$ is located near the center of the forbidden gap. The electrical conductivity of an insulator is very low because an ordinary electric field does not provide sufficient energy to excite electrons from the VB into vacant energy states of the CB. An *intrinsic semiconductor* has a band structure similar to an insulator, except a small forbidden gap permits thermal agitation to excite electrons from the top of the VB to the bottom of the CB (Figure 3.11c). At ordinary temperatures some electrons reside in the conduction band of a semiconductor and vacant states or *holes* occupy the top of the valence band.

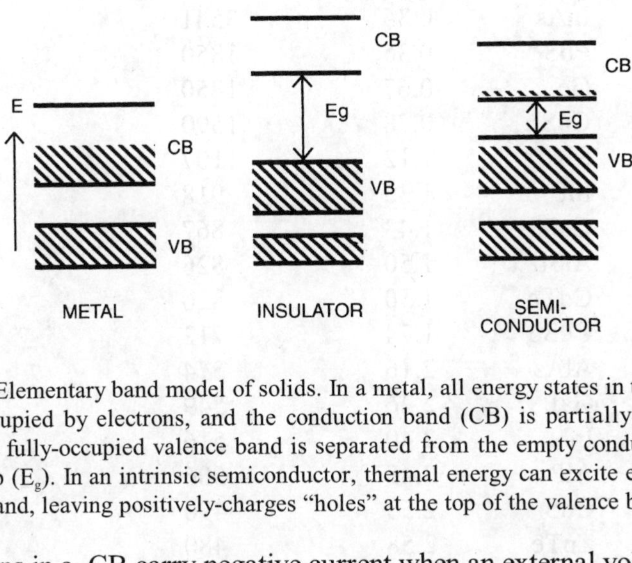


**Figure 3.11** Elementary band model of solids. In a metal, all energy states in the valence band (VB) are occupied by electrons, and the conduction band (CB) is partially occupied. In an insulator, the fully-occupied valence band is separated from the empty conduction band by a forbidden gap ($E_g$). In an intrinsic semiconductor, thermal energy can excite electrons into the conduction band, leaving positively-charges "holes" at the top of the valence band.

The electrons in a CB carry negative current when an external voltage or "bias" is applied. The holes in the VB act as positive charge carriers because the occupation of a hole by an electron is equivalent to migration of the hole in the opposite direction.[7] The electron and hole currents are additive because the motion of negative charges in one direction is equivalent to the motion of positive charges in the opposite direction.

The fabrication of an intrinsic semiconductor requires very pure crystals which were not available until the 1950s. The crystalline lattice of an intrinsic semiconductor has tetrahedral symmetry in which each atom makes covalent bonds with its four nearest neighbors. Intrinsic semiconductors can be fabricated from elements belonging to Group-IV of the periodic table (Si, Ge), from 1:1 alloys between elements of Group-III (Al, Ga, In) and Group-V (N, P, As, Sb), and from 1:1 alloys between elements of Group-II (Zn, Cd) and Group-VI (O, S, Se, Te). Values of $E_g$ and the equivalent band gap wavelengths for intrinsic semiconductors are given in Table 3.3. The electrical conductivity of an intrinsic semiconductor is enhanced by "doping" with impurities that introduce additional

**Table 3.3 Some values of semiconductor band gap**

| Material | Band Gap (eV) | Band Gap Wavelength (nm) |
|---|---|---|
| InSb | 0.18 | 6885 |
| PbSe | 0.27 | 4590 |
| PbTe | 0.30 | 4131 |
| InAs | 0.35 | 3541 |
| PbS | 0.35 | 1850 |
| Ge | 0.67 | 1850 |
| GaSb | 0.78 | 1590 |
| Si | 1.12 | 1107 |
| InP | 1.35 | 918 |
| GaAs | 1.43 | 867 |
| AlSb | 1.50 | 826 |
| CdTe | 1.50 | 826 |
| CdSe | 1.74 | 712 |
| AlAs | 2.16 | 574 |
| GaP | 2.26 | 549 |
| InN | 2.40 | 516 |
| AlP | 2.45 | 506 |
| CdS | 2.53 | 490 |
| ZnTe | 2.58 | 480 |
| SiC | 3.00 | 413 |
| GaN | 3.40 | 365 |
| ZnS(β) | 3.60 | 344 |
| ZnS(α) | 3.80 | 326 |

charge carriers. A Group-V element acts as an electron donor in a Group-IV base material leading to an *n-type* (for negative) semiconductor. Four of the five valance electrons of the donor atom are used for covalent bonding in the lattice and the extra electron remains localized near the positively-charged donor ion ($D^+$) where it may be excited by thermal agitation into the CB. A Group-III element acts an acceptor in a Group-IV-Group-IV base material leading to a *p-type* (for positive) semiconductor. The fourth electron required for covalent bonding of the acceptor atom is abstracted from the base material assisted by thermal excitation, leaving a hole in the CB. An acceptor can be viewed as a hole localized in the vicinity of a negatively- charged impurity ion ($A^-$) (Figure 3.12).

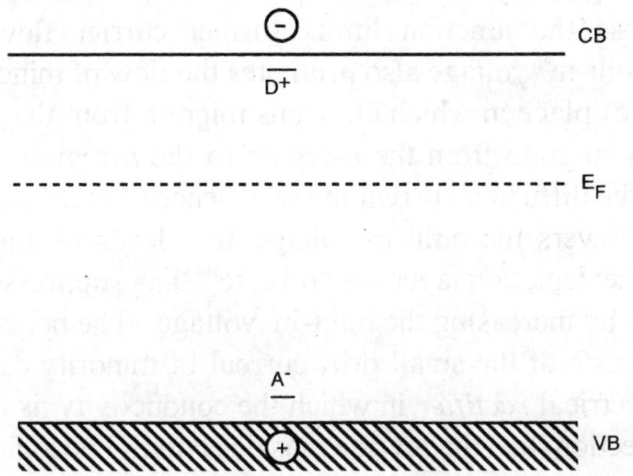


**Figure 3.12** Donor and acceptor levels in an extrinsic semiconductor. Ionized donors ($D^+$) contribute electrons to the conduction band; ionized acceptors ($A^-$) contribute holes to the valence band. The location of the equilibrium Fermi level ($E_F$) depends on the band structure of the semiconductor and its impurity concentrations. In a heavily doped n-type semiconductor, $E_F$ lies near $D^+$, and in a heavily doped p-type semiconductor, $E_F$ lies near $A^-$.

Group VI elements are donors and Group II elements are acceptors in a Group III-Group V semiconductor. Equivalent amounts of donors and acceptors can be added to the same base material leading to a *compensated semiconductor* having low electrical conductivity. The location of the $E_F$ in an extrinsic semiconductor depends on the relative concentrations of donors and acceptors. In a heavily doped n-type material $E_F$ is located near the top of the forbidden gap near the $D^+$ level and in a heavily-doped p-type material $E_F$ is located near the bottom of the forbidden gap near the $A^-$ level. The *minority carriers* make only a small contribution to the electrical conductivity. Specifically, *n*-type semiconductors have a low concentration of holes in the valence band and *p*-type semiconductors have a low concentration of electrons in the valence band. However, minority carriers are essential for the operation of semiconductor junctions.

### 3.6.3.2 Semiconductor Devices

Many useful devices can be constructed from junctions between a semiconductor and a metal and between different semiconductors. A *p-n junction* is fabricated from the same base material in which one side of the junction is n-type and the other side is p-type. The fermi level must equilibrate at the same energy after the junction is formed. To accomplish this, electrons flow "downhill" (like ball bearings) from the n-region to the p-region and holes flow "uphill" (like bubbles) from the *p*-region to the *n*-region. This *diffusion current* of the majority carriers leaves a narrow *depletion layer* containing positively-charged donors in the n-region and negatively-charged acceptors in the p-region. A "built-in" voltage established across the junction limits further carrier flow (Figure 3.13a). However, the "built-in" voltage also promotes the flow of minority carriers and a *drift current* takes place in which electrons migrate from the *p*-region to the *n*-region and holes migrate from the *n*-region to the *p*-region. The drift current exactly cancels the diffusion current in the absence of external bias and light. A "forward" bias lowers the built-in voltage and leads to injection of majority carriers across the depletion layer. A "reverse" bias suppresses the diffusion of majority carriers by increasing the built-in voltage. The net current flow under reverse bias consists of the small drift current of minority carriers. Thus, a p junction is an electrical *rectifier* in which the conductivity is much higher in the forward bias direction than the reverse bias direction. The logic elements in early digital computers were semiconductor diodes, in which the on-off status of a given element is determined by the direction of the applied bias.

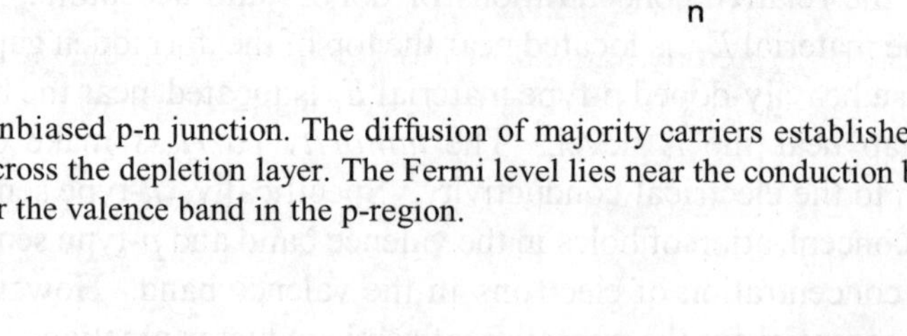


**Figure 3.13** Unbiased p-n junction. The diffusion of majority carriers establishes a "built-in" voltage ($V_{bi}$) across the depletion layer. The Fermi level lies near the conduction band in the n-region and near the valence band in the p-region.

The recombination of electrons and holes in a *p-n* junction semiconductor can lead to emission of light (Figure 3.14). A *light emitting diode* (LED) is a *p-n* junction operated in forward bias. The flow of electrons from the *n*-region to the *p*-region leads to excess electrons in the CB of the p-region, and the flow of holes leads to excess holes in the VB of the n-region. Steady-conditions are restored by recombination of electrons and holes resulting in light emission with the photon energy corresponding to the band gap (Table 3.3). Direct interband transitions are not feasible for some semiconductors owing to the electronic structure, for example, Si, Ge, GaP, and the III-V alloys of Al in Table 3.3. Electron-hole recombination in these materials requires energy transfer with the lattice. They are referred to as "indirect" band gap semiconductors.

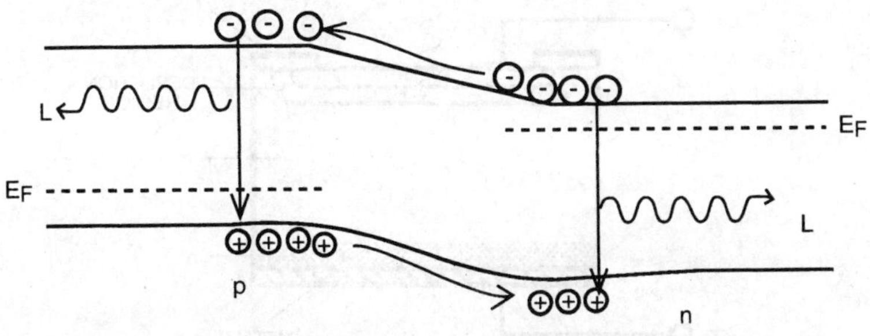


**Figure 3.14** Energy levels in a p-n junction light emitting diode (LED). Forward bias promotes the flow of majority carriers across the junction. Recombination of conduction band electrons and valence band holes generates photons corresponding to the energy of the optical band gap. The equilibrium Fermi level splits into an electron Fermi level and a hole Fermi level.

Semiconductor diodes can also respond to external light. The basic design of a photodiode consists of a thin semiconductor film located on an insulating substrate with attached electrodes. In a *photoconductive detector*, light absorption generates electron-hole pairs across the band gap which are converted to a photocurrent by a bias voltage. A *p-n* junction photodiode can be operated also as a *photovoltaic detector* without external bias. Figure 3.15 illustrates a modern type of silicon *p-n* junction photodiode. The upper p-type region ($p^+$) and the lower *n*-type region ($n^-$) are heavily doped for high conductivity. Electron-hole pairs are generated by light absorption in the $p^+$ layer and the central and n-type. The "built-in" voltage accelerates electrons further into the *n*- region and holes into the $p^+$ region, leading to a charge separation that can be measured with a sensitive voltmeter. A photocurrent is generated under reverse bias. A *photodiode array* consists of a chain of photodiodes fabricated as an integrated circuit. The optical energy absorbed by each subunit generates an accumulated charge

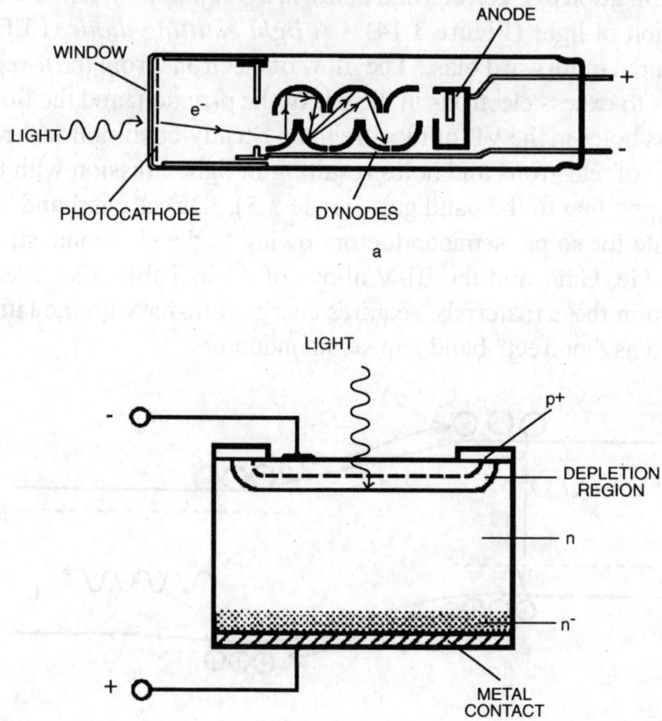


a

b

**Figure 3.15** Typical construction of photodetectors. (a) An end-window vacuum photomultiplier. Electrons emitted by the illuminated photocathode are accelerated by a series of dynodes, each of which amplifies the photocurrent. (b) A silicon p-n junction photodiode. Light absorption in the $p^+$, depletion, and n regions generates electrons and holes that are accelerated by the built-in voltage into the $n^-$ and $p^+$ regions, respectively. In unbiased operation, the charge separation is measured with a sensitive voltmeter. Application of reverse bias generates a photocurrent. Adapted from Oriel Corporation catalog *Light Soursecs, Monochromators & Spectrographs, Detectors & Detector Systems, Fiber Optics,* Vol. II, 1994, Stratford, CT. With permission.

that is read sequentially and converted to a voltage signal by the measurement system. The use of a photodiode array in a spectrograph instead of a photographic film makes it possible to record spectral data in a fraction of a second. The properties of different photodetectors are summarized in Table 3.4.

### 3.6.4 Semiconductor Lasers

A semiconductor laser is essentially an LED operated with population inversion. This is accomplished by employing very heavy doping such that $E_F$

**Table 3.4 Properties of photodetectors**

| Type | Spectral Range (μm) | Time Response (s) | Typical Gain[a] | D*[b] |
|---|---|---|---|---|
| Photomultiplier | 0.16-1.2 | $10^{-9}$ | $10^5$ A/W | $10^{15}$ |
| Thermopile | 0.2-50 | 1 | 10 V/W | $10^9$ |
| Pyroelectric | 0.2-15 | $10^{-3}$ | $10^2$ V/W | $10^8$ |
| Silicon PV | 0.6-1.1 | $10^{-8}$ | 0.5 A/W | $10^{14}$ |
| Germanium PV | 0.9-1.9 | $10^{-9}$ | $10^3$ A/W | $10^{11}$ |
| Lead Sulfide PV | 1.1-3.1 | $10^{-3}$ | $10^3$ V/W | $10^{11}$ |
| Lead Selenide PV | 1.5-4.9 | $10^{-5}$ | $10^3$ V/W | $10^9$ |
| Indium Antimonide PV | 1.8-3.8 | $10^{-6}$ | 0.7 A/W | $10^{10}$ |
| Indium Arsenide PV | 1.5-3.5 | $10^{-6}$ | 1 A/W | $10^{16}$ |
| IR Viewing Card | 0.75-1.5 | - | c | - |

[a]The typical output in volts or amperes per watt of radiant energy; [b]D* is the standard figure of merit for solid state photodectors and is based on the ratio of the output signal to the random noise per unit bandwidth, scaled to the area of the detector. The units of D* are $cm\sqrt{(Hz/W)}$; [C] Approximately 0.5 mW/cm$^2$ is required for a weak emission.

lies within the CB in the *n*-region and the VB lies in in the *p*-region (Figure 3.16a). Population inversion is achieved under forward bias in a narrow zone referred to as the "active region" leading to laser emission (Figure 3.16b). A basic "homojunction" GaAs semiconductor diode laser is depicted in Figure 3.6.

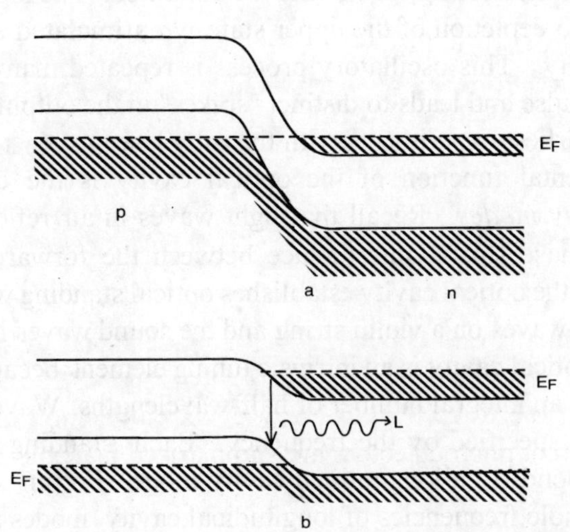


**Figure 3.16** Energy levels in a semiconductor diode laser. (a) Heavy doping shifts the equilibrium Fermi level in to the conduction band in the n-region, and into the valence band in the p-region. (b) Forward bias leads to population inversion in a thin "active region" and emission of laser radiation (L).

Laser radiation is emitted in the 1 μm - 3 μm active region located between *n*-type and *p*-type layers (A). The typical emissions of commercial GaAs laser range from 700-850 nm Semiconductor diode lasers with polished outer surfaces do not require external mirrors because the high refractive index of semiconductor materials (for example, $n_r$ = 3.6 for GaAs) leads to a high specular reflection coefficient. The threshold current density and optical properties of semiconductor lasers are improved in "heterojunction" devices utilizing stacked layers of different semiconductors. The advantages of semiconductor diode lasers include high power output for small size (up to hundreds of watts), high efficiency, very long useful lifetime, and limited wavelength tuning by varying the operating temperature.

### 3.6.5 Properties of the Laser Optical Cavity

An optical cavity confines laser radiation to the space between the end mirrors to permit the buildup of high optical power levels. Each round trip of light waves induces two stages of power amplification by the active medium. The *loop gain* of an optical cavity (GL) is defined as the fractional change of power in the beam after each round trip allowing for losses. Shortly after a CW laser is turned on, the radiation in the optical cavity attains an average power determined by the pumping level corresponding to GL = 1. Optical pumping by a pulsed light source leads to "overshooting" of the GL = 1 condition. The high gain cannot be sustained owing to depletion of the upper state *via* stimulated emission and GL falls to below unity. This oscillatory process is repeated many times during a single pumping pulse and leads to distinct "spikes" in the output pulse, that were observed in the earliest experiments with the ruby laser.

A fundamental function of the optical cavity is the establishment of *longitudinal cavity modes*. Recall that light waves in air reflected at a mirror undergo a λ/2 phase shift. Interference between the forward and backward running waves in the optical cavity establishes optical standing waves, analogous to the mechanical waves on a violin string and the sound waves in a closed-ended organ pipe. An optical cavity is an intrinsic tuning element because the sustained waves must have an integral number of half-wavelengths. Wave motion in laser science is usually specified by the frequency. Each standing wave of a given frequency corresponds to a longitudinal cavity mode. For an optical cavity of length *L*, the possible frequencies of longitudinal cavity modes are given by

$$f_p = p(c/2n_r L) \tag{3.7}$$

where $p$ is an integer referred to as the *mode number* and $n_r$ is the relative refractive index of the active medium. The mode number is very high owing to the short wavelength of light compared to the physical length of a laser cavity. The frequency separation between cavity modes $\Delta f$ is an important determinant of laser action. For a cavity with a path length $L_m$ in the active medium and an air path length $L_a$, the mode spacing is given by

$$\Delta f = c/2(n_r L_m + L_a) \tag{3.8}$$

**Example 3.2** Longitudinal cavity modes in a laser

A He-Ne laser emitting at 633 nm is 50 cm long. Calculate: (a) the mode number; (b) the spacing between longitudinal cavity modes.

(a) The frequency of the He-Ne laser is: $f$ = 3.00 x $10^8$ (m/s)/633 x$10^{-9}$ m = 3.74 x $10^{14}$ $s^{-1}$ or 474 THz. From Equation (3.7): p = 3.74 x 1014 $s^{-1}$/[3.00 x $10^8$ (m/s)/(2 x 0.5 m)] = 1.5 x $10^6$.

(b) From Equation (3.8), $\Delta f$ = 3.00 x $10^8$ (m/s)/2 x 0.5 m = 3 x 108 $s^{-1}$ or 300 MHz.

Although a laser cavity can accommodate millions of longitudinal modes, the sustained modes must lie within the spectral bandwidth of the laser emission line. The total bandwidth of a multimode He-Ne laser line is about 1500 MHz (0.002 nm). Therefore, the actual number of modes present during laser operation is about 1500/300 $\approx$ 5 for the parameters of Example 3.2. Figure 3.17 depicts the spacing of longitudinal cavity modes and the spectral power output at GL = 1.0, where the dashed line is the gain of the broadened laser emission line. The output of this hypothetical multimode He-Ne laser consists of 4 very narrow lines, typically of 1 MHz bandwidth.

The finite diameter of a laser optical cavity leads to an intensity pattern across the beam referred to as *transverse electromagnetic modes* (TEM). The TEM resemble the tranverse modes of an optical fiber (Section 2.6.3). Idealized irradiance profiles for an optical cavity formed by two plane mirrors are depicted for several TEM modes in Figure 3.18. The first subscript is the number of intensity minima in the horizontal direction and the second subscript is the number of intensity minima in the vertical direction. The $TEM_{00}$ mode has a gaussian (bell-shaped) intensity distribution. This mode has the smallest beam divergence. A pinhole in the optical path can force an optical cavity to oscillate in the $TEM_{00}$ mode. Clinical lasers are usually operated multimode to achieve higher power

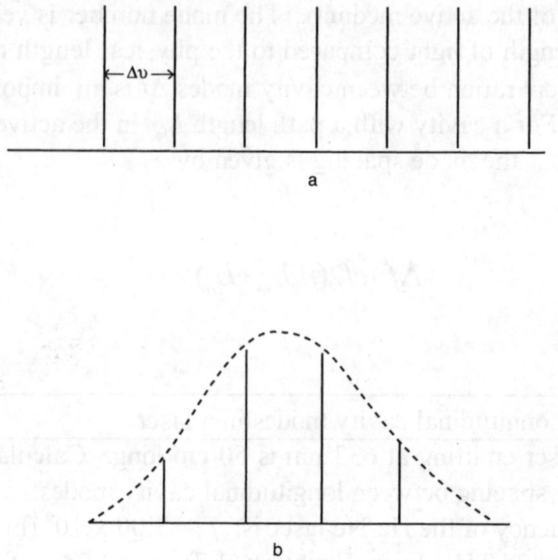


**Figure 3.17** Longitudinal modes in an optical cavity. (a) The frequency spectrum of axial modes is established by the length of the optical cavity. (b) The power output is modulated by the bandwidth of laser emission line, as indicated by the dashed line.

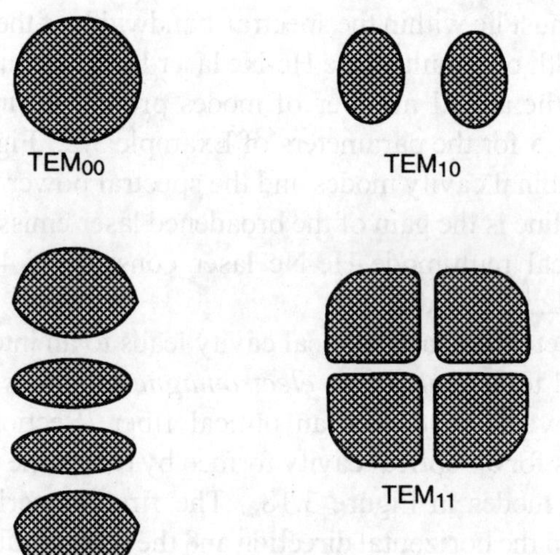


**Figure 3.18** Projection of some low-order transverse modes of a laser cavity. The $TEM_{00}$ mode has a gaussian irradiance profile.

levels. The irradiance profile of a gaussian laser beam is depicted in Figure 3.19a. The radial irradiance distribution is given by:

$$E(r)=(2P/\pi w^2)\exp(-2r^2/w^2) \tag{3.9}$$

where $P$ is the total power in the beam and $w$ is a width parameter referred to as the "1/e² radius". Direct substitution shows that the irradiance falls to 0.135 of the maximum at $r = w$. A laser beam with a constant irradiance in a circular spot is referred to as a "rectangular" or "top hat" beam. A top hat beam which has the same total power and peak power as a gaussian beam has a diameter of $\sqrt{2}w$.

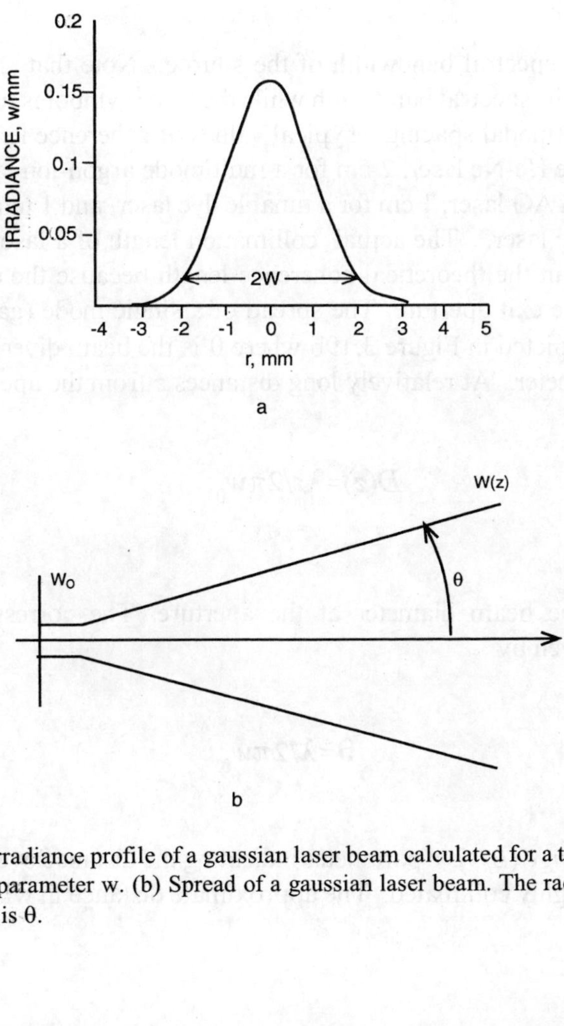


**Figure 3.19** (a) Irradiance profile of a gaussian laser beam calculated for a total power of 1 watt and 1 mm width parameter w. (b) Spread of a gaussian laser beam. The radius is w(z) and the divergence angle is θ.

### 3.6.6 Properties of Laser Beams

The coherence of laser light results from the narrow bandwidth and a small number of active axial modes. An operational definition of coherence is based on the ability of light to form interference fringes. In the two-slit experiment of Thomas Young (Figure 2.12), fringes are not observed if the distance between the single slit and the double slit is such that the wave trains do not overlap. The largest distance at which fringes are formed is a measure of the *longitudinal coherence length* ($L_c$). An approximate expression for $L_c$ is

$$L_c = c/\Delta f \tag{3.10}$$

where $\Delta f$ is the spectral bandwidth of the source. Note that $\Delta f$ in Equation (3.10) refers to the spectral bandwidth while the same symbol is used in Equation (3.8) for the intermodal spacing. Typical values of coherence length are 100 m for a single mode He-Ne laser, 2 cm for a multimode argon-ion laser, 1 cm for a multimode Nd:YAG laser, 1 cm for a tunable dye laser, and 1 mm for a GaAlAs multimode diode laser. The actual collimation length of a laser beam may be much shorter than the theoretical coherence length because the emitted light is diffracted by the exit aperture. The spread of a single mode (gaussian profile) laser beam is depicted in Figure 3.19b where θ is the beam divergence and $D(z)$ is the beam diameter. At relatively long distances $z$ from the aperture

$$D(z) = \lambda z / 2\pi w_0 \tag{3.11}$$

where $2w_0$ is the beam diameter at the aperture. The corresponding beam divergence is given by

$$\theta = \lambda / 2\pi w_0 \tag{3.12}$$

The *Rayleigh range* ($L_R$) is a parameter indicating the distance over which a laser beam remains highly collimated. The approximate distance in which a gaussian beam widens by $\sqrt{2}$ is given by

$$L_R = 4\pi w_0^2/\lambda \tag{3.13}$$

**Example 3.3** Properties of a gaussian laser beam

A He-Ne laser operating in the $TEM_{00}$ mode has a spectral bandwidth of 1 MHz and a beam diameter of 1.0 mm. Calculate: (a) the coherence length; (b) the beam divergence; (c) the Rayleigh range; (d) the beam diameter at 100 m from the laser.

(a) The coherence length from Equation (3.10) is: $L_c = 3.00 \times 10^8$ (m/s)/1 x $10^6$ $s^{-1}$ = 300 m.

(b) The beam divergence from Equation (3.12) is: $\theta = 633 \times 10^{-9}/(0.5 \times 10^{-3}\pi) = 3.03 \times 10^{-4}$ rad or 0.40 mrad.

(c) The Rayleigh range from Equation (3.13) is: $L_R = (0.5 \times 10^{-3})^2\, \pi/(633 \times 10^{-9}) = 1.24$ m. This distance is much shorter than the coherence length because of diffraction effects at the small aperture.

(d) The beam diameter at 100 m according to Equation (3.11) is: $D = 633 \times 10^{-9} \times 100/(0.5 \times 10^{-3}\pi) = 3.03 \times 10^{-2}$ m or 40.3 mm.

The collimation length of a laser beam can be lengthened by a "beam expander" which is essentially a telescope in reverse. If the magnification of the telescope is $M$, the beam divergence is reduced by $M$ and $L_R$ is increased by $M^2$. If the laser beam of Example 3.3 is expanded by a 20-power telescope, $L_R$ increases to 500 m. As early as 1962 a laser beam directed at the moon through a 48- inch diameter telescope in reverse had an estimated diameter of 4 miles at the surface of the moon. In many medical applications a laser beam is focused with a lens system to a small, high-intensity spot. The approximate diameter of a focused $TEM_{00}$ laser beam ($D_F$) is given by

$$D_F = \lambda F/D_0 \tag{3.14}$$

where $F$ is the focal length of the lens and $D_0$ is the laser beam diameter at the lens. For the He-Ne laser of Example 3.3, a 100 mm focal length lens would focus the 1 mm diameter beam to a spot diameter 63 μm. This calculation assumes a single mode laser beam and a perfect lens. The focused diameter of a multimode laser beam may be much larger.

### 3.6.7 Laser Operating Characteristics

The power output of a laser determines the rate at which radiant energy is delivered. A CW laser provides an approximately constant average output. Typical CW power levels range from a few milliwatts for a small He-Ne laser up to several hundred watts for surgical $CO_2$ laser. Some types of lasers can be operated in either CW or pulsed modes while other lasers require pulsing in order to achieve population inversion (Table 3.2). The designation "pulsed" is used for two different modes of operation. In a repetitively pulsed laser, the time interval between individual pulses is short compared to time constants of the induced effects. Such lasers are essentially CW for those applications. However, it cannot be assumed that CW and rapidly pulsed laser are always equivalent because a slow process may have fast intermediate steps that respond to laser pulsing. The *duty cycle* (*DC*) of a pulsed laser is the fraction of operating time occupied by light pulses: $DC = \delta t \times PRR$, where $\delta t$ is the duration of a single pulse and *PRR* is the pulse repetition rate. The *average power* ($P_{avg}$) must be distinguished from the *peak power* ($P_{max}$). If $\delta E$ is the energy in one pulse, $P_{avg} = \delta E \times PRR$ and $P_{max} = P_{ave}/DC$. The average power is relevant for slow processes relative to $\delta t$ such as tissue heating, in which case repetitively pulsed operation is equivalent to CW. The peak power may be relevant for fast photophysical and photochemical processes. Short-pulse operation of a laser is required for many applications. For example, *ablation* of tissue by a surgical laser, requires the delivery of high-energy and short-duration pulses.

---

**Example 3.4** Pulsed lasers

Consider a repetitive pulsed laser with an average power of 100 W, a pulse duration of 0.5 ms, and a pulse repetition rate of 500 Hz. Calculate the duty cycle, the peak power, and the energy per pulse.

The duty cycle is $0.5 \times 10^{-3}$ s x 500 $s^{-1}$ = 0.25; the peak power is 100 W/0.25 = 400 W; the average energy per pulse is 100 W/500 $s^{-1}$ = 0.2 J.

---

The basic pulsed mode of a laser is generated by a using a pulsed pumping source, for example, a gas-filled flash lamp for optical pumping and a high-voltage pulse for gaseous discharge pumping. Typical peak power levels are the order of kilowatts at pulse durations in the millisecond range. Each output pulse has a substructure consisting of many short spikes resulting from oscillations of the loop gain and a multiplicity of axial modes. This effect is illustrated for a flash lamp-pumped ruby laser in Figure 3.20a. Innovative techniques for generating shorter and higher power pulses were devised shortly after the invention of lasers. A *Q-switch* is a fast shutter that blocks the optical path during pulsed pumping until population inversion has been achieved.[8] Opening the Q-switch at this instant leads to a rapid depletion of the stored energy in the form of a single high-intensity light pulse. The first Q-switch was a rapidly rotating prism used as an end mirror. More efficient techniques are now employed. In a "passive" *saturable absorber* technique, the output mirror is blocked by the absorption of an organic dye solution. Population inversion in the active medium induced by an pulsed optical pumping continues to increase while the optical path is blocked by the dye absorption. Partial bleaching of dye

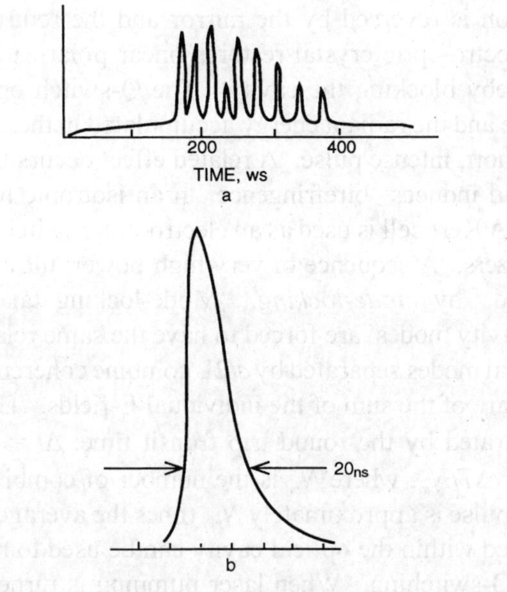


**Figure 3.20** Typical output characteristics of pulsed lasers. (a) The output pulse of a flash-lamp-pumped ruby laser has a sub-structure consisting of sharp spikes. (b) A flash-lamp pumped, Q-switched Nd-YAG laser provides a single "giant" pulse. Typical pulse widths range from 10-100 ns, depending on the flash lamp input energy. (c) The output of a mode-locked pulsed laser with a 60-cm optical cavity and 5-GHz linewidth.

at a sufficiently high power density initiates an intense Q-switched laser pulse in the nanosecond range at power levels of megawatts Figure 3.20b depicts a Q-switched pulse of about 20 ns duration as produced by a flash lamp-pumped Nd-YAG laser. An electro-optic crystal can be employed for "active" Q-switching. This technique is based on the *Pockels effect*, in which a change of refractive index in a crystal is induced by a strong external electric field. In one type of Pockels cell Q-switch, the electro-optic crystal is located in the laser cavity between a linear polarizer and a mirror. The application of a high voltage pulse to the crystal produces a temporary quarter-wave plate that converts the normally linearly polarized laser light into circularly polarized light. The direction of circular rotation is reversed by the mirror and the return passage of the light through the electro-optic crystal restores linear polarization in the transverse direction, thereby blocking the cavity. The Q-switch opens at the end of the electrical pulse and the radiant energy accumulated in the cavity is released in the form of very short, intense pulse. A related effect occurs in a *Kerr cell* in which an electric field induces birefringence in an isotropic medium such as liquid nitrobenzene. A Kerr cell is used as an electro-optic switch by locating it between crossed polarizers. A sequence of very high-power, ultra-short-duration pulses are generated by *mode-locking*. Mode-locking takes place when many longitudinal cavity modes are forced to have the same relative phase difference. The longitudinal modes separated by $c/2$L combine coherently and the total power equals the square of the sum of the individual E-fields. The result is a sequence of pulses separated by the round trip transit time: $\Delta t = 2L/c$ and with pulse duration: $\delta t = \Delta T/N_m$, where $N_m$ is the number of combining modes. The peak power in each pulse is approximately $N_m$ times the average power.[9] A saturable absorber located within the optical cavity can be used to mode-lock a laser with simultaneous Q-switching. When laser pumping is turned on, a strong noise spike is selectively amplified until it "burns" through the absorption band of the dye and establishes single mode operation. The output consists of a sequence of very short, high power mode-locked pulses with the duration of the Q-switched pulse. Active mode-locking is achieved by locating an amplitude or frequency modulating device, such as an electro-optic crystal, within the cavity activated at the round trip frequency $2L/c$. Figure 3.20c depicts a mode-locked pulse sequence from a Q-switched laser with a 60 cm long optical cavity and a 5 gHz line width. Approximately 20 pulses of 200 ps width are generated separated by 4 ns. The gain is higher near the pulse center where more cavity modes can combine. Mode-locked dye lasers can emit pulses as short as 20 fs.

**Example 3.5** Mode-locked laser

A 1.0 m mode-locked Nd:YAG laser emits at 1064 nm with 1.0 x $10^{11}$ Hz line width. Calculate the pulse width and pulse separation.

The bandwidth equals is given by: $\Delta f = c/2L$ = 3 x$10^8$/(2 x 1.0) = 1.5 x $10^8$ Hz. The number of modes is given by: $N_m$ = 1.0 x $10^{11}$/1.5 x $10^8$ = 667 modes. Therefore, the separation of mode locked pulses is given by: $\Delta t = 2L/c$ = 2 x 1.0/3 x $10^8$ = 6.67 x $10^{-9}$ s or 6.7 ns; the pulse width is given by $\delta t$ = 6.67 x $10^{-9}$/667 = 10.0 x $10^{-12}$ s or 10 ps.

### 3.6.8 Wavelength Shifting

The emission wavelengths of a laser depend on the active medium, physical construction, and operating conditions. Not infrequently the wavelengths of standard lasers are not optimal for the clinical applications. Unwanted wavelengths can be minimized by blocking filters and the spectral response of the laser mirrors. The spectral output of a semiconductor laser can be varied over a small range by controlling the operating temperature. Some lasers can be tuned over a limited range by locating a wavelength selection element in the optical cavity such as a prism, diffraction grating, or a variable thickness interference filter. The output of a tunable dye laser can be varied over a broad range by changing the dye. *Frequency doubling* (the same as wavelength halving) is achieved by passing laser light through an electro-optic crystal having a structure without a center of symmetry, for example, potassium dihydrogen phosphate (KDP), potassium trihydrogen phosphate (KTP), and ammonium dideuterium phosphate (AD*P). For example in a "KTP-dye laser" used for photodynamic therapy, the green light output of a frequency-doubled Nd-YAG laser is used to pump a red-light emitting dye laser. Frequency doubling originates from interactions of the higher harmonics of the lattice vibrations of the crystal with the E-field of the light. The strength of the interaction depends on the square of the E-field and, therefore, it is significant only at high optical power levels. From the quantum viewpoint, frequency multiplication can be viewed as the combination of input photons to give a higher energy output photons. Thus, two 1064 nm photons emitted by an Nd-YAG laser having a quantum energy of 1.165 eV can be "doubled" to give a green light photon at 532 nm (3.33 eV).

### 3.6.9 Lasers Safety

Lasers in the U.S. are classified by the American National Standards Institute and under Federal regulations issued by the United States Center for

Devices and Radiological Health (CDRH) . Similar classifications exist in many industrialized countries. Retinal hazard is the primary basis of laser risk analysis. According to present ANSI standards, *Class 1* or "exempt" lasers are "totally safe" and emit "levels at which biological damage has not been established" according to CDRH. This category is limited to very low power semiconductor diode lasers and certain enclosed lasers incorporated into consumer products. *Class 2* lasers are "low-power" devices. The CDRH classification applies to visible lasers (400-700 nm) that have insufficient output power to injure a person accidentally, but which cause "eye damage from chronic exposure". This group includes He-Ne lasers (< 1 mW) used as pointers. *Class 3* applies to "medium-power" lasers. The CDRH classification includes lasers that emit "at levels at which biological damage to human tissue is possible from acute direct exposure". The classification of *Class 3A* applies for lasers emitting visible light at less than 5mW for CW operation. *Class 3B* applies to laser emitting between 200 nm and 1 μm with output power limited to 500 mW for CW operation. This group of lasers can cause retinal injury on direct viewing of the laser beam or from specular reflections. *Class 4* lasers operate from 200 nm to 1 μm and the output power exceeds 500 mW. This group includes high-power lasers that can produce ocular damage from specularly reflected and diffusely reflected light and also skin damage from the direct beam and reflected light. *Class 4* includes most clinical and industrial lasers. Laser classification is a complex subject requiring specialized instrumentation. The necessary documentation for a given laser should be provided by the supplier for any clinical laser marketed in the U.S. The hazards associated with surgical lasers are considered in Section 9.3.

## NOTES

1. A black body absorbs all incident radiation. A large cavity with a small aperture is an approximate black body because external radiation cannot enter and radiation generated inside cannot escape.

2. A plasma is a unique state of matter in which electrons and ions interact collectively though long-range electromagnetic forces. Nearly all visible matter in the universe exists in the plasma state, including the stars, auroras, and lightning.

3. Equation (3.3) is derived in thermal physics by considering the exchange of energy between a small quantized system in thermal equilibrium with a large heat reservoir at absolute temperature *T*.

4. Temporary population inversion can be achieved in magnetic crystals such as LiF by applying a strong magnetic field at low temperature and then rapidly reversing the direction of the field. The original lower-energy state having the higher population becomes the higher-energy state for a short time until the magnetic moments of the nuclei attain equilibrium with the crystal lattice. This situation is referred to as "negative temperature". Inserting a negative temperature in Equation (3.3) leads to $N_2 > N_1$. The negative temperature decays when thermal equilibrium is restored.

5. The Einstein *A*-coefficient is calculated from the experimental fluorescence lifetime. The Einstein *B*-coefficient is calculated from the integrated area under the absorption spectrum of the transition. The *B*- coefficient is equivalent to a key property of an optical transition, referred to as the "transition dipole strength", that measures the separation of electric charge induced by the transition.

6. The conditions required for population inversion in a three-level laser are easily derived for CW operation. The power balance between levels C and G can be expressed as

$$BIN_g = BIN_c + N_c/\tau_{cf} \tag{3.15}$$

where $I$ is intensity of the pumping light, $B$ is the Einstein $B$ -oefficient for absorption by level G and stimulated emission by level C, and $\tau_{cf}$ is the spontaneous decay lifetime of level C. The power balance for the intermediate level F is given by

$$N_c/\tau_{cf} = N_f/\tau_{fg} \tag{3.16}$$

where $\tau_{fg}$ is the decay lifetime of level F. The overpopulation of level F is specified by the parameter $(N_f - N_g)/N$, where $N = N_g + N_f + N_c$. A little algebra leads to

$$(N_f - N_g)/N = \frac{BI\tau_{fg}(1-\alpha)-1}{BI\tau_{fg}(1-\alpha)+1} \tag{3.17}$$

where a $= t_{cf}/t_{fg}$. The conditions leading to $(N_f - N_g)/N > 0$ are: a $< 1$ and $B\,\boldsymbol{I} > 1/t_{fg}(1$ - a). Thus, laser action requires a fast, non-radiative decay of level C compared to the decay lifetime of level F and a sufficiently high optical pumping rate to maintain the overpopulation of level F.

7. Consider a parking lot with a few empty spaces. The motion of a car from a filled space into an empty space is equivalent to the motion of the space in the opposite direction.

8. The term "Q-switch" is taken from the "quality factor" of cavity oscillations, defined as the peak frequency divided by the bandwidth.

9. The temporal reinforcement of modes in a mode-locked laser resembles the frequency reinforcement of waves by a diffraction grating. In a diffraction grating, coherent waves originating from each grating element have the same frequency and uniformly-spaced phase differences. In laser mode locking, the longitudinal cavity modes have the same phase and uniformly-spaced frequency separations.

# CHAPTER 4

# QUANTUM DESCRIPTION OF LIGHT INTERACTIONS WITH MATTER

The molecular stage of every phototherapy is initiated by a photochemical reaction. Physiological endpoints are attained in successive stages, starting with very fast photophysical events, followed by the primary photochemical reactions, and continuing with slower events at the physiological level (Figure 4.1). The reacting molecules are either endogenous chromophores or are introduced into the system as an exogenous drug. The first step at ordinary light intensities is the absorption of one light photon by a chromophoric molecule or molecular unit. The primary absorption act is very fast, typically requiring only about $10^{-15}$ s. A series of fast *relaxation* events follow, leading either to release of the excess energy as heat and light or photochemical products. Some complex reactions utilize light in more than one stage. Photosynthesis in green plants utilizes two coupled photochemical pathways, one of which leads to reduction of carbon dioxide and the other oxidizes water to oxygen. A photochemical reaction may be partially or entirely reversible. In the "blue light" phototherapy of neonatal jaundice, simultaneous light absorption by serum bilirubin and a photochemical product of bilirubin establishes a *photoequilibrium* between the two species that limits the yield of the useful product. The intrinsic efficiency of a photochemical process is measured by a *quantum yield* (or quantum efficiency) which is defined as the number of atoms or molecules of a given species produced or lost for each absorbed photon. The quantum yield can exceed unity in a *chain reaction* in which one of the photochemical intermediates reacts with a starting molecule to form a product while regenerating the same intermediate. This process can continue until the intermediate is removed by a different reaction. The presence of oxygen may be a key determinant of the reaction pathway. In conventional organic photochemistry, reaction mechanisms are described by the structural relationships between the initial reactants and the final products, for example, photofragmentation, photoaddition, photoisomerization, etc. This systematic approach is less useful for biological molecules owing to their complex structures and the variability of the environments in which the reactions take place. Furthermore, final photochemical products are not necessarily relevant to biological endpoints. This chapter emphasizes the photochemical mechanisms that are involved in important phototherapies and related light-activated processes of biomedical interest.

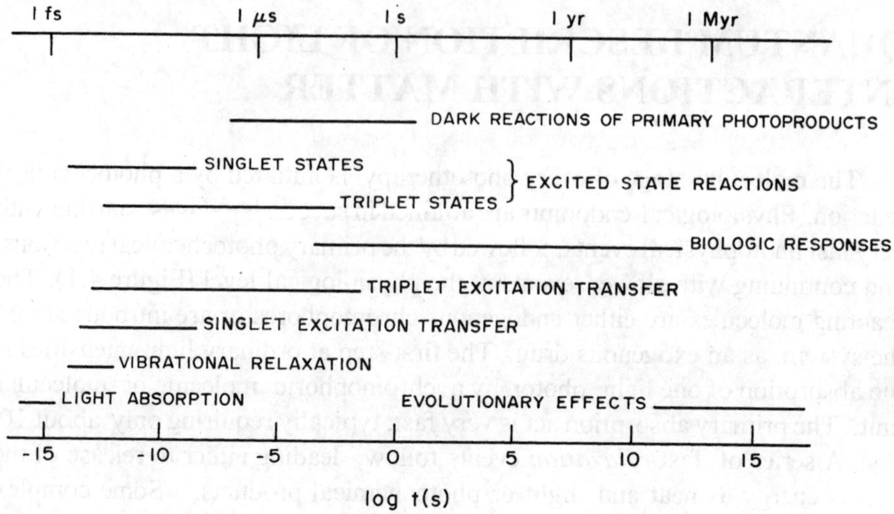


**Figure 4.1** The time scale of physical, chemical and physiological responses to photochemical radiations. The upper scale locates the time intervals in ordinary units. The lower scale is the logarithm of the tiome interval in seconds. From Grossweiner, L.I. and Smith, K.C., *The Science of Photobiology*, 2nd ed., Plenum Press, New York, 1989. With permission.

## 4.1 QUANTUM PHOTOPHYSICS

The photon picture of light is the basis of quantum photophysics. Einstein invoked the photon concept in 1905 to explain puzzling observations about the *photoelectric effect* - the process by which UV radiation ejects electrons from metals. According to classical physics, stronger illumination should generate more energetic photoelectrons. However, experiments show that light intensity has no effect on the energy of photoelectrons, although more photoelectrons are emitted per second at higher intensities. Furthermore, photoelectrons are not emitted for wavelengths exceeding a specific threshold, for example, approximately 375 nm for polished aluminum. Einstein explained the photoelectric effect by postulating that each photon transfers all of its energy to one electron inside the metal. The excess energy of the photon above the threshold appears as the kinetic energy of the photoelectron. He further assumed that all electrons within a given metal have the same binding energy, which is referred to as the *work function* . This energy balance is described by the famous Einstein photoelectric equation

$$K = hf - W \tag{4.1}$$

where $K$ is the kinetic energy of the ejected electron and $W$ is the work function of the metal. The confirmation of Equation (4.1) for many different metals was the first experimental proof of the photon concept.

---

**Example 4.1** Photoelectric effect

The photoelectric work function of polished silver is 4.73 eV. Calculate the longest UV wavelength that can eject photoelectrons.

The photon energy is given by Equation (2.2 ): $hf$ = 4.73 eV x 1.6 x $10^{-19}$ (J/eV) = 7.57 x $10^{-19}$ J. The equivalent light frequency is: 7.57 x $10^{-19}$ J/6.62 x $10^{-34}$ J-s = 1.14 x $10^{15}$ $s^{-1}$; the longest effective wavelength is 4.00 x $10^{8}$ (m/s)/1.14 x $10^{15}$ $s^{-1}$ = 2.63 x $10^{-7}$ m or 263 nm.

---

*Photoionization* is the atomic equivalent of the photoelectric effect. A similar energy balance applies for interactions of UV photons with electrons in atoms and molecules in which the *ionization potential* ($I$) of the ejected electron is the equivalent of the work function. Typical values of $I$ for valence electrons are the order of 5-15 eV. Short- wavelength UV photons have sufficient energy to eject inner shell electrons. The "hole" left by an ejected electron may be filled by a valence shell electron with simultaneous emission of a photon. This type of light emission is termed *fluorescence*. Alternatively, a second electron from the valence shell may be ejected. This is the *Auger effect*. The Auger effect is significant when photoionization is induced by energetic x-rays and gamma-rays.

UV interactions with molecules are more complicated than isolated atoms. The principal initial reactions are photoionization and *dissociation* by splitting of chemical bonds. UV photons with insufficient energy for electron ejection can generate electronically-excited molecules in which one electron exists in a less tightly-bound energy state.[1] Absorption of IR photons usually induces more energetic molecular vibrations leading to "hot" molecules. The lower photon energies of microwaves and radio waves interact with atoms and molecules by inducing temporary shifts of the electronic orbits relative to the heavier nuclei, which is the basis of *dielectric polarization*. In general, interactions of light with atoms and molecules resemble a "tuning" process in which the energy of the absorbed or emitted photon matches the energy change in the absorbing atom or molecule. This fundamental principle was first postulated by Bohr in his 1913 planetary model of atomic hydrogen. Bohr assumed that the single electron in a hydrogen atom rotates around the proton in a discrete circular orbit, that is unchanged until the atom absorbs a photon. The "allowed" orbits are calculated by the process of *quantization*. This is done by equating the angular momentum of the rotating electron to $nh/2\pi$, where $h$ is Planck's constant and $n$ is a positive integer. The emission spectrum of the hydrogen atom is predicted by the *Bohr frequency condition*

$$hf = E_n - E_m \tag{4.2}$$

where $E_n$ and $E_m$ are the energies of the higher and lower states, respectively.[2] Early refinements in the original Bohr model included the addition of elliptical as well as circular orbits and a relativistic correction to account for the increase of the electron mass at rotational velocities approaching the speed of light.

**Example 4.2** Bohr frequency condition

The total energy of the stable hydrogen atom relative to the separated electron and proton is -14.53 eV. The energy of the first excited state is 4.38 eV. Calculate the wavelength of the most efficient optical transition between these two states.

The energy difference is 14.53 eV - 4.38 eV = 10.15 eV. According to Equation (4.2) the photon frequency is: 10.15 eV x 1.6 x $10^{-19}$ (J/eV)/6.62 x $10^{-34}$ J-s = 2.45 x $10^{15}$ $s^{-1}$ ; the corresponding wavelength is: 3.00 x $10^{8}$ (m/s)/2.45 x $10^{15}$ $s^{-1}$ = 1.22 x $10^{-7}$ m or 122 nm in the far-UV.

The dual concepts of single-photon interactions and the Bohr frequency condition are the basis of molecular photophysics. Optical spectra of a complex molecule contain a wealth of information about the structure and properties. Quantum theory is required to extract this information.

## 4.2 ATOMIC AND MOLECULAR STRUCTURE

Attempts to extend the planetary model of atomic hydrogen to larger atoms were largely unsuccessful. A major conceptual advance occurred in 1924 when de Broglie proposed that material particles possess wave-like properties in the same sense that light displays particle-like properties. The *de Broglie wavelength* ($\lambda_D$) of a free particle depends on its mass ($m$) and velocity ($v$) according to:

$$\lambda_D = h/P \tag{4.3}$$

where $P = mv$ is the linear momentum. The matter waves associated with a free particle are represented by a group of waves or *wave packet* with the maximum amplitude located at the position of the particle. However, this localization can exist only for a very short time and the wave packet spreads out almost instantaneously. Unlike electromagnetic waves, matter waves are dispersive even in free space. It is the attractive force between an electron and a positively-charged nucleus that localizes electron waves to atomic dimensions. However, this localization has a certain "fuzziness" and electron waves can describe only the average position of the electron. Heisenberg postulated that this indeterminacy is a fundamental property of nature in his famous *uncertainty principle*. Any attempt to make a simultaneous measurement of position and momentum of a particle leads to small alterations in one or the other that cannot be predicted. A mathematical statement of the Heisenberg principle is

$$\Delta p \, \Delta x = h/2\pi \tag{4.4}$$

where $\Delta p$ and $\Delta x$ are the uncertainties in the position and momentum of the particle, respectively. The uncertainty principle has some interesting implications for simple mechanical systems. If a particle is confined to a one-dimensional "box" of length $Z$, then $\Delta x = Z$ and $\Delta p = h/2\pi Z$. Quantum theory predicts that the lowest energy state of an electron in a one-dimensional "box" cannot be zero. Inserting this result into the classical expression for the kinetic energy of a free particle gives : $KE = p^2/2m$ or $KE = h^2/8\pi^2 mZ^2$, where $KE$ is the lowest possible kinetic energy of an electron in a box. This result is utilized in the "free-electron model" for spectra of conjugated molecules (Section4.4.1). However, zero rotational kinetic energy is permitted for a spherical molecule spinning on its own axis. In this case, the position of the sphere is entirely unknown when the sphere is at rest and, therefore, the uncertainty principle is not violated.

The early wave mechanics proposed by de Broglie is limited to the motion of free particles. In 1926 the concept of wave-particle duality was extended by Schroedinger in his celebrated *wave equation* for matter waves. The Schroedinger equation is a partial differential equation that can be solved for specific systems by applying well-developed analytical techniques.[3] Unlike classical partial differential equations, the Schroedinger equation contains imaginary terms and, therefore, some solutions or *wave functions* consist of complex functions that cannot represent measurable physical quantities.[4]

According to the present interpretation, first postulated by Born, the wave function has a statistical relationship to the actual particle motion. In this context, the wave function is treated as a *probability amplitude* and its square amplitude or *probability density* is interpreted as the statistical distribution of position. This formalism resembles electromagnetic theory in which the wave amplitude is proportional to the strength of the E-field (or B-field) and the square of the amplitude is the intensity or power density. The probabilistic interpretation of wave mechanics has been the subject of much philosophical discussion and thus far it has not been replaced by a convincing fully-deterministic theory. The solutions of the Schroedinger equation for the hydrogen atom substitute "smeared out" electron density distributions for the discrete planetary orbits of the Bohr model. The Schroedinger equation is applicable, in principle, to essentially any non-relativistic physical system, including molecules and solids. However, exact solutions are not feasible for multi-electron systems and useful results require approximational techniques.

### 4.2.1 Atomic Orbitals

Attempts to extend the exact solutions of the Schroedinger wave equation beyond the helium atom are limited by the mathematical complexity associated with the multibody problem. A unified approach for larger atoms is based on the theory of *atomic orbitals*. An atomic orbital (AO) specifies a possible wave function for an atomic electron and its associated energy. The overall electronic state of a multi-electron atom is described by the occupation of the AOs. The scheme used to label AOs is based on solutions of the Schroedinger's equation for the hydrogen atom. Each AO is identified by an integer $n$ corresponding to the principal electronic shell and a letter $l$ specifying the angular momentum associated. In classical physics angular momentum is a perfectly-defined property of rotary motion. For example, the angular momentum of a rotating particle of mass $m$ in uniform circular motion is $mvR$, where $v$ is the orbital speed and $R$ is the radius. The quantum mechanical description of angular momentum is related to the symmetry properties of the wave function in space. Spherical symmetry corresponds to zero angular momentum and more complex symmetries correspond to higher angular momenta. Zero angular momentum is not meaningful in classical mechanics, although a particle oscillating along a line in space may be considered to have zero angular momentum. The equivalent concept in quantum theory is a motion that has no preferred direction. Unlike classical physics, in which a particle can have any angular momentum consistent with its energy, angular momentum in wave mechanics is quantized in integral or half-integral multiples of $h/2\pi$ . An arcane notation is used to label the quantum states of angular momenta. States of 0, 1, 2, 3, and 4 units of angular momentum are labeled *s, p, d*, and *f*, respectively. (These letters are abbreviations for systems of lines in atomic spectra.) Higher angular momenta are then labeled in alphabetical sequence: *g, h, i*, ... In the lowest energy or ground state of the hydrogen atom, the electron occupies a spherically symmetric orbital closest to the proton with zero units of angular momentum ( $l = 0$) which is labeled 1*s*. The shape is depicted in Figure 4.2a by a spherical surface that encloses approximately 90% of the electron density. The next higher energy AO labeled 2*s* has most of the electron density localized in a ring further out from the proton.[5] The $n = 2$ shell of the hydrogen atom accommodates a second orbital of exactly the same energy as the 2*s* orbital with one unit of angular momentum ( $l = 1$) labeled 2*p*. A 2*p* AO has two lobes of opposite sign with a node at the origin ( Figure 4.2 b ).

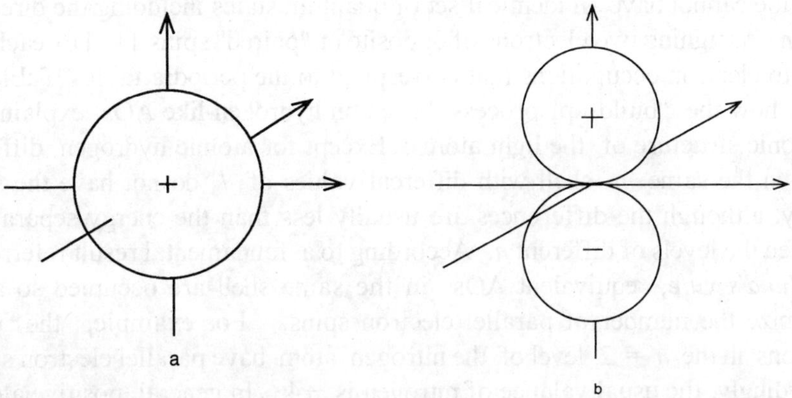


**Figure 4.2** Representation of atomic orbitals Aos by bounding surfaces enclosing 90% of the probability amplitude. (a) The *1s* AO is spherically symmetric. (b) The *2p* AO has equal positive and negative lobes, with a node at the origin.

There are three equivalent 2*p* orbitals directed along perpendicular axes, $2p_x$, $2p_y$, and $2p_z$. The complete solution of Schroedinger's equation for the hydrogen atom shows that the possible values of $l$ for a given $n$ range from 0 to $(n - 1)$. For a given $l$, the total number of equivalent orbitals is $(2l + 1)$. (These relations follow from the mathematics and are not intuitive.) Counting shows that the $n = 1$ shell provides one AO (1*s*), the $n = 2$ shell provides four AO (one 2*s* and three 2*p* ), the $n = 3$ shell provides nine AO (one 3*s* , three 3*p*, and five 3*d* ), etc. Remarkably, the total number of equivalent AO in each *n*-level of the hydrogen atom equals one-half of the number of elements in the principal groups of the periodic table of the elements which are 2, 10, 18, 36, etc.[6] For example, the first group ends with $^{2}$He, the second group ends with $^{10}$Ne, the third group ends with $^{18}$Ar, etc. The electronic structure of the elements larger than hydrogen is predicted by assigning two electrons to each hydrogen-like AO. The development of atomic structure from hydrogen atom energy states is referred to as the *build-up principle.*

The maximum occupation of an AO by two electrons is explained by a fundamental property of electrons referred to as *spin.* In 1925 Goudsmit and Uhlenbeck proposed that the electron possesses an intrinsic magnetism that can be viewed as originating from a charged sphere spinning around its own axis. Clockwise rotation generates a B-field pointing "down" [↓] and counter-clockwise rotation generates a B-field pointing "up" [↑]. The importance of electron spin for molecular structure originates from a fundamental principle of nature enunciated in 1925 by Pauli as the *exclusion principle.* The exclusion principle states that two electrons in the lowest energy state of an atom or molecule cannot have an identical set of quantum states including the direction of spin. Assigning two electrons of opposite or "paired" spins [↑↓] to each AO leads to electron occupations that correspond to the periodic table. Table 4.1 shows how the "build-up" process based on hydrogen-like AOs explains the electronic structure of the light atoms. Except for atomic hydrogen, different AOs in the same $n$ shell with different values of $l$ do not have the same energy, although the differences are usually less than the energy separations between the levels of different $n$. According to a fundamental result referred to as *Hund's rule*, equivalent AOs in the same shell are occupied so as to maximize the number of parallel electron spins. For example, the three electrons in the $n = 2$ level of the nitrogen atom have parallel electron spins. Accordingly, the usual valance of nitrogen is +4. In general, positive atomic valence equals the number of unpaired electrons in the outer shell for shells that are less than half filled. Negative atomic valence equals the number of "holes" in the outer shell for shells that are more than half filled. This elementary approach to chemical bonding does not explain the occurrence of multiple valence, such as found for boron and carbon. For example, the valence of carbon is +2 in CO and 4 in most organic compounds.

### 4.2.2 Molecular Orbitals

The solution of Schroedinger's equation for large molecules is complicated by the multiplicity of electronic and atomic motions. Two equivalent approaches to molecular calculations were developed in the late 1920s. In *valence-bond* theory, complete atoms are brought together to form molecules. Interactions between the outer electrons leads to new energy levels in the resultant molecule.

In the alternative *molecular orbital* (MO) theory, the atomic nuclei and inner shell electrons are brought together to form a molecular skeleton and the valence electrons are then allocated to MOs that extend over the groups of atoms. Chemical bonding does not affect the electrons in the inner shells that continue to occupy the original AOs. The essential properties of covalent bonding are explained by combining the outer shell AOs from two atoms. An MO can be *bonding* or *anti-bonding*. An MO is bonding if its occupation by one or two electrons stabilizes the molecule. An MO is anti-bonding if its occupation destabilizes the molecule.

**Table 4.1 Electronic states of the light elements**

| Element | Symbol | Occupation of Atomic Orbitals [a] | | | | | Valence [b] |
|---|---|---|---|---|---|---|---|
| Hydrogen | H | 1 $s^{\uparrow}$ | | | | | +1 |
| Helium | He | 1 | | | | | 0 |
| Lithium | LI | $s^{\uparrow\downarrow}$ | | | | | +1 |
| Beryllium | Be | 1 | 2 | | | | +2 |
| Boron | B | $s^{\uparrow\downarrow}$ | $s^{\uparrow\downarrow}$ | 2 $p_x^{\uparrow}$ | | | multivalent |
| Carbon | C | 1 | 2 | 2 $p_x^{\uparrow}$ | 2 $p_y^{\uparrow}$ | | +2, 4 |
| Nitrogen | N | $s^{\uparrow\downarrow}$ | $s^{\uparrow\downarrow}$ | 2 $p_x^{\uparrow}$ | 2 $p_y^{\uparrow}$ | 2 $p_z^{\uparrow}$ | multivalent |
| Oxygen | O | 1 | 2 | 2 | 2 $p_y^{\uparrow}$ | 2 $p_z^{\uparrow}$ | -2 |
| Fluorine | F | $s^{\uparrow\downarrow}$ | $s^{\uparrow\downarrow}$ | $p_x^{\uparrow\downarrow}$ | 2 | 2 $p_z^{\uparrow}$ | -1 |
| Neon | Ne | 1 | 2 | 2 | $p_y^{\uparrow\downarrow}$ | 2 $p_z^{\uparrow\downarrow}$ | 0 |
| | | $s^{\uparrow\downarrow}$ | $s^{\uparrow\downarrow}$ | $p_x^{\uparrow\downarrow}$ | 2 | | |
| | | 1 | 2 | 2 | $p_y^{\uparrow\downarrow}$ | | |
| | | $s^{\uparrow\downarrow}$ | $s^{\uparrow\downarrow}$ | $p_x^{\uparrow\downarrow}$ | | | |
| | | 1 | 2 | | | | |
| | | $s^{\uparrow\downarrow}$ | $s^{\uparrow\downarrow}$ | | | | |
| | | 1 | 2 | | | | |
| | | $s^{\uparrow\downarrow}$ | $s^{\uparrow\downarrow}$ | | | | |
| | | 1 | | | | | |
| | | $s^{\uparrow\downarrow}$ | | | | | |

[a]The arrows indicate the electron spin orientations; [b] The column indicates the ordinary valence for inorganic

Two bonding types of MOs are depicted in Figure 4.3 for the ethylene molecule $C_2H_4$. In a bonding *sigma* ($\sigma$) MO, the electric charge density is localized between the two atoms. In a bonding *pi* ($\pi$) MO, the electric charge density occupies two sausage-shaped lobes above and below the plane through the atoms. It is easy to see that an *s*-type AO and a *p*-type AO can combine to form a σ -type MO directed along a line between two atoms. A π – type MO is formed from two *p*-type AOs aligned perpendicular to the line between the two atoms. Other combinations of *s*-type and *p*-type AOs lead to higher-energy *antibonding* MOs labeled σ * and π *. A covalent single bond consists of one σ -type MO occupied by two electrons with paired (↑↓) electron spins. In a covalent double bond an additional π -type MO is occupied by two paired electrons, as exemplified by the C = C bond of ethylene. A second π–type MO perpendicular to the first is occupied by an electron pair in a covalent triple bond as in acetylene. Oxygen and nitrogen atoms also have *non-bonding* 2*s* AOs in the valence shell in which the electrons do not interact with other atoms. They are referred to as a *lone pair*.

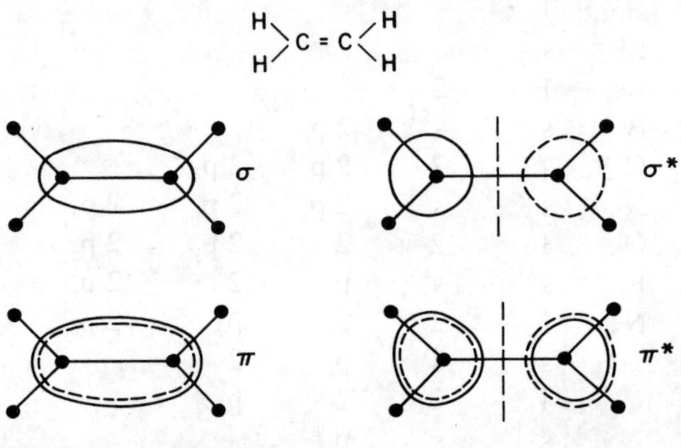


**Figure 4.3** Molecular orbitals for ethylene. The bonding σ MO lies along the C-C axis. The bonding π MO has a lobe above the C-C axis (dashed line), with a node between. The σ* MO and π* MO are higher energy, antibonding molelcular orbitals. From Grossweiner, L.I. and Smith, K.C., *The Science of Photobiology*, 2nd ed., Plenum Press, New York, 1989. With permission.

The localized negative charge of the lone pair attracts protons in water and in some crystals to form $H_3O^+$ and $NH_4^+$ ions. The electron configuration of a molecule is predicted by adding electrons to AOs and MOs in order of decreasing binding energy until all of the electrons are used. This "build-up" process requires information about the relative energies of the AOs and MOs which must be obtained by molecular calculations. In general, inner-shell AOs are filled first, followed σ -type MOs in the valence shell, π-type MOs in the valence shell, and then the lone-pair AOs. The electron spin orientations obey Hund's rule. For example, the configuration of ethylene is predicted by adding the 16 electrons to the AOs and MOs of decreasing binding energy: $C_2H_2 \Rightarrow$

$(1_{sC})^{\uparrow\downarrow}(1_{sC'})^{\uparrow\downarrow}(2_{sC})^{\uparrow\downarrow}(2_{sC'})^{\uparrow\downarrow}(\sigma_{CH})^{\uparrow\downarrow}(\sigma_{CH}')^{\uparrow\downarrow}(\sigma_{CC})^{\uparrow\downarrow}(\pi_{cc})^{\uparrow\downarrow}$. The subscript on the AO refers to the atom and the double subscript on the MO refers to the bond. Formaldehyde exemplifies a molecule with lone-pair electrons. The predicted configuration is: $H_2CO \Rightarrow K\,(\sigma_{CH})^2\,(\sigma_{CH}')^2\,(\sigma_{CO})^2(\pi_{cO})^2\,(n_O)^2$, where $K$ denotes the AO core of all four atoms and $n_O$ is the lone pair AO. on the oxygen atom. MO theory predicts that ethylene and formaldehyde have zero total spin in their lowest energy states. Molecular structure is determined by the symmetry properties of the MOs. For example, the valence electrons in the nitrogen atom occupy three mutually perpendicular $2p$ AOs which accounts for the pyramidal shape of ammonia. Two perpendicular $2p$ AOs are occupied in the oxygen atom, leading to an approximately 90° bend in the structure of water. (The electrostatic repulsion of the two partially unshielded protons stretches the angle to 105°). However, the two unpaired $p$ electrons in the carbon atom cannot explain the ubiquitous tetrahedral structure of carbon atoms in organic molecules. In his celebrated 1931 theory of chemical bonding, Pauling showed that two $2s$ and two $2p$ AOs in a carbon atom can combine to form four equivalent *hybrid* AOs referred to as $sp^3$ with the required tetrahedral symmetry. The further development of the Pauling theory explains chemical bonding in many inorganic and organic molecules. A different type of bonding occurs in conjugated aromatic molecules. According to the MO description of benzene, six $p$-type AOs form two rings of electric charge above and below the plane of the molecule (Figure 4.4). In the alternative valence-bond description of benzene, the six carbon atoms are connected by three single bonds and three alternating or *conjugated* double bonds. The two possible arrangements of the conjugated system correspond to physically-equivalent *resonance* structures that act to stabilize the molecule. Quantitative calculations based on the MO and VB models lead to equivalent results when all possible structures are considered in each approximation.

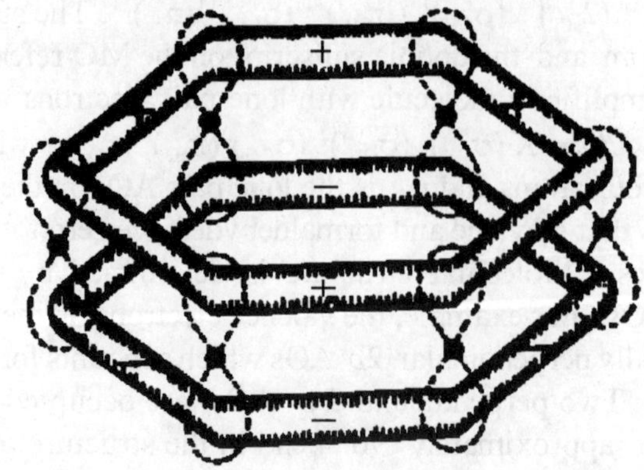


**Figure 4.4** In the Kekule benzene structure, six π molecular orbitals are localized in two parallel streamers lying above and below the plane of the ring..

### 4.2.3 Molecular Oxygen

The lowest energy MOs in almost all stable molecules are occupied by two electrons with no unpaired spins. This spin configuration is referred to as a *singlet state.*[7] However, molecular oxygen and nitric oxide are exceptions in which the lowest energy state is a *triplet state* with two unpaired electron spins in the ground state. A free radical with one unpaired electron spin is a termed a *doublet state*. Free radicals are usually, but not always, unstable. Melanin in epidermal melanocytes is a stable, high-molecular weight free radical. An early achievement of MO theory is the prediction that the ground state of molecular oxygen is a triplet. The electron configuration is determined by the six electrons in the four π - type MOs: $O_2 \Rightarrow K\,(\pi_x)^{\uparrow\downarrow}(\pi_y)^{\uparrow\downarrow}(\pi^*_x)^{\uparrow}(\pi^*_y)^{\uparrow}$, where the x and y subscripts refer to the directions of each MO in the plane perpendicular to the molecular axis. The two unpaired electrons in two different π* MOs explain the experimental fact that oxygen is a magnetic substance. Diatomic molecules such as $O_2$ are labeled by a spectroscopic *term symbol* that indicate the spin and symmetry properties of the molecule. The term symbol for the oxygen ground state is $^3\Sigma_g^-$ , where the superscript designates a triplet state and the other symbols refer to symmetry properties of the molecule. Molecular oxygen has three excited states above the triplet ground state. The configurations of two equivalent lower-energy excited states are: $O_2\,(^1\Delta_g^+) \Rightarrow K\,(\pi_x)^{\uparrow\downarrow}(\pi_y)^{\uparrow\downarrow}\,(\pi^*_x)^{\uparrow\downarrow}$ and $O_2\,(^1\Delta_g^+) \Rightarrow K\,(\pi_x)^{\uparrow\downarrow}(\pi_y)^{\uparrow\downarrow}(\pi^*_y)^{\uparrow\downarrow}$. These species are referred to as *singlet molecular oxygen* or "singlet oxygen". Singlet oxygen is an important intermediate in many photochemical reactions. The excitation energy of singlet oxygen is 0.98 eV (22.6 kcal/mole). and the radiative decay lifetime is 45 minutes at very low gas pressures. However, collisions with other molecules induce much shorter lifetimes, for example, 14 minutes in oxygen gas at 760 torr. The decay lifetime of singlet oxygen in liquids depends on the composition. Singlet oxygen is least stable in water with a decay lifetime about 4 μs. Light absorption by most dye-like molecules leads to excited states whose energy above the ground state exceeds 0.98 eV. The excess energy of the absorbing molecule can be transferred to oxygen molecules leading to the formation of singlet oxygen, which can then react with other substrates. This type of energy transfer process is more favorable for excited triplet states of the dye-like molecule than excited singlet states because triplet state are longer-lived and their spin orientations can conserve the total spin of the interacting species. Energy transfer from the triplet state of a dye-like molecule to molecular oxygen is an important mechanism in *photosensitization*. Molecular oxygen also has a higher energy excited state: $O_2\ (^1\Sigma_g^+)\ \Rightarrow\ K\ (\pi_x)^{\downarrow}(\pi_y)^{\uparrow\downarrow}\ (\pi^*_x)^{\uparrow}(\pi^*_y)^{\downarrow}$. The excitation energy of this species is 1.63 eV (37.6 kcal/mole) and the decay lifetime is 7 seconds in oxygen gas at 760 torr. $O_2\ (^1\Sigma_g^+)$ is converted rapidly to $O_2\ (^1\Delta_g^+)$ in condensed media and is seldom important in photobiology.

### 4.2.4 The Physical Significance of Spin

The concept of electron spin was originally postulated in connection with atomic structure. It is now known that other fundamental particles possess intrinsic angular momenta also referred to as spin, including protons, neutrons, many nuclei, and particles present in cosmic rays and generated by high-energy accelerators. Spin is one the most fundamental features of the physical universe. Spin angular momentum is closely related to intrinsic magnetism. In classical physics a rotating electric charge generates a magnetic field by virtue of being a current loop. This is the operating principle of the electromagnet. The magnetic field associated with a permanent magnet is attributed to the microscopic current loops of the atomic electrons. The fundamental element of magnetism is the *magnetic dipole*, consisting of separated magnetic north and south poles of equal strength. Magnetic lines of force leave at a north pole and enter at a south pole. The *magnetic dipole moment* is defined as the product of the pole strength and their separation. (The designation "moment" is borrowed from mechanics in which a turning moment is the product of the magnitude of a force and the perpendicular distance from the point of application of the force to the axis of rotation.) The magnetic field generated by a current loop is equivalent to a magnetic dipole oriented perpendicular to the plane of the loop. Ampere showed that the dipole moment of a current loop equals the product of the electric current and the area enclosed by the loop. If this concept is extended to a "classical" electron rotating in a circular orbit of radius $R$ at a speed $v$, the angular momentum is $mvR$ and the magnetic dipole moment is $-evR/2$. (The current is $-ev/2\pi R$ and the loop area is $\pi R^2$). The *gyromagnetic ratio* ($g$) is defined as the ratio of the magnetic moment to the angular momentum. Thus, the classical value of $g$ for the electron is $-e/2m$. Note that the gyromagnetic ratio does not depend on the speed or the orbit of the charged particle. Quantum theory shows that orbital angular momentum is quantized in integral units of $h/2\pi$. In order to obtain agreement of quantum theory with atomic spectra, the intrinsic spin angular momentum of the electron must be $h/4\pi$, which corresponds to a spin of 1/2 in units of $h/2\pi$. The corresponding magnetic moment of the electron is $- eh/4\pi m$. Therefore, $g$ for electron spin is $- e/m$, which is twice as large as orbital electron motions. The neutron and proton are also spin 1/2 particles. The corresponding gyromagnetic ratios of nuclear particles are about 2000 times smaller than that of the electron owing to the larger mass. Many atoms have an intrinsic nuclear magnetic moment including $^{1}H$, $^{13}C$, $^{14}N$, $^{19}F$, and $^{31}P$. Other atoms have zero nuclear spin owing to cancellation of the magnetism of the neutrons and protons, for example, $^{12}C$ and $^{16}O$. Particle spin has a profound effect on the macroscopic properties of matter. All particles in nature are divided into two fundamental classes. Those particles having half-integral spin are termed *fermions* , in honor of Enrico Fermi, including electrons, protons, and neutrons. Fermions obey the Pauli exclusion principle that limits the occupation of a quantum state to two particles. This property of electrons is responsible for atomic and molecular structure. The other major class of particles with zero or integral spin are termed *bosons* in honor of Subhas Chandra Bose. Photons and helium nuclei exemplify bosons. Bosons are not limited by the Pauli exclusion principle and any number of bosons can occupy the same quantum state. This property explains the key features of black body radiation in a cavity in which many photons can exist in the same energy state.[8] *Superfluid helium* resembles a quantum gas with all $^{4}He$ atoms in their ground state. Superconductivity has features in common with boson systems in which coupled pairs of electrons (*Cooper pairs*) are transported as spin- zero pseudoparticles.

## 4.3 OPTICAL TRANSITIONS IN MOLECULES

The electronic state of an entire molecule is determined by adding the orbital angular momenta ( $l$ ) and spin angular momenta ($s$) of the constituent electrons. This process must follow the rules of quantum physics such that the total orbital angular momentum $L$ and total spin angular momentum $S$ are integral or half-integral multiples of $h/2\pi$. The total angular momentum $J$ is then calculated using quantum rules. This can be done by adding the the total $L$ and the total $S$; this approach is referred to as L-S coupling. Alternatively, $l$ and $s$ can be added for each electron to give $j$ for each electron and then the total $J$ is calculated by adding the individual j values; this is j-j coupling. In general, L-S coupling is more accurate for light atoms and j-j coupling is more accurate for heavy atoms. The lines in atomic and molecular spectra are directly related to the $S$, $L$, and $J$ values of the absorbing or emitting species.

In an efficient optical process the energy of the absorbed or emitted photon matches the energy difference between the initial and final energy states of the entire molecule. *Selection rules* impose additional constraints on electronic transitions In an *allowed* optical transition one photon is absorbed or emitted without a change of electron spin direction. The corresponding transition with a simultaneous "spin flip" is much less likely; this process is referred to as *forbidden*. Another selection rule limits the change of $L$ an allowed optical transition to $\pm 1$. The application of selection rules to molecular spectra assists in sorting out the energy levels. Instead of specifying the occupation of all MOs before and after light absorption or emission, it is more convenient to identify the initial and final energy states of the entire molecule. In spectroscopic notation, singlet states are labeled $S_0$, $S_1$, $S_2$, etc., in order of increasing energy. The corresponding triplet states are labeled $T_0$, $T_1$, $T_2$, etc. The occupation of lower-energy MOs is shown in Figure 4.5 for ethylene In the $S_0$ ground state, two electrons having paired spins occupy the lowest energy $\sigma$ and $\pi$ MOs. One electron is elevated from the highest-energy occupied $\pi$ MO to the next higher-energy empty $\pi$ * MO in the $S_1$ excited state without a change of spin direction. The equivalent optical transition with a spin flip leads to the $T_1$ excited state. The $S_1$ state is short-lived because transitions back to the $S_0$ ground state are allowed. However, the $T_1$ state is relatively long-lived or *metastable* owing to the forbidden nature of the transition to $S_0$.

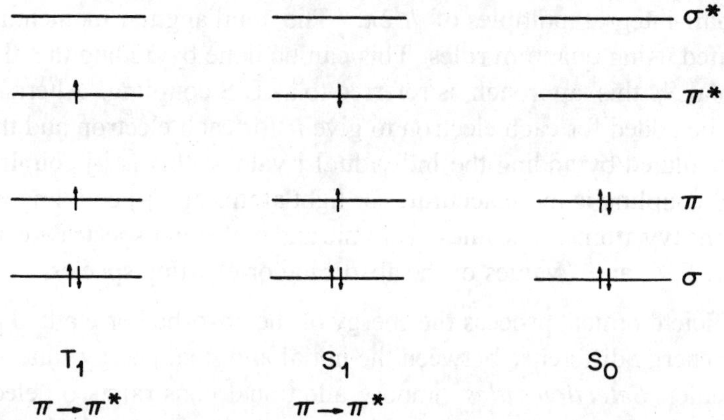


**Figure 4.5** Occupation of C-C molecular orbitals in ethylene. In the ground state So, the lowest energy σ MO and π MO are each occupied by two electrons of opposite spin. In the first excited singlet state (S1), one electron is elevated from the occupied π MO to the π* MO, with a simultaneous reversal of spin direction. This process is "forbidden" by the spin selection rule and its occurrence has low probability. From Grossweiner, L.I. and Smith, K.C., *The Science of Photobiology*, 2nd ed., Plenum Press, New York, 1989. With permission.

## 4.3.1 The Free Electron Model

MO theory has been very successful in predicting the energy levels of conjugated molecules. The approach can be illustrated with a minimum of mathematical complexity using the *free electron model* (FEM). The molecular backbone of a conjugated molecule is treated as a one-dimensional "box" of length $Z$ that contains $N$ mobile π electrons. According to Equation (4.3), the de Broglie wavelength of a free electron equals $h/mv$. The electron wavelengths that "fit" into a one-dimensional box of length $Z$ are given by $\lambda_D = 2Z/p$, where $p$ in an integer. For example, $p = 1$ corresponds to a state having de Broglie wavelength $\lambda_D/2$, $p = 2$ corresponds to $\lambda_D$, etc. Thus, the allowed values of linear momentum are quantized according to: $mv = ph/2Z$. The corresponding kinetic energy of the electron is given by the classical relation: $KE = (1/2)Mv^2$. Substituting the quantization expression leads to

$$KE = \frac{p^2 h^2}{8mZ^2} \tag{4.5}$$

Equation (4.5) was derived for $p = 1$ using only the uncertainty principle in Section 4.2.0. Each integral value of $p$ specifies an allowed energy state. For a conjugated molecule with $N$ *pi* electrons, two *pi* electrons of opposite spins occupy each of the lowest energy MO up to a maximum value of $p = N/2$. Absorption of a photon elevates one electron from the highest filled MO to the lowest vacant MO corresponding to: $p = (N/2 + 1)$. Setting the energy difference between the two states equal to the photon energy according to Equation (4.2) leads to:

$$\lambda = \frac{8mZ^2 c}{h(N+1)} \tag{4.6}$$

Typical results for a group of linear cyanine dye analogs with *n* vinyl groups are given in Table 4.2 . The value of $Z$ is based on a spacing of 0.139 nm per C - C bond in benzene. Each carbon atom of a CH = CH group contributes one mobile *pi* electron and two nitrogen atoms contribute three mobile π electrons. The total number of *pi* electrons is (2n + 4). The agreement between the observed and calculated values of λ is excellent.

**Table 4.2 Application of the free electron model to cyanide dye analogs**
**$(CH_3)_2N^+ = CH\text{-}(CH{=}CH)_n\text{-}N(CH_3)_2$**

| n | λ ( calculated ) nm | λ( observed ) nm |
|---|---|---|
| 1 | 309 | 309 |
| 2 | 409 | 409 |
| 3 | 509 | 511 |

A similar approach is employed for a cyclic conjugated molecule by assuming that the *pi* electrons are confined to a circular "box" of radius $R$. Porphyrins are important class of conjugated biomolecules. The typical absorption spectrum of a metal- free porphyrin consists of a very strong band near 400 nm referred to as the *Soret band* or B band and four decreasingly weaker bands between 450 and 600 nm named Q bands. The actual conjugated structure is replaced by the inner ring of 16 atoms for the FEM calculation. The 12 carbon atoms contribute 12 *pi* electrons and 4 nitrogen atoms contribute 6 *pi* electrons. Rotational angular momentum is quantized for circular geometry. Equation (4.4) applies with $p$ replaced by an integer $l$ proportional to the angular momentum of the electron and $Z$ replaced by the circumference $2\pi R$, where $R$ is the radius of the porphyrin ring. This result is obtained by quantizing the angular momentum: $mvR = l\,(h/2\pi)$. The possible values of $l$ are 0, ± 1, ± 2, ..., where $+l$ refers to an electron circulating in a clockwise direction and $-l$ refers to clockwise rotation. Allowing for motion in both directions, the eighteen *pi* electrons occupy the lowest five energy levels corresponding to $l = 0$, ±1, ±2, ±3, and ± 4. The ground state energy levels are depicted in Figure 4.6a. The lowest-energy excited state corresponds to the elevation of an electron from the $l = 4$ level to the $l = 5$ level. The numerical calculation for $R = 0.35$ nm for a typical porphyrin leads to $\lambda = 444$ nm, which is the median wavelength of the B and Q bands. The FEM predicts more detailed features of porphyrin spectra by applying the quantum mechanical rules for addition of angular momenta.[9] In the FEM calculation for metal-free porphyrins it was assumed that 18 mobile *pi* electrons are confined to an inner circular "box" of constant size. More exact calculations with the *linear-combination-of-atomic orbitals* (LCAO) method include all 24 *p*-type AOs of the skeleton structure and the resultant MOs are sorted out according to their symmetries. The results are similar to Figure 4.6b.

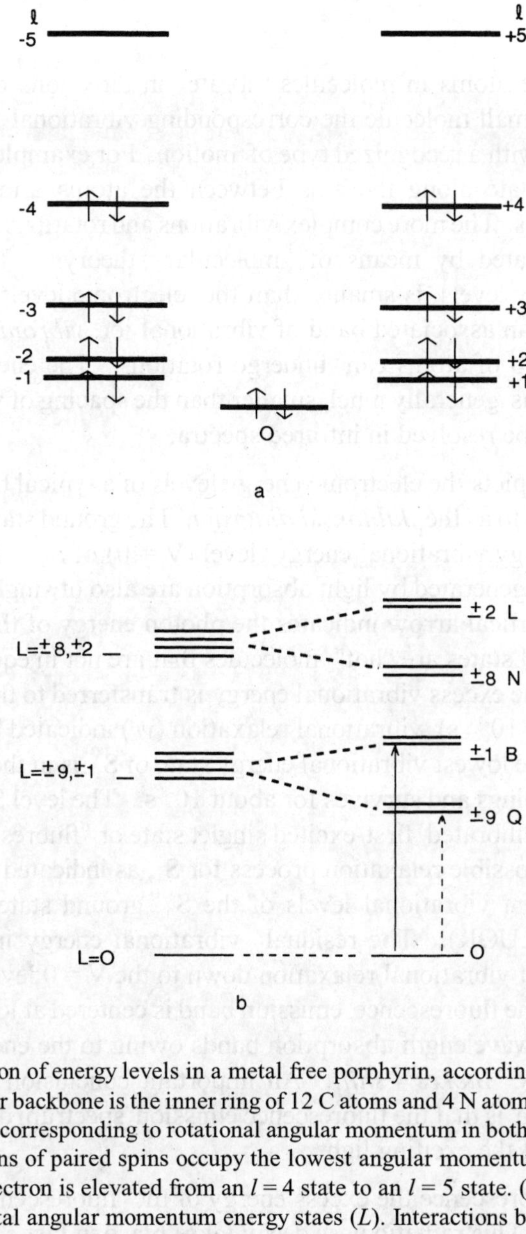


**Figure 4.6** Occupation of energy levels in a metal free porphyrin, according to the free electron model. The molecular backbone is the inner ring of 12 C atoms and 4 N atoms. The 18 $\pi$ electrons occupy energy states corresponding to rotational angular momentum in both directions. (a) In the ground state, electrons of paired spins occupy the lowest angular momentum states. In the first excited state, one electron is elevated from an $l = 4$ state to an $l = 5$ state. (b) The energy levels on the left are the total angular momentum energy staes ($L$). Interactions between the electrons lead to the energy separations on the right. The allowed transition (solid line) corresponds to the intense B band (Soret band) and the forbidden transition (dashed line) is responsible for the weak Q bands. The higher energy transitions corresond to the shorter wavelength N and L absorption bands.

### 4.3.2 The Jablonski Diagram

Groups of atoms in molecules vibrate in directions determined by the structure. In a small molecule the corresponding vibrational energy levels are easily identified with a recognized type of motion. For example, the two C atoms in ethylene oscillate along the axis between the atoms and rotate around a perpendicular axis. The more complex vibrations and rotations in large molecules must be calculated by means of molecular theory. The separation of vibrational energy levels is smaller than the electronic levels. Each electronic energy level has an associated band of vibrational or *vibronic* levels. In some molecules a group of atoms can undergo rotations The energy separation of rotational levels is generally much smaller than the spacing of vibrational energy levels. They can be resolved in infrared spectra.

Figure 4.7 depicts the electronic energy levels of a typical biological pigment molecule referred to as the *Jablonski diagram.* The ground state of the molecule is the lowest energy vibrational energy level (V = 0) of $S_0$ . The higher energy states $S_1$ , $S_2$ , ... generated by light absorption are also of singlet character. The length of each vertical arrow indicates the photon energy of the transition. The higher vibrational states are "hot" molecules that are not in equilibrium with the environment. The excess vibrational energy is transferred to the medium as heat *via* a very fast ( $\approx 10^{-11}$ s) vibrational relaxation (*vr*) indicated by the downward wavy arrow. The lowest vibrational energy state of $S_1$ is in thermal equilibrium with the surroundings and survives for about $10^{-8}$ s. The level $S_1$ is referred to as the thermally-equilibrated first-excited singlet state or "fluorescent" state. Light emission is one possible relaxation process for $S_1$, as indicated by the downward arrows to different vibrational levels of the $S_0$ ground state. This process is *fluorescence* (FLUOR). The residual vibrational energy in the $S_0$ state is released by a fast vibrational relaxation down to the V = 0 level. It is evident in Figure 4.7 that the fluorescence emission band is centered at longer wavelengths than the longest wavelength absorption bands owing to the energy lost as heat. This effect is the *Stokes' shift.* An important conclusion derived from the Jablonski diagram is that the fluorescence emission spectrum does not depend on the wavelength of the exciting light.

Instead of fluorescence the excess energy of the fluorescent state $S_1$ may be released as heat. This radiationless decay takes place in fast consecutive stages. The first step is an energy-conserving "horizontal" transition to a high vibrational level of $S_0$ termed *internal conversion* (*ic* ), followed by a fast vibrational cascade to the ground state. Another and very important possible fate of the fluorescent state is its spontaneous conversion to a triplet state *via* a change of an electron spin direction. This process of *intersystem crossing* (*isc*) is shown by a wavy horizontal line. The probability of *isc* is promoted by the presence of heavy atoms and paramagnetic ions in the molecular structure and in the environment. Intersystem crossing is followed by stepwise radiationless transitions to the lowest energy vibrational level of the $T_1$ state. Triplet states have longer decay lifetimes than the corresponding singlet state because transitions from $T_1$ to $S_0$ are forbidden. A typical triplet state in solution may survive for milliseconds to seconds and even longer in rigid media at low

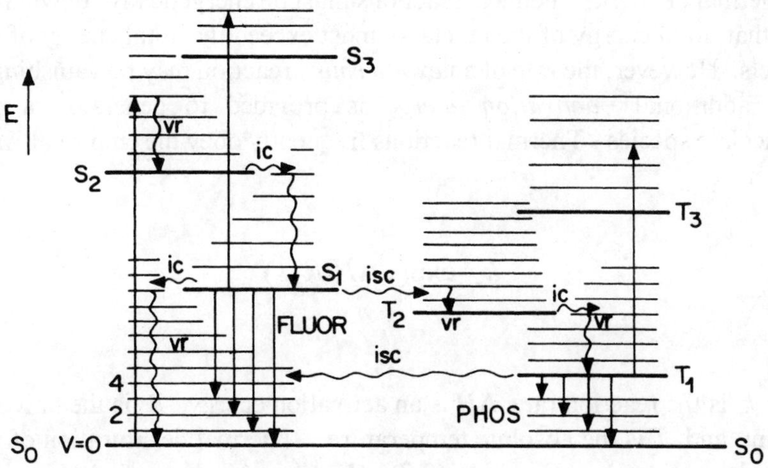


**Figure 4.7** A Jablonski diagram representing the lower energy levels of a conjugated dye or pigment molecule. $S_0$, $S_1$, $S_2$, $S_3$ are singlet states; $T_1$, $T_2$, $T_3$ are triplet states. The vibronic levels (light horizontal lines) are shown for only one vibrational mode of each electronic energy state (heavy horizontal lines). The solid vertical arrows indicate radiative transitions. The *absorption* events depicted are the transition from the thermally-equilibrated first excited singlet state $S_1$(v=0) to the third excited singlet state $S_3$(v=3); and from the lowest triplet state $T_1$(v=0) to a higher triplet state $T_3$(v=3). *Fluorescence* (FLUOR) is an allowed optical transition from $S_1$(v=0) to h igher vibrational levels of $S_0$. *Phosphorescence* (PHOS) is the forbidden optical transition from $T_1$(v=0) to higher vibrational levels of $S_0$. The non-radiative transitions shown (wavy arrows) are vibrational relaxation (vr), or "vibrational cascade"; internal conversion (ic); and intersystem crossing (isc). From Grossweiner, L.I. and Smith, K.C., *The Science of Photobiology*, 2nd ed., Plenum Press, New York, 1989. With permission.

temperature. Triplet states have been observed in dry irradiated silk after many days. In the absence of interactions with other molecules, triplet states release their excess energy as light and heat. Light emission by a triplet state may be observed as a *phosphorescence* or "afterglow" after the exciting light has been removed. In summary, light absorption by a highly conjugated aromatic or heterocyclic molecule leads to two types of electronically excited states: a short-lived excited singlet state and a longer-lived triplet states. The subsequent reactions of these energy-rich molecular species accounts for many photosensitized biochemical reactions.

## 4.4 OVERALL PHOTOCHEMICAL REACTIONS

Thermal or "dark" chemical reactions must be energetically "downhill" in the sense that total energy of the reactants must exceed the total energy of the final products. However, the rate of an *exothermic* reaction may be vanishingly slow unless additional *activation energy* is provided to generate a required intermediate species . Thermal reactions frequently obey the empirical Arrhenius equation

$$k_r = \exp(-\Delta H / R_g T) \tag{4.7}$$

where $k_r$ is the reaction rate, $\Delta H$ is an activation energy, $R_g$ is the universal gas constant, and $T$ is the absolute temperature. Thermal denaturation of proteins and nucleic acids obey Equation (4.7). A typical values of $\Delta H$ is about 25 kcal/mol (105 kJ/mol) for a reaction that take place in some hours at room temperature. Raising the temperature to 100°C would speed up the reaction rate 5000 times.[10] Photochemical reactions are more selective than thermal reactions because the optical energy is channeled into specific pathways. This effect is exemplified by some important one-step photochemical reactions of biological importance.

The photochemical generation of thymine dimers from thymine is a one-step reaction initiated by UV-C

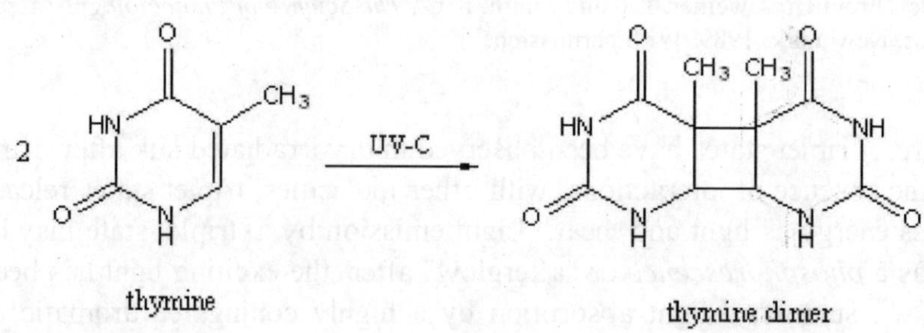


(4.8)

Reaction (4.8) takes place in frozen thymine solutions and between adjacent pyrimidine bases on the same chain of DNA. The actual structure is U-shaped with the thymine moieties perpendicular to the four-member cyclobutane ring. Generation of pyrimidine dimers in epidermal cells is the first step in the induction of skin cancer by sunlight. Many cells contain enzymic systems under genetic control that remove thymine dimers (Section 6.2.1).

Retinal (vitamin A aldehyde) is the chromophore in the visual pigment rhodopsin. The first photochemical step in retinal vision is the *photoisomerization* of *all-trans* retinal to the *11-cis* form

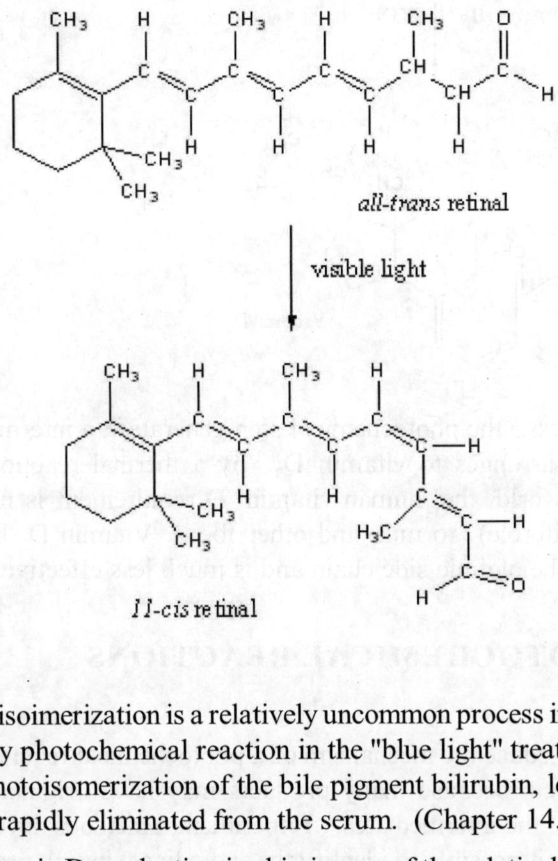


(4.9)

Photoisoimerization is a relatively uncommon process in photobiology. However, the key photochemical reaction in the "blue light" treatment of neonatal jaundice is a photoisomerization of the bile pigment bilirubin, leading to an isomer that is more rapidly eliminated from the serum. (Chapter 14.).

Vitamin D production in skin is one of the relatively few beneficial reactions induced by sunlight exposure in humans. The overall reaction is a *photorearrangement* of 7-dehydrocholesterol (ergosterol) in skin to form vitamin $D_3$.

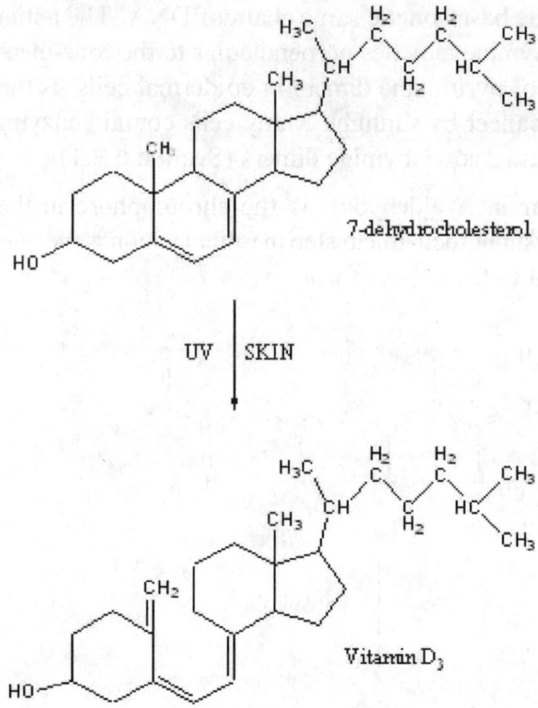


(4.10)

In the physiological process, the photochemical step generates an intermediate previtamin $D_3$ which re-arranges to vitamin $D_3$ by a thermal reaction. In developed regions of the world the human vitamin D requirement is met by adding vitamin $D_2$ (calciferole) to milk and other foods. Vitamin $D_2$ has an unsaturated C-C bond in the olefinic side chain and is much less effective.

## 4.4 PRIMARY PHOTOCHEMICAL REACTIONS

It is often difficult to deduce the mechanism of a photochemical reaction by analyzing the final products because many products may have originated by photochemical reactions of the intermediates. Experimental methods developed since the 1950s have made it possible to identify the short-lived initial products before they undergo further reactions. This approach has shown that most photochemical reactions are initiated by relatively few types of primary reactions. Short-lived photochemical intermediates can be stabilized in a rigid medium and identified by means of optical and magnetic resonance. Alternatively, a high concentration of the initial reaction products can be generated and detected by applying pulse excitation. *Flash photolysis* is the most widely used approach in which the intermediates are identified and followed in real time using optical techniques. In related methods, the reactions are started by applying a pulse of heat (temperature-jump), ionizing radiation (pulse radiolysis), or ultrasound.

The first products of most photochemical reactions are short-lived electronically-excited molecules and reactive free radicals. In *photolytic bond cleavage* free radicals are generated by splitting a stable molecule

$$R_1 - R_2 + \text{light} \Rightarrow R_1^{\bullet} + R_2^{\bullet} \qquad (4.11)$$

where $^{\bullet}$ denotes a free radical with an unpaired electron. The absorbed photon energy must exceed the strength of the chemical bond in the gas phase and hydrophobic liquids. However, Reaction (4.11) may be driven by less energetic (longer wavelength) photons if additional energy is provided by stabilization of the primary products in the medium.

---

**Example 4.3.** Photolytic bond cleavage

The typical energy of a C-C single bond is 85 kcal/mol. (a) Calculate the longest wavelength of light that can split a C-C bond in the gas phase. (b) Repeat the calculation for a liquid system in which solubilization energy of the products is 20 kcal/mol. Convenient conversions between thermochemical bond energy and photon energy are: (1) A photon energy of 1 eV corresponds to a chemical bond energy of 24.06 kcal/mole. (2) A photon wavelength $\lambda$ (nm) corresponds to a photon energy of 1239/ $\lambda$ eV.

(a) The bond energy per molecule is: 85 kcal/mol x (1 eV/24.06 kcal/mol) = 4.69 eV. The equivalent photon wavelength is: 1239/4.69 = 335 nm.

(b) The overall energy requirement is 65 kcal/mol in the liquid. The photon wavelength is 1239/(65/24.06) = 439 nm.

---

*Photoionization* is a primary photochemical reaction in which an electron is ejected from a molecule leaving a positively-charged free radical. The ionization energies of organic molecules in the gas phase are the order of 7-10 eV, corresponding to far-UV wavelengths. However, photoionization may be favored over bond splitting in a polar solvent assisted by the additional hydration energy of ionic products that can contribute 1-2 eV. Some inorganic anions and aromatic molecules are photoionized at relatively low photon energies in water. The *hydrated electron* ($e^-_{aq}$) is an initial product of this reaction:

$$R + \text{light} \Rightarrow R^{\bullet +} + e^-_{aq} \qquad (4.12)$$

where $R^{\cdot +}$ is a positively-charged free radical. A simple model of $e^-_{aq}$ consists of a mobile electron surrounded by a sheath of oriented water dipoles.[11] This unique species has a strong red absorption band and characteristic chemical reactions. $e^-_{aq}$ can be considered as the anion of the aqueous hydrogen atom with $pK_a = 9.7$

$$H(aq) \rightleftharpoons H^+(aq) + e^-_{aq} \tag{4.13}$$

$e^-_{aq}$ is the strongest known reducing agent with a reduction potential $E_0' = -2.9$ V at pH 7.[12] Aqueous halide ions are photoionized by UV-C to give halogen atoms and $e^-_{aq}$. The resultant halogen atoms exist in equilibrium with halide ions to form *dihalide radical anions*: $I_2^-$, $Br_2^-$, and $Cl_2^-$. Photoionization of aromatic amino acids is significant in the UV photolysis of proteins (Section 6.1.1). This reaction is involved in photochemical inactivation of enzymes and also in tissue ablation by the UV excimer laser. The reactions of $e^-_{aq}$ with oxygen generates the *superoxide* radical-anion ($O_2^-$) in an oxygenated system:

$$e^-_{aq} + O_2 \Rightarrow O_2^-(aq) \tag{4.14}$$

Superoxide is a strong oxidizing agent and the precursor of hydrogen peroxide. Superoxide has been implicated in coronary artery disease, aging, and cancer induction. Erythrocytes and other cells contain *superoxide dismutase* and other enzymes that convert superoxide to water and oxygen.

## 4.5 QUENCHING REACTIONS

The term *quenching* describes any process in which the presence of an additive inhibits a photophysical or photochemical process. For example, a fluorescence quencher reduces the efficiency of a fluorescent emission. Quenching can take place *via* different pathways. In *physical quenching* the excitation energy of a short-lived intermediate is dissipated harmlessly as heat and the original species is restored. The products of *chemical quenching* are different chemical species. Many types of quenching mechanisms have been identified. In a *photoredox* reaction an electron (or H atom) is transferred from an excited "donor" (D*) to an unexcited "acceptor" (A) or *vice versa*:

$$D^* + A \Rightarrow D^+ + A^- \tag{4.15}$$

$$D^* + A \Rightarrow D^- + A^+ \qquad (4.16)$$

Reactions (4.15) and (4.16) requires"downhill" energetics. Specifically, the excitation energy of D* must exceed the difference between the ionization energy of D and the electron affinity of A or *vise versa*, less any solvation energy gained by the production of ionic species. The products on the right side of Reaction (4.15) and Reaction (4.16) are usually free radicals and undergo additional dark reactions. Triplet states of some dye-like molecules may act as oxidizing or reducing agents. The halogenated fluorescein dyes such as eosin Y (tetra-bromofluorescein) and rose bengal (tetra-iodo-tetra-chlorofluorescein) have this property. The fluorescein dye triplet states react with ferricynanide ion according to Reaction (4.15) to form a half-oxidized dye free radical and react with tryptophan according to Reaction (4.16) to give a half-reduced dye free radical. These reactions correspond to extracting an electron from the highest-energy filled MO of the triplet state or adding an electron to the lowest-energy unfilled MO of the triplet state, respectively.

Physical quenching may lead to loss of excitation energy

$$D^* + A \Rightarrow D + A + \text{heat} \qquad (4.17)$$

The acceptor in Reaction (4.17) may be an unexcited donor molecule. This process is termed *self-quenching*. Both physical and chemical quenching may occur for some D-A systems at different rates. *Intermolecular energy transfer* is a mode of physical quenching process in which the excitation energy of a donor is transferred to an acceptor:

$$D^* + A \Rightarrow D + A^* \qquad (4.18)$$

Reaction (4.18) requires both "downhill" energetics and *spin conservation.*[13] Several different mechanisms of energy transfer have been identified. "Long-range" singlet-singlet energy transfer is especially important because excitation energy can "hop" across intermolecular distances the order of several nanometers in solution and longer distances in ordered systems. The theory of Förster shows that the energy transfer efficiency (*Eff*) decreases with the sixth power of the intermolecular separation (*r*)

$$Eff = \frac{R_0^6}{R_0^6 + r^6} \tag{4.19}$$

where $R_0$ is the distance at which the rate of energy transfer equals the fluorescence decay rate of the donor in the absence of the acceptor. The detailed theory shows that $R_0$ depends on the fluorescence efficiency and fluorescence lifetime of the donor, the absorption strength of the acceptor, and the overlap between the donor fluorescence spectrum and the acceptor absorption spectrum. Intermolecular long-rang energy transfer is feasible between separated regions of a large molecule.

---

**Example 4.4** Long-range energy transfer in rhodopsin

Rhodopsin is the photoreceptor protein in the disk membranes of retinal rods. The chromophoric moiety of rhodopsin is 11-*cis* retinal that absorbs maximally at 500 nm. In a laboratory experiment, exogenous fluorescent molecules were bound to rhodopsin at different distances from 11-*cis* retinal and the efficiency of energy transfer was measured. In one such experiment the energy transfer efficiency was 36% when the donor was 5.7 nm distant from the 11-*cis* retinal molecule.

(a) Calculate $R_0$ for this donor. (b) Calculate the energy transfer efficiency if the separation is increased to 7.0 nm.

(a) From Eq. (4.16): $0.36 = R_0^6/(R_0^6 + 5.7^6)$; $R_0 = 5.2$.nm.

(b) For r = 7.0 nm, $Eff = 5.2^6/(5.2^6 + 7.0^6) = 0.14$.

This example shows that the long-range energy transfer efficiency is very sensitive to the intermolecular separation.

---

Intermolecular energy transfer may take place between like species.

$$D^* + D \Rightarrow D + D^* \tag{4.20}$$

The net effect of Reaction (4.20) is the transfer of excitation energy from the initial site of light absorption. The *exciton* is a particle-like description of this process. An exciton may be viewed as an energetic particle that hops from one entity to another until the energy is lost by emission of light or heat or by quenching. Exciton migration may be viewed as a collective excitation of an entire system. This description is most useful for molecules are arranged in an ordered array. New optical absorption bands in aggregates of dye molecules are attributed to excitons. Similar particle-like concepts have been applied to other types of excited systems. An *excimer* is a excited dimeric species that is stable only in the excited state. The active substance in an *excimer laser* is a mixture of two gases that react only in the excited state. For example, the active substance in an ArF excimer laser is a mixture of argon and fluorine gases. An *exciplex* is an excited molecular complex that is stable only in the excited state and dissociated in the ground state. New emission bands in mixtures of two aromatic hydrocarbons in solution are attributed to exciplex formation.

## 4.6 PHOTOSENSITIZATION

Photosensitization is a process in which the presence of a light-absorbing *photosensitizing agent* activates reactions in a *substrate* that otherwise would not take place. Virtually all natural photoprocesses utilize photosensitization, including photosynthesis, photoperiodism, photomorphogenesis, and retinal vision. Conventional photography is a practical application of photosensitization in which light absorbed by organic dyes extends the spectral response of silver salt emulsions. Some chemicals and drugs photosensitize human and animal skin to visible sunlight leading to erythema and other deleterious effects. Most photosensitized biological processes require, or are augmented by, the presence of molecular oxygen. A photochemical reaction in which an optically-excited photosensitizer molecule reacts first with a major component of the system other than molecular oxygen is referred to as *Type 1*. Oxygen may be involved in subsequent reactions. Photosynthesis and PUVA therapy are examples of *Type 1* photosensitizations. In the alternative *Type 2* pathway, the optically-excited photosensitizer molecule reacts first with molecular oxygen. This process generates a form of "active" oxygen that initiates further reactions with a substrate. *Type 2* photosensitizations in biology are referred to as *photodynamic action*. The formation of "active oxygen" was demonstrated in 1931 by Kautsky, an assistant professor at Heidelberg University, who proceeded to investigate its properties for another 25 years. The photosensitizer in Kautsky's original experiments was trypaflavine dye adsorbed on silica gel and illuminated in an enclosed vessel. Kautsky found that the presence of oxygen quenched the normal blue fluorescence of trypaflavine, and in the same system, the colorless form of the dye malachite green (the *leuco base* of the dye), also adsorbed on silica gel, was converted to the normal bright green form. He correctly deduced that energy was transferred from optically-excited trypaflavine molecules to molecular oxygen and the resultant "active oxygen" oxidized leuco malachite green to malachite green. The unstable, energy-rich form ofoxygen identified by Kautsky is singlet molecular oxygen ($^1O_2$). Some writers limit *Type 2* processes to those reactions in which singlet oxygen is the intermediate. This definition is unnecessarily limiting because both $^1O_2$ and other active oxygen intermediates may participate in some reactions by an extent that depends on the experimental conditions. Figure 4.8. depicts the overall scheme of *Type 1* and *Type 2* photosensitization reactions.

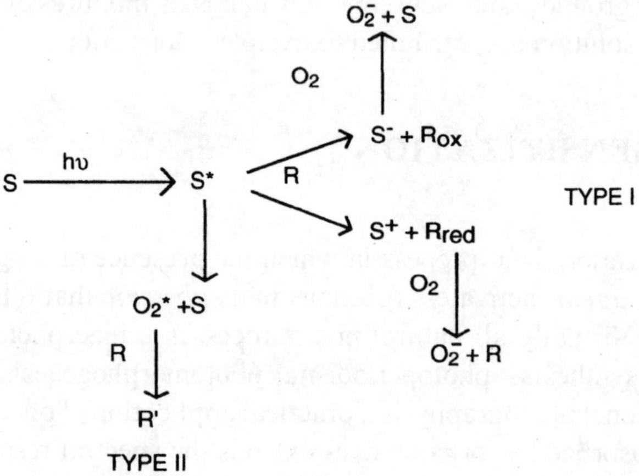


**Figure 4.8** General pathways of photosensitization. In the *Type II* pathway, an excited state of the photosensitizer (S*) transfers energy to molecular oxygen, leading to singlet oxygen ($O^*_2$); this singlet oxygen ($O^*_2$) in turn reacts with a substrate molecule (R), leading to product R'. S* may be either an excited singlet state or a triplet state. In the *Type I* pathway, S* oxidizes or reduces a substrate molecule (R). The reduced substrate ($R_{red}$) and reduced sensitizer ($S^-$) may be restored by reacting with molecular oxygen, leading to superoxide ($O^-_2$). The oxidized substrate ($R_{ox}$) and oxidized sensitizer ($S^+$) may further react.

## 4.6.1 *TypeI* Photosensitization of Membrane Lipid Peroxidation

Biological membranes are essential for life. Membranes regulate the flow of materials and information between different regions of cells and between a cell and its environment. Photosensitization of membrane lipid peroxidation is an important reaction that is involved in the initial stages of PDT. Lipid structures are described in standard biochemistry texts. Briefly, a *fatty acid* is the simplest type of lipid consisting of a carboxylate group attached to a long hydrocarbon tail. A fatty acid is *unsaturated* if the tail contains one or more double bonds.

For example, the saturated fatty acid stearic acid has a chain of 18 carbon atoms.

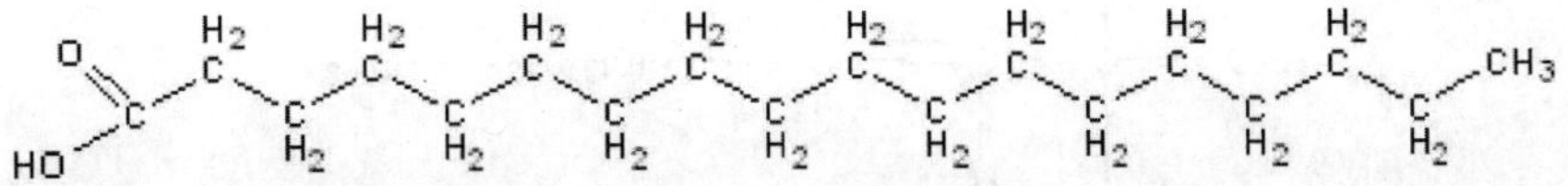

Fatty acids are stored in organisms in the form of *triglyceroles*. This structure is exemplified by the fat tristearin.

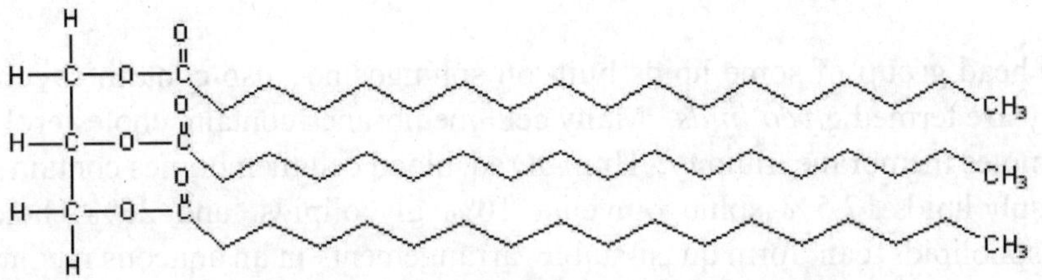

Natural fats contain a mixture of saturated and unsaturated triglyceroles. For example, olive oil contains 20% saturated fats ranging from 4 to 18 C atom chains and 80% unsaturated triglyceroles ranging from 16 to 18 C atom chains. Butter and beef fat are 40% and 46 % unsaturated, respectively. *Glycerophospholipids* are the major building blocks of biological membranes. The basic structure consists of a glycerole-3-phosphate with two acyl side chains and a substituted "head group".

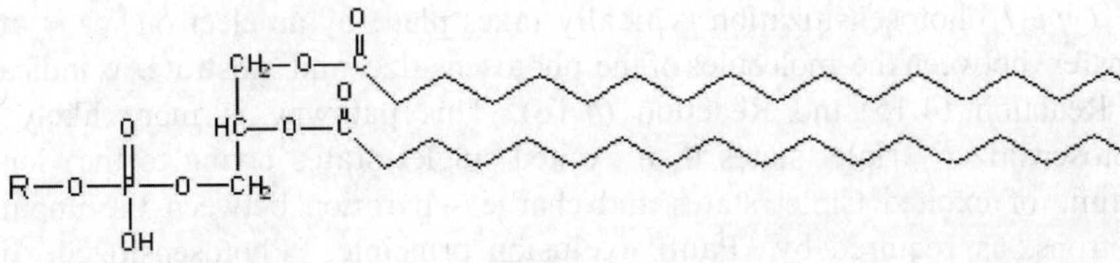

The most common head groups (R) in natural phosopholipids are phosphatidylethanolamine, phosphatidylcholine, and phosphatidylserine. Sphingolipids are a second major class of membrane constituents in which the basic structure is the long-chain amino alcohol sphingosine.

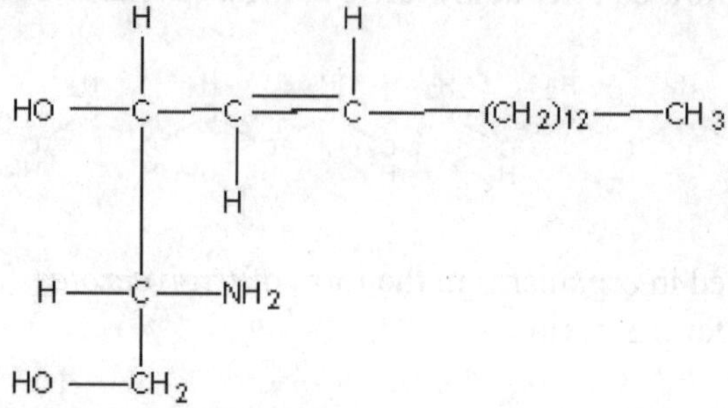


The head group of some lipids built on sphingosine also contain saccharides. They are termed *glycolipids*. Many cell membranes contain cholesterol which promotes membrane fluidity. Human red blood cell membranes contain 47.5 % phospholipids,17.5% sphingomyelin, 10% glycolipids, and 25% cholesterol. Phospholipids can form quasi-stable arrangements in an aqueous medium. In a phospholipid *monolayer* the hydrocarbon head groups are aligned in a parallel array in the aqueous region and the fatty acid tails extend into the air. A *bilayer* is a more stable structure consisting of two anti-parallel monolayers. The polar head groups of a bilayer are aligned on each exterior face and the hydrocarbon tails project into the interior (Figure 4.9). Membrane structure leads to the classification of bacteria into *gram positive* and *gram negative* depending on their coloration after treatment with crystal violet amd iodine, followed by destaining with an organic solvent. Gram-positive bacteria remain colored while gram-negative bacteria are decolorized.[14]

*Type 1* photosensitization typically takes place by an electron or H atom transfer between the molecules of the photosensitizer and substrate, as indicated by Reaction (4.15).and Reaction (4.16). This pathway is more likely for photosensitizer triplet states than excited singlet states owing to the longer lifetime of excited triplet states and charge separation between the unpaired electrons, as required by Pauli exclusion principle. Photosensitized lipid peroxidation is an important example of a *Type 1* process that is initiated by photosensitizer triplet state and stabilized by oxygen. The initial step in the photooxidation of a phospholipid linoleoyl group (LH) can be represented as:

$$T_1 + LH \rightarrow SH^{\cdot} + L^{\cdot} \qquad (4.21)$$

where $T_1$, $SH^{\cdot}$, and $L^{\cdot}$ indicate the triplet state of the photosensitizer, the

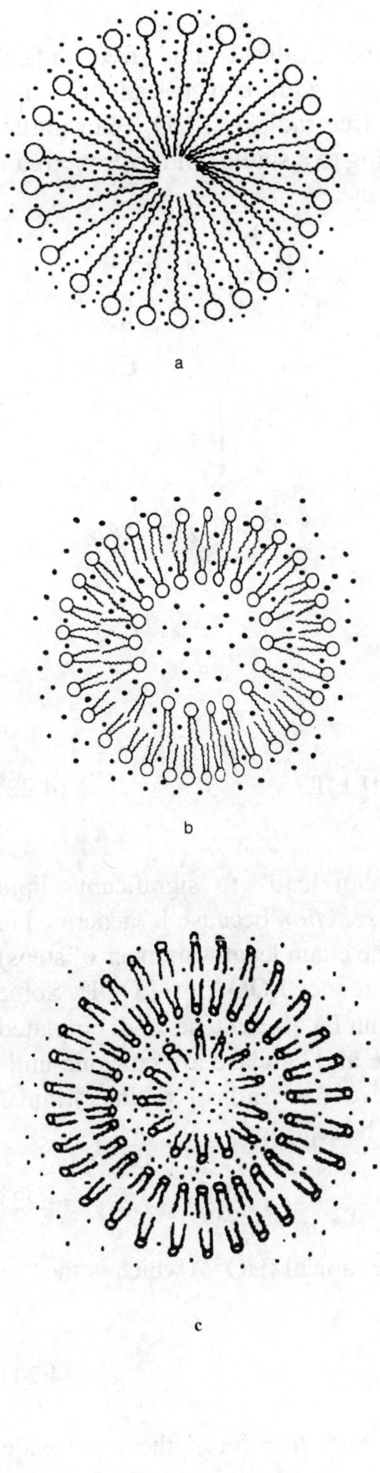


**Figure 4.9** Structural arrangements of amphipathic molecules. (a) A *micelle* is an ordred arrangement of amphipathic molecules, with the polar head groups on the outside and the hydrophobic tails on the interior. The dots depict water molecules. (b) In a *bilayer vesicle*, the polar head groups are located in the interior and exterior aqueous regions. (c) A *multilamellar liposome* is aspherical vesicle comprising two-chain phospholipids.

reduced photosensitizer free radical, and the oxidized lipid free radical, respectively. The reaction of the oxidized lipid with oxygen leads to a lipid peroxyl radical (LOO $^{\bullet}$). The intermediate free radical LOO $^{\bullet}$ can initiate additional oxidation *via* a chain reaction leading to a stable lipid hydroperoxide (LOOH). This mechanism can be represented as

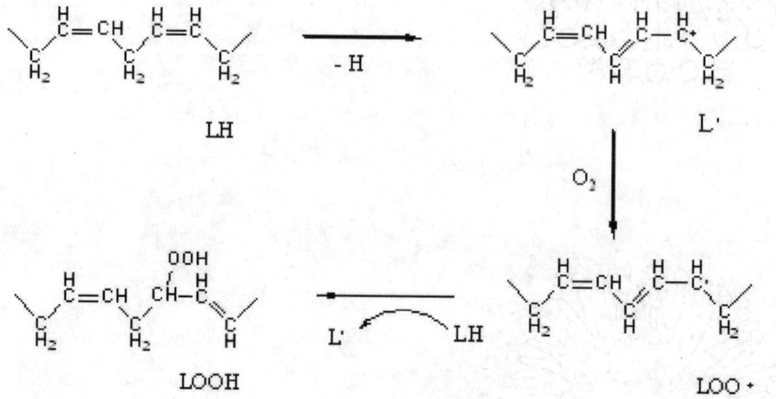


$$\begin{aligned} LH &\Rightarrow L^{\bullet} + H \\ L^{\bullet} + O_2 &\Rightarrow LOO^{\bullet} \\ LOO^{\bullet} + LH &\Rightarrow LOOH + L^{\bullet} \end{aligned} \tag{4.22}$$

Thus, a small amount of lipid free radical can lead to significant lipid peroxidation. This system constitutes a *chain reaction* because $L^{\bullet}$ generated in the first step is regenerated in the third step. The chain length (number of steps) is reduced by the presence of antioxidants that reduce LOO $^{\bullet}$ to LOOH. Some common anti-oxidants are α–tocopherol (vitamin E), β-carotene, and butylated hydroxytoluene (BHT). Superoxide dismutase and catalase are enzymic anti-oxidants. Oxygen may restore the reduced dye free radical to the original photosensitizer and the reaction sequence may be repeated

$$SH^{\bullet} + O_2 \Rightarrow S + HO_2^{\bullet} \tag{4.23}$$

Reaction (4.23) generates the hydroperoxy free radical ($HO_2^{\bullet}$) which is the acidic form of superoxide

$$HO_2^{\bullet} \rightleftharpoons O_2^{-} + H^{+} \tag{4.24}$$

The $pK_a$ of the hydroperoxy radical is 4.4 and, therefore, the superoxide radical-anion dominates at physiological conditions. Additional lipid peroxidation may be initiated by superoxide. This process is inhibited by superoxide dismutase

$$2\,O_2^{\cdot -} + 2\,H^+ \Rightarrow H_2O + O_2 \qquad (4.25)$$

The enzymes catalase and peroxidase can remove the resultant hydrogen peroxide:

$$2\,H_2O_2 \Rightarrow 2\,H_2O + O_2 \qquad (4.26)$$

In 1934 Haber and Weiss suggested that highly reactive hydroxyl free radicals ($OH^{\cdot}$) are generated by the reaction of superoxide and hydrogen peroxide

$$O_2^{\cdot -} + H_2O_2 + H^+ \Rightarrow OH^{\cdot} + OH^- + O_2 \qquad (4.27)$$

The Haber-Weiss reaction is very slow. However, the process is catalyzed by the presence of oxidizing metal ions:

$$O_2^{\cdot -} + Fe^{3+} \Rightarrow O_2 + Fe^{2+} \qquad (4.28)$$
$$H_2O_2 + Fe^{2+} \Rightarrow OH^{\cdot} + OH^- + Fe^{3+} \qquad (4.29)$$

Reaction (4.29) is termed a *Fenton reaction* . The iron-catalyzed Haber-Weiss pathway is probably involved in lipid peroxidation catalyzed by enzymes that generate $O_2^{\cdot -}$ in the dark An important example of this process is the co-oxidation of a lipid and xanthine by the enzyme xanthine oxidase. Xanthine is a purine derivative, discovered in 1817, and the metabolic precursor of uric acid. A hereditary deficiency of xanthine oxidase can lead to accumulation of xanthine kidney stones.

## 4.6.2 Properties of Singlet Molecular Oxygen

Considerable evidence indicates that singlet oxygen is the dominant intermediate in many biological photosensitizations including photohemolysis of red blood cells and photodynamic therapy (PDT). In view of its importance, it is appropriate to review the relevant information about the production, identification, and chemical properties of singlet oxygen. Molecular oxygen is a unique species because the most stable form is a triplet state ($^3O_2$) and the lowest energy electronically-excited state is a singlet state. The lowest excitation energy of $^1O_2$ is 0.98 eV, corresponding to a wavelength of 1265 nm. The triplet energy of many dye and natural pigments exceed 0.98 eV favoring "downhill" energy transfer

$$T_1(\downarrow\downarrow) + \ {}^3O_2\,(\uparrow\uparrow) \Rightarrow S_0\,(\uparrow\downarrow) + {}^1O_2\,(\uparrow\downarrow) \tag{4.30}$$

The electron spin directions show that total spin is conserved by Reaction (4.30). The singlet oxygen *quantum yield* ($\Phi_\Delta$) is an important property of a photosensitizing agent. This quantity is defined as the number of $^1O_2$ molecules generated for each photon absorbed by the photosensitizer. The production of $^1O_2$ by photosensitization involves four steps: (1) Light absorption by the photosensitizer. (2) Formation of the photosensitizer triplet state; the quantum yield of this process is the *intersystem crossing efficiency* or *triplet yield* ($\Phi_T$). (3) Trapping of the photosensitizer triplet state by molecular oxygen within its lifetime; the fraction of trapped triplet states in a given system is designated by $f_T$. (4) Energy transfer from the photosensitizer triplet state to molecular oxygen; The probability of this energy transfer is $S_\Delta$. The experimental value of $S_\Delta$ is usually unity for those agents in which the fluorescence is not quenched by oxygen. Overall, $\Phi_\Delta = \Phi_T f_T S_\Delta$.

Virtually all measurements of $\Phi_\Delta$ are scaled to a reference substance. Frequently employed water-soluble standards include rose bengal, methylene blue, and fullerene $C_{60}$. The published values of $\Phi_\Delta$ show considerable variations with the solvent, reaction conditions, and the measurement technique. Some representative results are given in Table 4.3. The decay lifetime of $^1O_2$ in the absence of chemical reactions ($\tau_\Delta$) depends on the solvent. Some experimental values of the solvent-induced decay lifetime of $^1O_2$ are approximately 4 μs in $H_2O$, 30 μs in $D_2O$, 12 μs in ethanol, 700 μs in $CCl_4$, and $\approx$0.2 μs in living cells. There is a correlation between $\tau_\Delta$ and the strength of the infrared absorption of the solvent near 1250 nm. This result suggests that $^1O_2$ decays by transferring its electronic excitation energy to vibrational energy of the solvent.

#### 4.6.2.1 Identification of Singlet Oxygen

The techniques of $^1O_2$ detection and measurement depend on the method of generation. Optical spectroscopy is widely employed for gaseous systems. Light emission by $^1O_2$ has been identified in *airglow* originating 40 to 180 miles above the surface of the Earth and also from the atmospheres of Venus and Mars. Luminescence is the "gold standard" of singlet oxygen identification in

**Table 4.3 Triplet state and singlet oxygen quantum yields** [a]

| **Molecule** | **Medium** | $\Phi_T$ | $\Phi_D$ |
|---|---|---|---|
| Anthracene | $C_6H_6$ | 0.78 | 0.61 |
| Methylene Blue | $H_2O$ | 0.52 | 0.52 |
| Rose Bengal | $H_2O$ | 0.86 | 0.75 |
| Riboflavin | $CH_3OH$ | 0.61 | 0.48 |
| Hematoporphyrin | $CH_3OH$ | 0.82 | 0.53 |
| Uroporphyrin | $H_2O$ | 0.93 | 0.71 |
| Bacteriochlorophyll a | benzene | 0.32 | 0.32 |
| Aluminum Phthalocyanine | $H_2O$ | 0.38 | 0.36 |
| Tetrasulfonate octaethyl-purpurin | benezene | 0.82 | 0.67 |
| Hypericin | ethanol | 0.72 | 0.72 |
| 8-Methoxypsoralen | benezene | 0.011 | 0.0044 |

[a] The results in this table were taken from different sources

solution. The measurements are based on the 1269 nm luminescence emitted by the radiative decay of $^1O_2$ and the 634 nm *dimol emission*. In the latter process, two $^1O_2$ molecules combine to form two molecules of the higher-energy excited singlet state of molecular oxygen and the excess energy is emitted as a 634 nm red light photon. An essentially unambiguous test for $^1O_2$ in aqueous systems is based on the increase of the decay lifetime from $\approx 4$ μs in $H_2O$ to $\approx 30$ μs in $D_2O$. Thus, the rate of a photochemical reaction mediated by $^1O_2$ is accelerated in perdeuterated solvents. Indirect tests for singlet oxygen are based on the inhibiting effects of an additive on the rate of a photochemical reaction mediated by $^1O_2$. Azide ion ($N_3^-$) is a useful water-soluble probe. The protective effect of azide is attributed to physical quenching

$$^1O_2 + N_3^- \Rightarrow {}^3O_2 + N_3^- \qquad (4.31)$$

Other frequently employed water-soluble quenching agents include 1,4-diazabicyclo-[2.2.2]octane (DABCO), *p*-nitrosodimethylaniline plus histidine, 2,5-dimethylfuran (DMF), 9,10-dimethylanthracene (DMA), 1,3-diphenyl-isobenzofuran (DPBF), ascorbic acid, and tryptophan. Some quenchers of $^1O_2$ used in organic solvents and lipid membranes are cholesterol, α-tocopherol and β–carotene. Cholesterol provides a useful diagnostic test that distinguishes free radical reactions from those of $^1O_2$. The reaction of $^1O_2$ with cholesterol leads to a stable product 3β-hydroxy-5α-hydroperoxy-$\Delta^6$-cholestene, while reactions of cholesterol with free radicals produce the corresponding 7α– hydroperoxide that can separated from the 5α-hydroperoxide by chromatography

### 4.6.2.2 EPR Spin Label Techniques

An unambiguous test for $^1O_2$ is based on *electron paramagnetic resonance* (EPR). EPR is a non-optical technique in which energy transfer between the intrinsic magnetism of unpaired electrons and an external magnetic field is measured with a sensitive microwave detection system. The interaction energy between an unpaired electron and a B-field corresponds to the energy difference between the spin-up and spin-down states

$$\Delta E = 2\mu_B B \tag{4.32}$$

where $\mu_B$ is the magnetic moment of the electron (the *Bohr magneton*) and $B$ is the strength of the effective B-field. In an EPR measurement a small sample is located in a microwave cavity that lies between the poles of a strong electromagnet. A transverse microwave field generated by a *klystron* is applied to the microwave cavity which induces spin transitions between the two spin states. The maximum power absorption takes place at the exact resonance condition

$$hf = 2\mu_B B \tag{4.33}$$

where $f$ is the microwave frequency. In practical units, the resonance B-field of a free electron is 1.25 tesla (12,500 gauss) at the Q-band microwave frequency of 35 Ghz. Various interactions induce shifts of the resonance B-field for unpaired electrons in compounds. An EPR spectrum is recorded by measuring the strength of the microwave signal when the B-field is swept over a small range. The spectrum of a free electron consists of a narrow absorption band corresponding to transitions between the spin-up and spin-down states. (Each resonance "line" comprises an adjacent "hill" and "valley" because the measurement technique leads to the first derivative of the absorption band.) The perturbing effects of nearby magnetic nuclei induce a substructure to EPR spectra termed the *hyperfine interaction*. The B-field of the adjacent nuclei add or subtract from the external B-field depending on the orientation of the nuclei in the external B-field. $^1O_2$ is a non-magnetic molecule and, therefore, it cannot be detected directly by EPR. However, the reaction of $^1O_2$ with a stable molecule can generate a long-lived free radical, referred to as a *spin label*, whose structure as determined by EPR provides an unambiguous identification. The compound 2,2,6,6-tetramethyl-4-piperidone (TEMP) has been employed as a spin label probe for $^1O_2$. The reaction of $^1O_2$ with TEMP with leads to the free radical 2,2,6,6-tetramethyl-4-piperidone-N-oxyl (TEMPO).

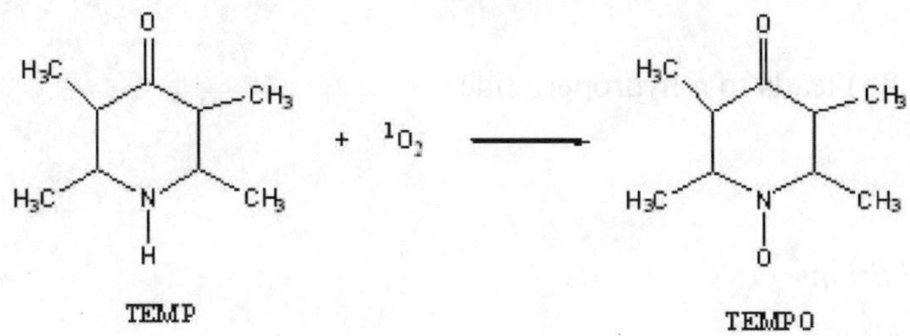


(4.34)

The EPR spectrum of TEMPO in ethanol consists of three equal intensity lines characteristic of a nitroxide radical owing to $I = 1$ for the $^{14}N$ atom. Spin labels have been employed for detection of other active oxygen intermediates. Superoxide forms an unstable free radical with 5,5-dimethyl-1-pyrroline-1-oxide (DMPO) which has a characteristic four-line EPR spectrum. DMPO also reacts with hydroxyl radicals to give a different four-line spectrum.

### 4.6.3 Singlet Oxygen Reactions

$^1O_2$ is *electrophilic* and reacts with molecules having electron-rich functionality. Classes of compounds that are known to react with $^1O_2$ include furans, electron-rich olefins, phenols, dialkylsulfides, tryptophan, and lipids. Some important types of $^1O_2$ reactions include

(1) *ene reaction with olefins*

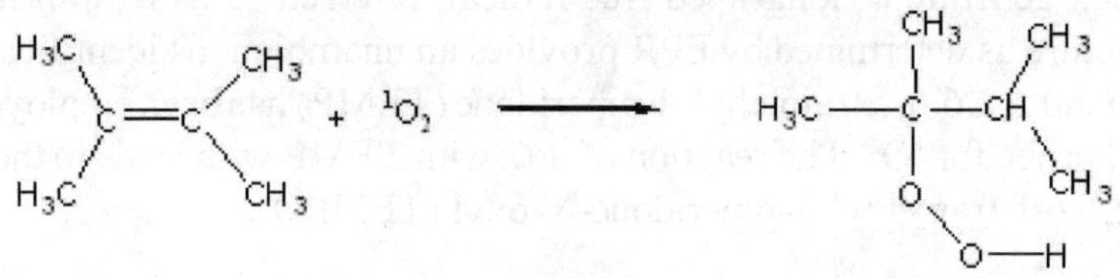

(4.35)

Reaction (4.35) leads to a hydroperoxide

(2) *cycloaddition*

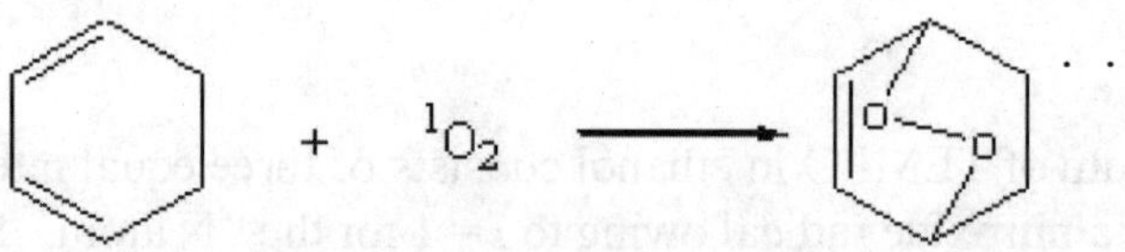

(4.36)

The product in Reaction (4.36) is an endo-peroxide.

(3) *oxidation of sulfides*

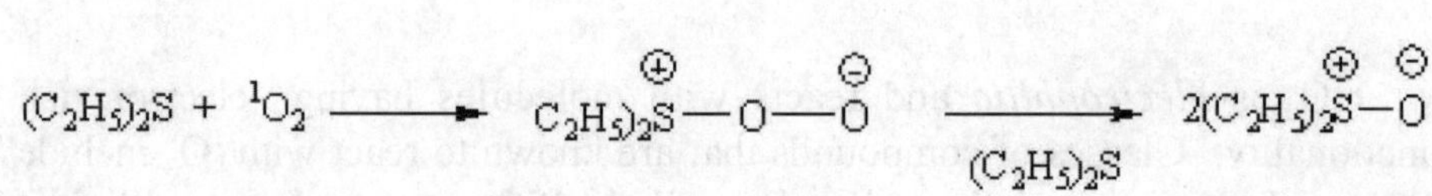

(4.37)

Reaction (4.37) corresponds to the photooxidation of the amino acid methione to a sulfoxide.

(4) *photooxidation of phenols*

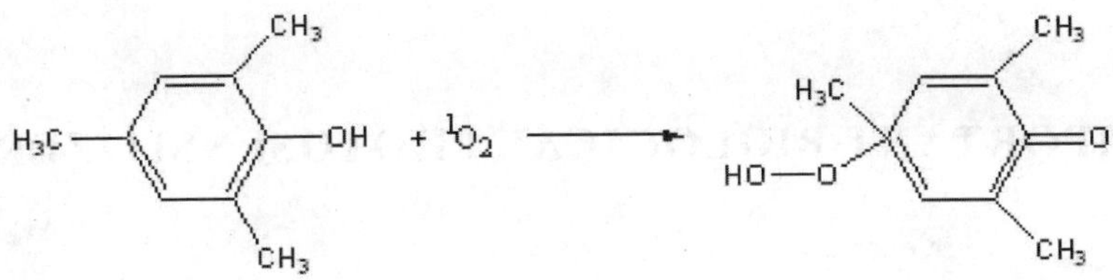


(4.38)

Reaction (4.38) indicates the photooxidation of 2,6-di-tert-butylphenol to a quinone hydroperoxide.

### 4.6.3.1 *Type 2* Photosensitization of Lipid Peroxidation

Photodynamic damage to cells has been identified with the attack of $^1O_2$ on the plasma membrane and intracellular membranes. The likely reactive sites are unsaturated phospholipids, cholesterol, and membrane proteins. Lipid peroxidation mediated by $^1O_2$ can be represented as

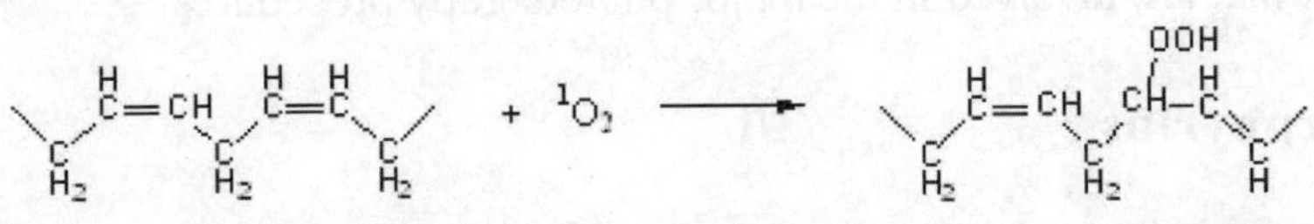


(4.39)

Reaction (4.39) is another example of the *ene* reaction. Generation of $^1O_2$ in the lower atmosphere by the action of sunlight on *polycyclic aromatic hydrocarbons* (PAH) are involved in atmospheric pollution and smog. PAHs are a large class of organic pollutants released in the atmosphere by many types of natural and artificial sources, including volcanos, forest fires, automobile exhausts, electric power generators, and incinerators. More than a hundred PAHs have been identified as pollutants. PAHs absorb sunlight in the UV-A region, leading to metastable triplet states which are favorable for the generation of $^1O_2$. PAHs also react with $^1O_2$ to form dioxetanes and endoperoxides which are potential candidates for induction of biological damage.

## 4.7 IMPORTANT BIOLOGICAL PHOTOSENSITIZERS

Known photosensitizers of biological systems number in the thousands. However, extensive investigations have been carried out only for those relatively few compounds that are involved in natural photobiology and important phototherapies. The research objective is to relate the photophysical and photochemical reactions at the molecular level to physiological endpoints. This goal is not easily achieved because photosensitization pathways are highly sensitive to the environment in which they take place. A comprehensive research program includes studies on isolated molecular systems, microorganisms, and experimental plants and animals. This information plus judicious guessing may lead to a reasonable description of the action mechanism. However, the application of this ambitious program to a phototherapy is limited by the difficulties of correlating results from different laboratories, biological variability, possibly the lack of an animal model (for example, psoriasis), as well as the high cost and time expenditure. The last essential step consists of controlled clinical trials on human patients which requires regulatory approvals. It is useful at this point to review the basic molecular properties of those classes of molecules that are involved in the major phototherapy procedures.

### 4.7.1 Porphyrins

Porphyrins are derivatives of the porphine molecule with a characteristic tetrapyrrole ring structure. Thousands of natural and synthetic porphyrins have been identified. The important protoporphyin IX (PpIX) and the closely related hematoporphyrin (HP) and hydroxyethylvinyldeuteroporphyrin (HVD) are metal-free substituted porphines

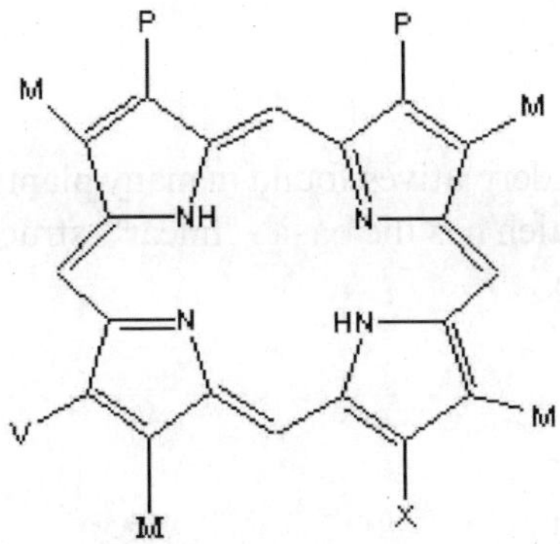

(4.40)

with M = $CH_3$ and P = $CH_2CH_2COOH$. In HP, X = Y = $CHOHCH_3$ ; in HVD, X = CH=$CH_2$, Y = $CHOHCH_3$ ; in PpIX, X = Y = CH=$CH_2$ . Ferrous protoporphyin IX is the red chromophore in hemoglobin and myoglobin. Human porphyrias result from genetic or chemical induced deficiencies in the enzymatic pathways leading to an excess of metal-free porphyrins in the circulation. The uncommon genetic disease erythropoietic protoporphyria (EPP) results from a deficiency of ferrochelatase, the enzyme that catalyzes the insertion of ferrous iron into protoporphyrin. EPP leads to abnormally high PpIX levels in RBC, feces, and plasma, accompanied by enhanced photosensitization of skin to sunlight, and a high risk of hepatic failure. It is speculated that vampires were victims of EPP, which explains their bizarre appearance, fluorescent teeth, preference for the dark, and propensity for drinking blood. Chlorophyll is a reduced porphyrin (a chlorin) with a central magnesium atom and a long phytyl tail. Porphyrins can act as *Type 1* and *Type 2* photosensitizers, depending on the chemical structure and central metal atom. Aggregation of the more hydrophobic porphyrins in an aqueous medium leads to spectral shifts and a marked reduction of the photosensitizing efficiency. A hydrophobic porphyrin may be good photosensitizer when bound to a serum protein or incorporated in a biomembrane. The first successful PDT drug was *hematoporphyrin derivative* (HPD), a porphyrin mixture synthesized from commercial HP. The active constituent in HPD consists of covalently-linked porphyrin dimers and small oligomers connected by ether and ester bridges. Photofrin® is a purified form of HPD approved for PDT in thc U.S and many other countries. Many "second-generation" PDT drugs are based on the tetrapyrrole macrocycle, including porphines, chlorins, phthalocyanines, purpurins, pheophobides, and texaphyrins Additional information is provided in Chapter 10 and Chapter 11.

## 4.7.2 Psoralens

Psoralens are furocoumarin derivatives found in many plants including parsley, rue, figs, and legumes. Psoralen has the basic "linear" structure and angelicin has the basic"bent" structure.

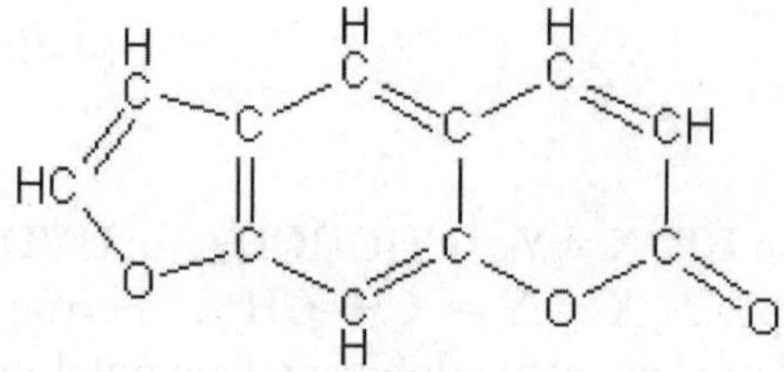

psoralen

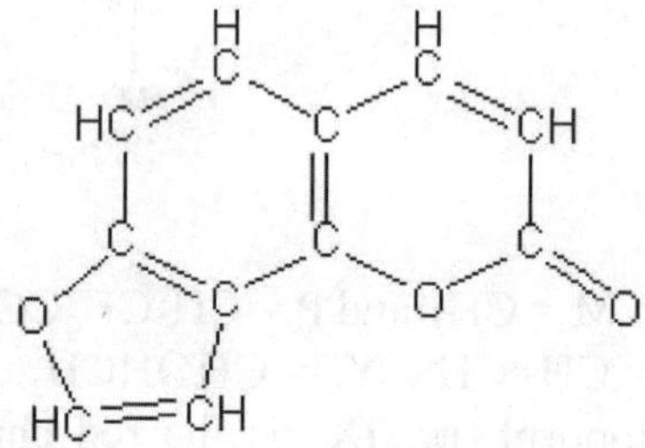

angelicin

(4.41)

Certain linear furocoumarins are described as being "active", indicating that they photosensitize human skin to UV-A. The active derivatives include 8-methoxypsoralen (8-MOP) and 4,5',8-trimethylpsoralen (TMP) both of which are used for PUVA phototherapy. The clinical responses in PUVA are attributed to *Type 1* photosensitization reactions with DNA. Psoralen derivatives have low single oxygen quantum yields which may be involved in *Type 2* reactions with substrates other than DNA.

## 4.7.3 Hypericin

The anthraquinoid pigment hypericin (HY) is one of the oldest known natural photosensitizing pigments.

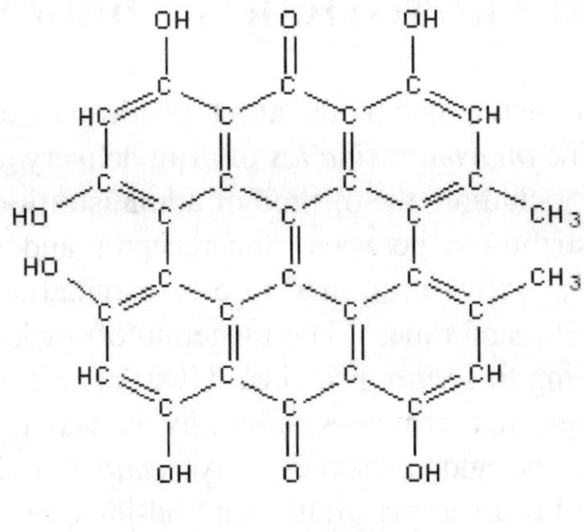


(4.42)

hypericin

Exposure of grazing animals to sunlight after ingestion of plants containing HY leads to inflammation, ulceration, and infection, and in severe cases, convulsions and death. This disease is *hypericism*. Over 200 species of the plant genus *Hypericum* contain HY and/or the closely related pseudohypericin (pHY), in which one of the methyl groups is replaced by a hydroxymethylene group. *H. perforatum* (klamath weed, St. John's wort) found in the United States and Australia contains HY and pHY. Buckwheat (*Fagopyrum esculentum*) contains the HY derivative fagopyrin and the mold *Penicilliopsis clavariaeformis* contains pencilliopsin that can be photooxidized to HY. The ciliate *Blepharisma undulans* contains a light-sensitive polymer blepharismin (or zoopurpurin) which is similar to hypericin. HY is an acidic compound having a p$K_a$ about 4.5. Alcoholic solutions of HY are bright red with the longest wavelength absorption band at 590 nm and a strong red fluorescence emission band at 600 nm. HY can be dispersed in water by vigorous stirring, leading to a violet non-fluorescent material of very high molecular weight. Medical properties of HY have been of interest for centuries. The red juice of *Hypericum* was recommended by Paracelsus for treatment of melancholia. St. John's wort is currently marketed as an over-the-counter tranquilizer. HY is a potent *Type 2* photosensitizer. The classic study of Blum in 1941 showed that HY photosensitizes hemolysis of red blood cells *in vitro* and that oxygen is required. HY also photosensitizes human skin, plant tissues, insect larvae, protozoa, cells, and viruses. The mechanism of action depends on the specific target and reaction conditions. Recent interest in HY derives from its ability to inhibit retroviral infection *in vitro* and *in vivo*.

## 4.8 TARGETING OF PHOTOTHERAPY DRUGS

A drug must reach and be retained by its target in adequate concentrations in order to be effective. The *pharmacokinetics* of drug delivery and utilization depends on many factors, including the method of administration, absorption into the bloodstream, distribution between the receptor and other tissues, metabolic alterations of the active drug, and rate of elimination. Lipophilic drugs can pass through cell membranes. The targeting of hydrophilic drugs may be facilitated by binding to serum proteins. Hydrophobic agents may be administered *via* a carrier that promotes solubility in water, such as an emulsion, surfactant micelles, and liposomes. An *emulsion* is a colloidal dispersion of two liquids. For example, milk is an oil-in-water emulsion and butter is a water-in-oil emulsion. Detergent *surfactants* are *amphiphiles* with a hydrophilic head group attached to long hydrophobic tail. A *micelle* consists of a loose arrangement of amphiphiles in which the tails form a spherical interior region with the head groups located in the aqueous medium (Figure 4.9a). Micelles can be formed by anionic detergent molecules (for example, sodium dodecyl sulfate, sodium cholate), uncharged molecules (for example, polyoxyethylene ethers), and cationic molecules (for example, cetyltrimethylammonium bromide). The *critical micelle concentration* (CMC) is the minimum detergent concentration required for micelle formation. The CMC ranges from ranges from 2 to about 150 depending on the detergent and its concentration. A spherical bilayer is another possible arrangement of amphiphiles. A soap bubble consists of a head-to-head arrangement of amphiphiles with the tails on the outside in air. (A soap is a sodium or potassium salt of a fatty acid.) A bilayer vesicle is a more stable tail-to-tail arrangement of amphiphiles with the polar head groups located in the interior aqueous region and the exterior water region (Figure 4.9b). *Liposomes* are bilayer vesicles formed by phospholipids. Drug targeting by liposomes has evolved as a useful technology during the past three decades.

### 4.8.1 Preparation and Properties of Liposomes

Liposomes are increasingly being used for drug delivery. The composition and method of preparation determines the liposome size distribution, the membrane properties, the substances that can be encapsulated, and the loading efficiency. In a simplest method of liposome preparation pioneered by Bangham, an aqueous buffer solution is added to the dry lipid film and stored for a few hours with occasional stirring. The suspension turns cloudy and microscopic examination reveals the presence of *multilamellar vesicles* (MLV) with a wide size distribution. A water-soluble compound such as an enzyme can be encapsulated in MLV by dissolution in the buffer prior to liposome formation. Water-soluble lipophilic additives are loaded into preformed liposomes by partitioning from an external aqueous phase. Sonication converts MLV to *small unilamellar vesicles* (SUV) with a narrower size distribution, typically 20 nm - 40 nm in diameter. In an alternative "detergent exclusion" method, the lipid is first solubilized by an aqueous detergent and then the SUV are formed by passing the dispersion through a gel column that retains the smaller detergent molecules. SUV can be formed from MLV also by sieving and pressing through a membrane filter. *Large unilamellar vesicles* (LUV) can be formed by injecting an organic solvent containing the lipid into the aqueous phase through a fine needle. Phospholipid membranes including lipsomes exist in an ordered *gel* phase below a phase transition temperature $T_c$ and in a more fluid *liquid crystal* phase above $T_c$. The value of $T_c$ depends on the phospholipid structure. For example, $T_c$ = + 41°C for saturated (16:0/16:0) phosphatidylcholine (PC) liposomes and $T_c$ = - 36 °C for unsaturated (16:1/16:1) PC liposomes. In this notation the first integer is the total length of each chain and the second integer is the number of unsaturated bonds in that chain. Liposome preparations usually incorporate additional constituents. The presence of 20-50% cholesterol increases the membrane fluidity, an ionic compound confers a positive or negative surface charge, and antioxidants inhibit lipid peroxidation. The pharmacokinetics of liposomes depend on the size distribution, lipid composition, and mode of administration. Circulating liposomes localize predominantly in the liver, spleen, and bone marrow, and in circulating macrophages. Liposome targeting applications include enzyme replacement therapies, stimulation of macrophage-mediated host defense mechanisms against tumor cells and infective agents, and delivery of metal-chelating agents. Another potential application involves drug targeting to circulating blood cells and vascular endothelial cells. Coupling of liposomes to monoclonal antibodies can be employed for selective targeting of anticancer drugs. Temperature-sensitive liposomes been used to deliver PDT drugs to vascular epithelia for treatment of age-related macular degeneration (AMD). In this procedure, liposomes incorporating the PDT drug are injected into the bloodstream where they are lysed by heating with blue-green light from an argon-ion laser to release the PDT drug. The retina is then exposed to laser-delivered red light which initiates the PDT process.

## NOTES

1. The lowest energy state of a molecule corresponds to the most stable or "ground" state. Except at ultra-high optical power levels, absorption of one photon excites one electron of the molecule to a higher electronic state. Electronic excitation also affects the vibrational motions of the molecule owing to the changes in the potential energy of the entire molecule.

2. The Bohr model is applicable also to one-electron ions such as $He^{+}$ and $Li^{++}$.

3. A total differential equation is the relationship between a function y(x) and total derivatives of y(x). Differential equations can be solved by integration and numerical techniques. A partial differential equation includes partial derivatives of two or more variables. Many of the important equations of classical physics are partial differential equations including the equations of wave motion, Maxwell's equations of the electromagnetic field, and the diffusion equation. Solutions of partial differential equations are a major part of mathematical analysis.

4. A complex number consists of a "real" term and an "imaginary" term, for example $a + i\, b$ where $i = \sqrt{-1}$. (In electrical engineering $j = \sqrt{-1}$). The amplitude of a complex number is $\sqrt{(a^2 + b^2)}$.

5. The wave function of an *s* state is spherically symmetric and corresponds to zero angular momentum.

6. A periodic table arranges the stable elements in a rectangular array in order of increasing atomic weight. In a typical periodic table, the first row has 2 elements, the next two rows have 8 elements, the next row has 18 elements, etc. The elements in each column or "group" have similar properties. For example, Group I consists of the alkali metals Li, Na, K, Rb, Cs, and Fr, and Group 8 consists of the rare gases He, Ne, Ar, Kr. Xe. and Rn.

7. In spectroscopic notation, a singlet state has no unpaired electrons, a doublet state has one unpaired electron, and a triplet state has two unpaired electrons. The quantum-mechanical multiplicities indicate the number of unique orientations an atom or molecule can acquire in an external magnetic field. The spin multiplicity is $(2S+1)$, where s is the electron spin angular momentum in units of (h/2p). Therefore, $S = 0$ for a singlet state, $S = 1/2$ for a doublet state, and $S = 1$ for a triplet state. The electron spin quantum number $S$ should be distinguished from an *s*-type state which corresponds to zero orbital angular momentum.

8. The spin of a photon is unity because electromagnetic radiation has two unique directions of polarization.

9. Quantum states having the same energy are referred to as degenerate. For example, the $n = 2$ state of atomic hydrogen is two-fold degenerate, consisting of the states with $l = 0$ and $l = 1$. In the FEM model for a porphyrin the lowest energy state $l = 0$ is non-degenerate. There are four equivalent excited states in which one electron occupies an $l = 4$ level and one electron occupies a $l = 5$ level. The total orbital angular momentum ($L$) of each state is calculated by the quantum mechanical rules of addition according to: $|5 \pm 4| = \pm 9, \pm 8, \pm 7, ..., \pm 1$. Two of the 18 *pi* electrons electrons can occupy the $L = 0$ state, one electron for each rotation direction. The remaining 16 *pi* electrons occupy two of the four-fold degenerate excited states, with two electrons for each rotation direction. The detailed calculation shows that these states correspond to $L = \pm 9, \pm 1$ and $L = \pm 8, \pm 2$ (Figure 4.5 left side). Electron interactions split the four-fold degenerate states into pairs of two-fold degenerate states (Fig. 4.5, right side). The strong B band (Soret band) corresponds to allowed optical transitions from $L = 0$ to $L = \pm 1$. The weak Q bands correspond to forbidden transitions from $L = 0$ to $L = \pm 9$. The total angular momentum of the degenerate energy states is labeled L in the left side of Figure 4.5b.

10. The theory of absolute reaction rates identifies $\Delta H$ and $A$ with thermodynamic properties of the unstable "activated complex" formed by the reactants. The dissociation of the activated complex restores the original reactants or leads to the products.

11. An ionic species is stabilized in a fluid medium by dielectric polarization. In the first step of this process, the electrons of adjacent solvent molecules shift towards a positively charged species and away from a negatively charged species. This process stabilizes the product because energy must be expended to restore the original state of the solvent. In a polar medium such as water, the water dipoles undergo rotations that further stabilize a dissolved charged species.

12. Reduction-oxidation (redox) potentials indicate the ability of a single chemical species to accept electrons. The numerical values $E_0$(volts)are scaled to the free energy required for electrochemical reduction of hydrogen ions to hydrogen gas at standard conditions. Half-cell reactions with positive values of $E_0$ involve reactants that are more readily reduced than hydrogen ions and *vice versa* for negative values of $E_0$. Therefore, a species with a large negative $E_0$ is a strong reducing agent. Values of $E_0'$ used in biochemistry are scaled to aqueous solutions at pH 7.

13. Spin conservation means that the total spin of the products must equal the total spin of the reactants. All electrons in a singlet state are paired (↑↓). Two electrons in a triplet state have parallel spins (↑↑). Only a fraction of the reactants leading to products may satisfy the spin conservation rule. Three cases can be distinguished. (1) Both D* and A* are excited singlet states in "singlet-singlet" (S-S) energy transfer. This process always conserves total spin. The addition of an acceptor may convert the normal fluorescence of the donor to that of the acceptor. This effect is *sensitized fluorescence*. (2) Both D* and A* are triplet states in "triplet-triplet" (T-

T) energy transfer. This process conserves total spin only for those interactions in which the spin orientations are parallel in D* and A*. (3) In "triplet-singlet" (T-S) energy transfer D* is a triplet state and A* is an excited singlet state. This process cannot conserve total spin. However, this "forbidden" interaction may take place if D* is a long-lived triplet state.

14. Microscopic examination shows that the cell wall of gram -negative *E. coli* consists of an outer lipid bilayer, followed by a thin interior glycopeptide layer, a gap, and another phospholipid bilayer. Gram-positive bacteria such as *Staphylococcus aureus* remain stained deep blue-black after treatment with Gram stain. The *S. aureus* cell wall consists of a thick and highly cross-linked glycopeptide layer outside a phospholipid membrane bilayer. The distinction between the two types of cell wall structures has a profound effect on the responses of bacteria to antibiotics and photosensitization.

15. An adjacent atom of nuclear spin $I$ leads to $2I + 1$ EPR lines. For example, the EPR spectrum of the unpaired electron on a free H atom consists of two equal intensity lines corresponding to $I = ½$. N equivalent adjacent protons lead to (N +1) EPR lines.

16. The *ene* reaction consists of hydrogen abstraction by oxygen followed by the addition of -O-O-H at a double bond.

# CHAPTER 5

# TISSUE OPTICS

## 5.1 CONCEPTS OF TISSUE OPTICS

Tissue optics is a sub-field of photophysics that describes light propagation and utilization in biological materials. The applications of tissue optics to photobiology are based on idealized models of tissues and modes of light delivery. Physical systems that simulate the optical properties of tissues are referred to as *tissue phantoms* and the term "models" is reserved for analytical treatments. A major objective of tissue optics in phototherapy is to relate the delivered light dose to the physiological endpoints. Modeling is useful for treatment planning, analysis of data, and optimization of clinical protocols. Non-invasive optical diagnosis and optical imaging are recent applications of tissue optics. Light propagation in a biological tissue differs from homogeneous substances.

Tissues are optically "cloudy" or *turbid* materials owing to the presence of microscopic entities, including macromolecules, cells, water pools, layer structures, and light-absorbing substances. Multiple light scattering is primarily responsible for the weak penetration of light in non-pigmented tissues. A collimated light beam incident on a tissue spreads out and loses its directionality. A significant fraction of the radiant flux exits at the illuminated face. The emitted light is termed *diffuse reflection* to distinguish it from specular reflection at a polished surface. The alternative term *remission* is applicable to non-layer structures. The radiation within tissue has the properties of diffuse light in which the rays (or photon motions) are almost non-directional or *isotropic*. Some light interactions with tissue are depicted in Figure 5.1. Each trajectory can be identified either with an individual light ray or a single photon.

## 5.2 TISSUE OPTIC THEORIES

The classical theory of *radiative transfer* leads to quantitative information about the distribution of radiant flux in a light scattering material. Radiative transfer was developed by Chandrasekhar to explain light propagation in stellar atmospheres. Radiative transfer simplifies the electromagnetic theory description of light that requires rigorous solutions of Maxwell's equations.

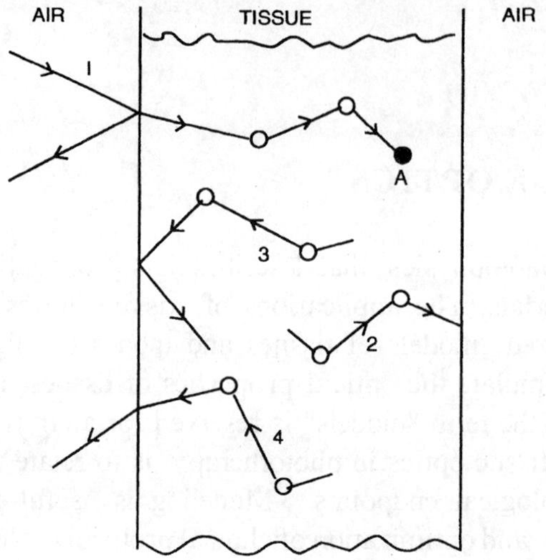


**Figure 5.1** Propagation of light rays in tissue. (1) Absorption after several scattering events. (2) Transmission after several scattering events. (3) Internal reflection after several scattering events. (4) Diffuse reflection after several scattering events. From Grossweiner. L.I., *Photodynamic Therapy of Neoplastic Disease,* Vol. I, Kessel, D., Ed., CRC Press, Boca Raton, FL, 1990.

Instead, the radiant flux density is calculated for an idealized medium consisting of discrete scattering and absorbing centers. Radiative transfer resembles a "pinball machine" in which the balls are photons, the bumpers are scattering centers, and the holes are absorbing centers. The pinball analogy resembles an alternative statistical approach to light propagation that does not require solving equations. In *Monte Carlo* (MC) simulation, a trajectory for each photon is determined by using random numbers to predict the probability of each microscopic event, for example, the direction and length of a photon jump and the fate of the photon after a jump. The superposition of many thousands of photon paths approaches the actual flux density distribution in the medium. MC is considered as the "gold standard" of tissue optic calculations. However each case must be calculated *ab initio*, whereas radiative transfer leads to analytical expressions applicable to similar systems, for example, light transmission by a thick layer with plane parallel faces. In recent tissue optic research the results of many MC calculations are combined for a limited range of parameters to give empirical algebraic expressions.

The *transfer equation* is derived out by equating the radiant flux that enters and leaves a small region in an idealized medium. The radiance ($L$) is the appropriate radiometric quantity for this calculation because radiance specifies the flux density in different directions. The detailed analysis requires three factors: (1) A *source function* must be introduced to account for internal and external light sources and sinks. (2) The angular distribution of light scattering must be specified. This is done by assuming a *phase function* that indicates the scattering intensity in all directions of solid angle. (3) If the external refractive index is lower than that of the medium, then total internal reflection (TIR) takes place for those internal rays that reach an interface at angles exceeding the Snell's law critical angle $\phi_c$. This situation is referred to as *refractive index mismatch*. The equation of transfer is an integral-differential equation that is quite difficult to solve even for simple geometries. The classic papers of Chandrasekhar provide accurate numerical calculations for plane parallel layers with matched boundaries based on the standard method of "finite ordinates". The availability of digital computers facilitated the use of more convenient computational procedures. "Multi-flux" methods divide the interior flux into two or more streams moving in different directions. In the "adding-doubling" method, the optical constants are calculated for a very thin layer and the layer depth is repetitively doubled until the desired layer thickness is reached. These methods are applicable to the "direct problem" of calculating the flux density distributions for input values of the optical constants. MC simulation has the same limitation. However, experimental measurements on tissues lead to macroscopic optical parameters such as the reflection coefficient and transmission coefficient. An approximate form of radiative transfer theory referred to as the *diffusion approximation* (DA) is better suited to the *"indirect (or inverse) problem"* in which the optical constants of a given material are calculated from experimental optical measurements. Prior to radiative transfer theory, Kubelka and Munk developed a two-stream model that has been employed extensively in the paper and pigment industries.[1] The Kubelka-Munk model is not used often in tissue optics owing to the higher accuracy and versatility of the DA and its applicability to cylindrical and spherical geometries. However, it does have applications to strongly absorbing and weakly scattering tissues for which the DA is inaccurate.

## 5.3 MICROSCOPIC AND MACROSCOPIC OPTICAL CONSTANTS

Microscopic optical constants describe optical interactions at the level of individual events. Direct experimental measurements of microscopic optical constants are feasible only for isolated entities, for example a very thin layer or a dilute suspension of cells. Macroscopic optical properties extending over large regions of a medium can be measured experimentally. For convenience, microscopic parameters are referred to as "optical constants" and macroscopic parameters are referred to as "optical coefficients". The coefficients applicable to a planar layer include the *total transmission coefficient* ($T_{tot}$) - the fraction of incident flux exiting in all directions at the rear surface; the *total reflection coefficient* ($R_{tot}$) - the fraction of incident flux exiting in all directions at the front surface; the *optical penetration depth* (δ) - the depth at which the flux density decreases by a factor of $1/e$ (0.3679) of the initial value. For a thick uniformly-illuminated layer, the flux density deep within the layer is attenuated by approximately 63 % at a depth δ, by 86 % in 2δ, by 95 % in 3δ, etc. Table 5.1 lists some experimental values of δ for red light in excised tissues.

**Table 5.1 *In vitro* optical properties on human tissues [a]**

| Tissue | $\mu_a$ ($cm^{-1}$) | $\mu_s(1-g)$ ($cm^{-1}$) | δ [b] (cm) |
|---|---|---|---|
| Aorta | 0.52 | 41 | 0.17 |
| Bladder | 1.4 | 3.5 | 0.45 |
| Brain | | | |
| white matter | 1.6 | 2.0 | 0.52 |
| gray matter | 1.6 | 7.2 | 0.15 |
| Breast | 0.3 | 16 | 0.26 |
| Fatty | 0.7 | 9 | 0.22 |
| Liver | 2.3 | 100 | 0.037 |
| Lung | 8.1 | 81 | 0.022 |
| Skin | | | |
| epidermis | 1.5 | 24 | 0.093 |
| dermis | 2.7 | 36 | 0.056 |
| Tumor | 1.5 | 5 | 0.18 |
| Whole blood | 1.3 | 6.1 | 0.19 |

[a] Values for 633-635 nm. [b] Calculated with Equation (5.15) as $\delta = 1/\mu_{eff}$.

The magnitude of a microscopic optical constant is specified by a *cross section* (σ). The concept of a cross section is used throughout physics to quantify the probability of an individual encounter between microscopic entities. The cross sections used in radiative transfer are the *absorption* cross section ($\sigma_a$), *total scattering cross section* ($\sigma_s$), and the *attenuation cross section* ($\sigma_{att}$) They are related by

$$\sigma_{att} = \sigma_s + \sigma_a \qquad (5.1)$$

The optical absorption cross section of a molecule is equivalent to the *molar extinction coefficient* (ε) according to

$$\sigma_a(cm^2)=4.34x10^{-21}\epsilon(l/mol-cm) \tag{5.2}$$

The value of $\sigma_a$ is comparable to the geometrical area for a strongly absorbing molecule. For example, $\epsilon$ = 1 x $10^5$ (l/mol-cm) corresponds to $\sigma_a$ = 4 x $10^{-16}$ $cm^2$ and a molecular diameter of 0.2 nm. Cross section calculations are more complicated for entities that both scatter and absorb light. Mie scattering theory is applicable for isolated spheres, cylinders, and other simple geometric forms (Section 2.5). An exact calculation is not feasible for the poorly characterized absorbing and scattering centers in biological tissues. For most applications the cross section $\sigma$ and corresponding particle concentration ($N$) occur as the product $N\sigma$ with units of reciprocal distance. This quantity is a *linear attenuation coefficient*. Accordingly, there is a linear attenuation coefficient for total scattering ($\mu_s$), absorption ($\mu_a$) and attenuation ($\mu_{att}$) where

$$\mu_{att}=\mu_s+\mu_a \tag{5.3}$$

The linear scattering coefficient has a straightforward physical interpretation. $1/\mu_s$ is the average penetration depth of non-scattered photons that enter a turbid medium from the outside. For example, entering photons travel an average distance of 0.01 cm before they are scattered for $\mu_s$ = 100 $cm^{-1}$.

## 5.3.1 More Details about Phase Functions

A phase function specifies the angular distribution of scattering by a single center. This terminology has no relation to the phase of a wave or "phases" of the moon. The choice of a phase function is arbitrary for a composite material such as a biological tissue owing to the diversity and unknown nature of the scattering particles. The relevant property of the phase function is described by its average value in the forward direction. This number is the *asymmetry parameter* ($g$) or the average cosine of the scattering angle.[2] The value of $g$ ranges from -1 to +1, where $g$ = 0 corresponds to equal scattering probability in the forward and backward directions, $g$ = 1 corresponds to ideally forward scattering (scattering angle of zero degrees), and $g$ = -1 corresponds to ideally backward scattering (scattering angle 180 degrees). Typical experimental values of $g$ for biological tissues are 0.7 - 0.9 corresponding to an average scattering angle of 25 to 45 degrees. Radiative transfer calculations have been carried out with different assumed phase functions. Isotropic scattering, in which scattering is equally probable in all directions is referred to as the "P0" approximation in tissue optic jargon.[3] An anisotropic phase function can average to $g = 0$ if the forward and backwards components are the same, for example, Rayleigh scattering (Figure 2.16a). The phase function for Mie scattering by a large sphere is highly forward directed with $g$ close to 1 (Figure 2.16b). For example, $g \approx 0.99$ for red blood cells for visible light. The *Henyey-Greenstein phase function* (H-G) approximates the angular dependence of Mie scattering. The combination of an isotropic phase function with a forward scattering component proportional to cos θ is referred to as the "P1" or the Eddington approximation. The maximum value of $g$ for the P1 approximation is 1/3 which is significantly lower than measured for biological tissues. Enhanced forward scattering can treated by the mathematical device of adding to the P1 phase function a sharply-peaked scattering function or "delta function" in the forward direction which is referred to as the *delta-Eddingon* approximation.[4]

---

**Example 5.1** Calculation of the asymmetry parameter g of a phase function

A simple phase function employed for isotropic scattering with an additional anisotropic component is: $p(\theta,\phi) = 1 + a\cos\theta$ where $a \leq 1$. Calculate the value of $g$ for this phase function.

The value of g is calculated by integrating $f(\theta,\phi)\cos\theta$ over the total solid angle 4π sr.

$$g = \int_0^{2\pi} d\phi \int_0^{\pi} (1 + a\cos\theta)\cos\theta \, d\theta = 1/3$$

The maximum $g = 1/3$ is much smaller than typical values of $g$ for biological tissues.

---

## 5.3.2 More Details About Cross Sections

The cross section concept may be illustrated by considering balls striking a plate with holes in it. Assume that a ball may either bounce off the plate with a probability specified by the scattering cross section $\sigma_s$, stick to the plate with a probability specified by the absorption cross section $\sigma_a$, or pass through the plate without any change. The total attenuation cross section $\sigma_{att} = \sigma_s + \sigma_a$. specifies the probability of any interaction of a ball with a plate. Now consider an open window of total area $S$ partially covered with a random distribution of $N'$ similar plates per unit area.. Assume that a continuous stream of balls are directed towards the window. The blocked fraction of the window is $N'S_{att}/S = N'\sigma_{att}$. and, therefore, the fraction $1 - N'\sigma_{att}$ of the incident balls pass through the window. A more detailed calculation shows that the transmitted fraction is $\exp(-N'\sigma_{att})$ which is the same result for $N'\sigma_{att} << 1$. The two-dimensional calculation shows that the attenuation of the incident stream depends on the product of the total cross-section of one plate and the average concentration of plates. An equivalent result applies for propagation of photons in a light-scattering medium in three-dimensions in which $N'$ is replaced by the volume concentration of scattering centers.

## 5.4 LIGHT PROPOGATION IN TISSUES

The general features of tissue optics may be illustrated for a plane parallel layer illuminated by a collimated light beam. In this case, $N'\sigma_{att}$ is replaced by $N\sigma_{att}d$, where $N$ is the number of scattering centers per unit volume and $d$ is the thickness of the layer. Introducing the attenuation coefficient for photons $\mu_{att} = N\sigma_{att}$ gives for the transmission coefficient

$$T = \exp(-\mu_{att}d) \tag{5.4}$$

Equation (5.4) is applicable to a thin layer in which $\mu_{att}d << 1$. The transmission of an "optically thick" layer is higher owing to forward scattering. Radiative transfer treats the more general situation for layers of any thickness, ranging from the epidermis to interstellar space. The diffusion approximation is a very useful simplified approach to radiative transfer theory that is applicable to planar, cylindrical, and spherical geometries. The DA is readily adapted to the "indirect problem" of calculating microscopic optical constants from experimental coefficients and can account for refractive index mismatch at the boundaries.

### 5.4.1 The Diffusion Approximation

The diffusion approximation (DA) is an approximate form of the exact equation of transfer

$$\boldsymbol{s}\cdot L(\boldsymbol{r},\boldsymbol{s}) = -\rho\sigma_t L(\boldsymbol{r},\boldsymbol{s}) + (\rho\sigma_t/4\pi)\int_{4\pi\ sr} p(\boldsymbol{s},\boldsymbol{s}')L(\boldsymbol{r},\boldsymbol{s}')d\Omega + \epsilon(\boldsymbol{r},\boldsymbol{s})) \quad (5.5)$$

The left side of Equation (5.5) represents the change of the radiance *L(r,s)* at position ***r*** in the direction of unit vector ***s***. The first term on the right side represents the depletion of *L(r,s)* owing to absorption and scattering. The integral represents the increase of *L (r, s)* from photons scattered from all other directions ***s'*** into direction ***s***, where *p(s,s')* is the phase function. The last term specifies the contribution to *L(r, s)* of sources located within the medium. The non-scattered photons entering the medium from the outside are equivalent to sources. The equation of transfer can be solved by the method of "discrete ordinates" in which the integral is approximated by the Gauss's quadrature formula. This procedure leads to a set of simultaneous differential equations that can be solved numerically. Chandrasekhar has given rigorous solutions for isotropic scattering and the Eddington phase function in terms of special tabulated functions.

The DA is an approximate solution of the radiative transfer equation applicable for media in which light scattering dominates over absorption ($\mu_s >> \mu_a$) and in which the asymmetry parameter *g* is not too close to unity. The initial photon directions are randomized after a few scattering events in a highly-scattering and weakly- absorbing medium. An incident photon acquires random directions after approximately 1/(1 - *g)* scatters, which is only five scatters for *g* = 0.8. The integral-differential radiative transfer equation reduces to a diffusion equation in this limit, similar to the differential equations of heat flow and particle diffusion. The derivation from the equation of radiative transfer is straightforward but lengthy. The basic principles can be illustrated by a heuristic approach. The total radiance within the medium is divided into two components: $L = L_d + L_c$, where $L_d$ is the radiance of diffuse flux and $L_c$ is the radiance of collimated flux. Averaging $L_d$ over all solid angles leads to the energy fluence rate for diffuse flux

$$\Phi_d(\boldsymbol{r}) = \int_{4\pi\ sr} L_d(\boldsymbol{r},\boldsymbol{s})d\Omega \quad (5.6)$$

The net flow of diffuse flux in the forward direction is described by a flow vector $F_d$. The magnitude of $F_d$ is calculated by integrating the projection of $L_d$ in the forward direction over a hemisphere

$$F_d(\mathbf{r}) = \int_{2\pi\ sr} L_d(\mathbf{r},\mathbf{s})\cos\theta d\Omega \tag{5.7}$$

A generalized form of *Fick's first law* is assumed that relates the flow rate to the gradient of the driving force according to

$$F_d(\mathbf{r}) = -D\nabla\Phi_d(\mathbf{r}) \tag{5.8}$$

where $\nabla$ is the gradient operator, $D$ is the diffusion constant for diffuse flux, and the minus sign indicates that flux flows from high to low diffuse flux density.[5] An independent expression follows from conservation of energy. The net flow of flux out of a volume element equals the difference between the rate at which flux is absorbed in that volume element and the rate at which flux is scattered from other regions into that volume element. Using the well-known result from mathematical physics that the divergence of a flow vector equals the net outward flow leads to

$$\nabla \cdot F_d(\mathbf{r}) = -\mu_a \Phi_d(\mathbf{r}) + S(\mathbf{r}) \tag{5.9}$$

where $S(\mathbf{r})$ is a source term that accounts for the flux scattered into the volume element located at $\mathbf{r}$. Substituting Equation (5.8) in Equation (5.9 ) gives the time-independent diffusion equation

$$\nabla^2\Phi_d(\mathbf{r}) = (\mu_a/D)\Phi_d(\mathbf{r}) = -S(\mathbf{r})/D \tag{5.10}$$

When the medium does not contain actual light sources, the source term consists of the incident flux that has not been scattered. This contribution can be expressed by: $S(r) = \mu_s\Phi_{ri}$, where $\Phi_{ri}$ is the fluence rate of "reduced incident" flux. The contribution of $S(r)$ to the total flux density decreases with the distance from light entry and is negligible deep into a large volume. The detailed analysis leads an explicit expression for $D$

$$D=1/3[\mu_a+\mu_s(1-g)] \tag{5.11}$$

The quantity $\mu_s(1 - g)$ is referred to as the "reduced scattering coefficient".

### 5.4.2 Elementary Solutions of the Diffusion Equation

Exact solutions of the diffusion equation are straightforward for idealized planar layers and idealized boundary conditions at the interfaces. However, the results are still approximate owing to the assumptions made in deducing Eq. (5.10) from the equation of transfer. A useful further approximation can be employed by setting the right side of Eq. (5.10) equal to zero. This assumption ignores the contribution of the non-scattered incident flux to the total flux density within the medium. The resultant differential diffusion equation for a plane parallel layer is

$$\frac{d^2\Phi_d}{dz^2}-\mu_{eff}^2\Phi_d=0 \tag{5.12}$$

where $z$ is the depth into the layer and $\mu_{eff}$ is defined by

$$\mu_{eff}=\sqrt{[3\mu_a(\mu_a+\mu_s')]} \tag{5.13}$$

The elementary theory of differential equations leads to the following solution for a uniformly irradiated "semi-infinite" layer that extends from $z = 0$ to $z = \infty$

$$\Phi_d(z)=ME_0\exp(-\mu_{eff}z) \tag{5.14}$$

where $E_0$ is the irradiance of the incident beam and $M$ is a dimensionless constant that depends on the change of refractive index at the air-tissue boundary. Equation (5.14) shows that diffuse flux density falls off exponentially within the layer. The reciprocal of $\mu_{eff}$ is the optical penetration depth for diffuse flux

$$\delta_d = 1/\mu_{eff} \tag{5.15}$$

Values of $\delta_d$ for tissues at different wavelengths provide a useful indicator of their attenuation properties. For example, $\delta_d$ = 1 mm indicates that the flux within a layer decreases by 63% at an additional depth of 1 mm, by 86% at 2 mm, by 95% at 3 mm, by 98% at 4 mm, etc. Comparing $\mu_{eff}$ and $\mu_{att}$ shows that the diffuse flux has a much deeper penetration than the incident flux. Figure 5.2 shows a plot of Equation (5.14) for a relatively thick layer with mismatched boundaries taking $\delta_d$ = 2 mm. and $M$ = 5. The dashed line shows a more accurate Monte Carlo calculation that includes the incident flux density. The buildup of flux near the interface results from internal reflection

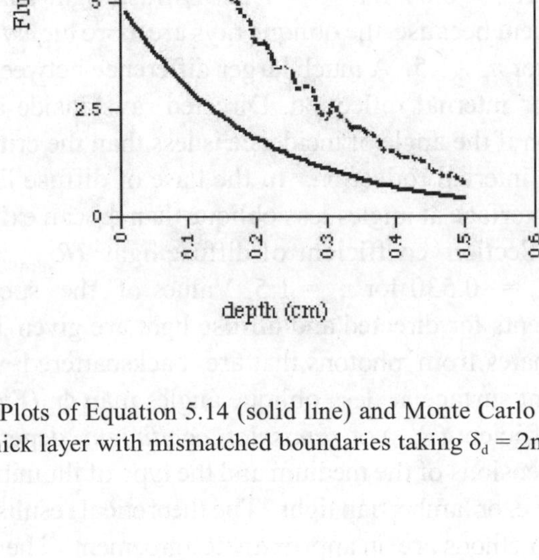


**Figure 5.2** Plots of Equation 5.14 (solid line) and Monte Carlo simulation (dashed line) for a relatively thick layer with mismatched boundaries taking $\delta_d$ = 2mm and M = 5

**Example 5.2** Light propagation in a thick tissue layer

For a thick layer of human breast tissue illuminated at 630 nm calculate (a) the attenuation coefficient; (b) the average distance traveled by incident photons before scattering; (c) the penetration depth of the incident beam; (d) the optical penetration depth for diffuse flux. Use the optical constants in Table 5.1 and assume $g = 0.8$.

(a) From Table 5.1, $\mu_a = 0.3\ \text{cm}^{-1}$ and $\mu_s(1 - g) = 16\ \text{cm}^{-1}$. The attenuation coefficient from Equation (5.3) is: $\mu_{att} = 0.3 + 16/(1 - 0.8) = 80.3\ \text{cm}^{-1}$.

(b) The incident photons travel an average distance of 1/80 = 0.013 cm = 0.13 mm before being scattered. This distance corresponds to an average of 1/(1 - 0.8) = 5 scattering events.

(c) The average penetration depth of the incident beam is 1/80.3 = 0.013 cm = 0.13 mm. This calculation shows that most of the attenuation within the layer results from light scattering.

(d) The optical penetration depth for diffuse flux from Equation (5.13) and Equation (5.15) is 0.26 cm = 2.6 mm. The average light penetration is much deeper than indicated by $\mu_{att}$ owing to multiple scattering of the incident photons before they are absorbed.

### 5.4.3 More Details about Diffuse Reflection

Reflection coefficients differ for directed and diffuse light. The specular reflection coefficient ($R_{sp}$) for directed light at an air-tissue interface is given by Eq. (2.6). For example, $R_{sp} = 0.04$ for $n_r = 1.5$. Diffuse light has a higher specular reflection coefficient because the oblique rays are more highly reflected, for example, $R_{sp} = 0.077$ for $n_r = 1.5$. A much larger difference between directed and diffuse light exists for internal reflection. Directed rays inside a medium undergo specular reflection if the angle of incidence is less than the critical angle $\phi_c$ or else undergo total internal reflection. In the case of diffuse light, only those rays that reach the interface at angles less oblique than $\phi_c$ can exit from the medium. The internal reflection coefficient of diffuse light ($R_{int, diff}$) is quite high, for example, $R_{int, diff} = 0.530$ for $n_r = 1.5$. Values of the specular and internal reflection coefficients for directed and diffuse light are given Table 5.2. Diffuse reflection originates from photons that are backscattered within the medium and reach the front surface at less oblique angles than $\phi_c$ (Figure 5.1). The diffuse reflection coefficient ($R_{diff}$) or *remission coefficient* depends on the optical properties and dimensions of the medium and the type of illumination, for example, collimated, diffuse, or lambertian light. The theoretical results obtained with different calculation methods are in approximate agreement. The following result is obtained with the DA for a uniformly illuminated semi-infinite layer with collimated incident light

$$R_d = \frac{N'}{[(1+N')+\sqrt{3(1+N')}][1+(2/3)b\sqrt{3/(1+N')}]} \qquad (5.16)$$

where $N' \equiv \mu_s(1 - g)/\mu_a$. and $b \equiv (1 + R_{int,diff})/(1 - R_{int,diff})$. Note that diffuse reflection depends on the ratio of the absorption coefficient to the reduced scattering coefficient and the relative refractive index at the air-medium interface. A less complicated result was calculated from MC simulations for typical tissues illuminated at 630 nm and assuming $n_r = 1.38$

$$R_d = \exp[-(7/3)\sqrt{3/(1+N')}] \qquad (5.17)$$

**Table 5.2 Reflection coefficients for diffuse light**[a]

| $n_{rel}$ | $R_{sp,dir}$[b] | $R_{sp,dif}$[c] | $R_{int,dif}$[d] |
|---|---|---|---|
| 1.00 | 0.000 | 0.000 | 0.000 |
| 1.10 | 0.0023 | 0.026 | 0.193 |
| 1.20 | 0.0083 | 0.045 | 0.336 |
| 1.30 | 0.017 | 0.061 | 0.445 |
| 1.40 | 0.028 | 0.077 | 0.530 |
| 1.50 | 0.040 | 0.092 | 0.597 |
| 1.60 | 0.053 | 0.107 | 0.651 |
| 1.70 | 0.067 | 0.121 | 0.696 |
| 1.80 | 0.082 | 0.134 | 0.733 |

[a] For unpolarized light. [b] Specular reflection coefficient for directed light at perpendicular incidence. [c] Specular reflection coefficient for diffuse light. [d] Internal reflection coefficient for diffuse light.

**Example 5.3** Diffuse reflection

(a) Calculate the diffuse reflection coefficient for a thick layer of human breast tissue illuminated with collimated light at 630 nm. Assume $n_r = 1.38$.
(b) Repeat the calculation using the K-M model.

From Table 5.1, $\mu_a = 0.3$ cm$^{-1}$ and $\mu_s(1 - g) = 16$ cm$^{-1}$. $N' = 16/0.3 = 53.3$. From Table 5.2, $R_{\text{int,diff}} = 0.513$ and $b = (1 + 0.513)/(1 - 0.513) = 3.11$. Substitution in Equation (5.16) gives: $R_{\text{diff}} = 0.32$. Equation (5.17) gives: $R_{\text{diff}} = 0.28$.

### 5.4.4 Experimental Measurements of Diffuse Reflection

The most accurate measurements of transmission and reflection coefficients are obtained with an integrating sphere. The total transmission measurement is depicted in Figure 2.22. The same device can be used to measure total reflection by locating the tissue layer behind a small aperture at the back of the sphere. A black painted plate or a cavity behind the layer acts as a "beam stop". A layer of a highly reflecting white powder or paint is a standard reference for 100% reflection. Some commercial spectrophotometers provide integrating sphere attachments for measurements of total reflection spectra. An alternative technique for measuring reflection spectra is depicted in Figure 5.3.

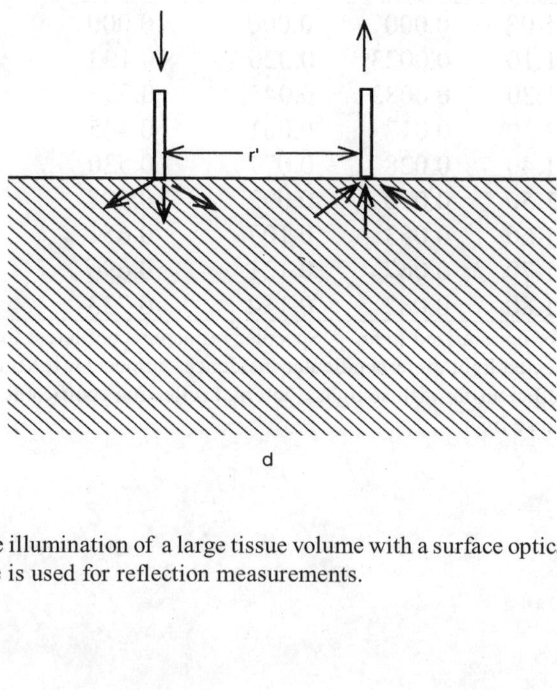


**Figure 5.3** Surface illumination of a large tissue volume with a surface optical fiber. The optical fiber surface probe is used for reflection measurements.

A turbid layer is illuminated with collimated white light and the reflected light is detected by an optical fiber bundle and analyzed by a spectrograph reading into an optical multichannel analyzer. The analysis for this arrangement leads to

$$R_d = z_0(\mu_{eff} + 1/r')\frac{\exp(-\mu_{eff} r')}{2\pi r^2} \tag{5.18}$$

where $r'$ is the distance between the source fiber and the probe fiber and $z_0 = 1/\mu_s(1 - g)$.

### 5.4.5 Transient Measurements

Transient measurements of tissue optic parameters are practical with fast pulsed lasers. The distance traveled by light in a homogeneous medium is $(c/n)t$ where $t$ is the *time-of-flight*. Thus for $n = 1.4$, light travels approximately 0.2 mm in 1 ps. The situation is more complicated in tissue owing to absorption and scattering. The *mean free path* (*mfp*) of a photon along the actual track is $1/(\mu_a + \mu_s)$. For typical values $\mu_s = 50\ \text{cm}^{-1}$ and $\mu_a = 5\ \text{cm}^{-1}$, the *mfp* is 0.18 mm and the time-of-flight is 0.9 ps. A scattering event occurs on the average of every $1/\mu_s = 0.2$ mm along the track. If a tissue is illuminated with a 1 ps light pulse, the first few photons that emerge have not been scattered; they are referred to as *ballistic*. The next group of emerging photons which have undergone a few scatters are termed *quasi-ballistic*. The photons that emerge after a relatively long time delay, say 10 ps, have been scattered on the average of ten times. These photons are diffuse and the diffusion approximation should be valid. The *transport mean free path* is defined in this regime: $mfp' = 1/[\mu_s(1 - g) + \mu_a]$. A typical value of *mfp'* is 0.7 mm for a non-pigmented tumor illuminated with red light, corresponding to a 7 ps time-of-flight.

Two experimental techniques are used for transient measurements on tissues. In a *time domain* measurement, an ultra-short laser pulse is incident on the tissue and the transmitted and reflected light signals are measured in real time. Mode-locked laser pulses in the picosecond range must be employed for these measurements. In a *frequency domain* measurement the tissue sample is illuminated continuously with very high-frequency, sinusoidally-modulated light. The reflected and transmitted light have the same modulation frequency as the incident light, but they are attenuated and phase-shifted by scattering and absorption within the tissue. Applications of frequency domain measurements in optical imaging are referred to as the method of *photon density waves*. The measurements can be carried out with a modulated diode laser at a fraction of the cost of the ultrafast pulsed laser and fast detection system required for time-domain measurements. The two types of measurements lead to equivalent information and can be inter-converted by means of Fourier transformation techniques (see footnote 1 in Chapter 2).

### 5.4.5.1 Time-Dependent Diffusion Equation

Equation (5.10) applies for steady-state conditions in which the photon density distribution is constant in time. The time-dependent DA requires an additional term expressing the increase or decrease of the flux density

$$-(1/c'D)\frac{\partial\Phi_d(\boldsymbol{r},t)}{\partial t}+\nabla^2\Phi_d(\boldsymbol{r},t)-(\mu_a/D)\Phi_d(\boldsymbol{r},t)=-S(\boldsymbol{r},t)/D \qquad (5.19)$$

where $c'$ is the light velocity in the medium. Equation (5.19) is the same as Equation (5.10) except for the first term on the left that expresses the loss of diffuse flux density at a time $t$ in a volume element locate at position $\boldsymbol{r}$. Optical image reconstruction techniques frequently utilize the time-dependent DA (see Chapter 7). An approximate solution to Equation (5.19) for the surface optical fiber geometry in Figure 5.3 leads to the time-dependent diffuse reflection coefficient

$$R_d(r',t)=z_0\frac{\exp(-\mu_a c't)}{(4\pi Dc't)^{3/2}t^{5/2}}\exp[-\frac{(r'^2+z_0^2)}{4Dc't}] \qquad (5.20)$$

where $R_d$ is the time-dependent diffuse reflection coefficient and $r'$ and $z_0$ are the same as in Equation (5.18). Equation (5.20) was applied to measurements on a 10 cm diameter cylindrical tissue phantom consisting of a lipid emulsion containing India ink. Mode-locked 6 ps laser pulses were delivered to the source fiber and the remitted photons were recorded with a multichannel plate photomultiplier and time-correlated single photon counting techniques.[6] The points in Figure 5.4 show the time course of the remission signal, corrected for the incident pulse, with the probes separated by 4 cm. The solid line in Figure 5.4 is a fit of the measurements to Equation (5.20) for $\mu_a = 0.053$ cm$^{-1}$ and $\mu_s(1 - g) = 15.5$ cm$^{-1}$. The data in Figure 5.4 show that the peak of the remitted light

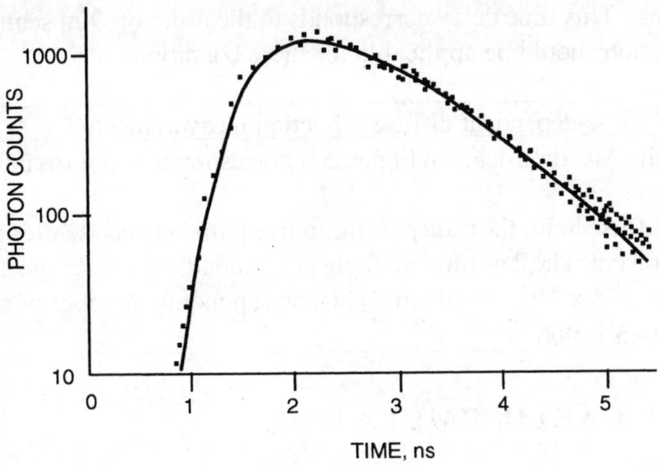


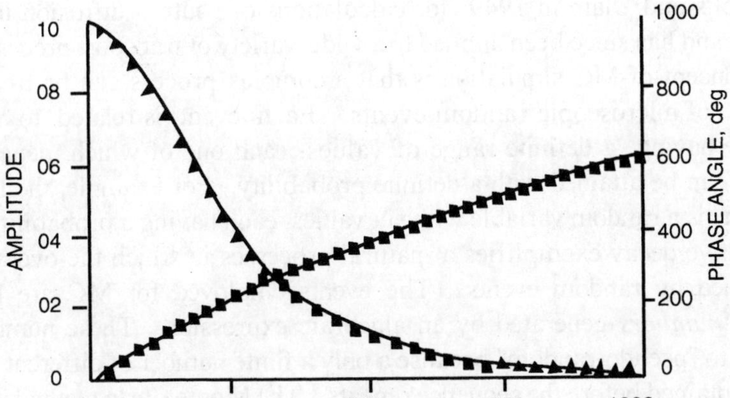


**Figure 5.4** (a) Time domain measurements of the transient diffuse reflection coefficient in a tissue model consisting of a lipid emulsion containing India ink. 6 ps laser pulses were delivered through a surface optical fiber and detected with a surface optical fiber separated by approximately 4 cm. The points are experimental. The line is the fit of the data to the time-dependent diffusion model, Equation 5.19, for $\mu_a = 0.053$ cm$^{-1}$ and $\mu_s' = 14.5$ cm$^{-1}$. (b) Simulated

frequency domain measurement of the same data. The points were calculated by Fourier trnasformation of the time-domain data. The lines are fitted to the theoretical expression. Adapted from Patterson, M.S., Moulton, J.D., Wilson, B. C., and Chance, B., *Int. Soc. Opt. Eng.*, Vol. 1203, Bellingham, 1990. With permission.

exited after 2 ns. This time delay corresponds to the order of 900 scatters. The DA approximation should be applicable for these conditions.

**Example 5.4** Time-dependent diffuse reflection measurements

Show that the 2 ns time delay in Figure 5.4 corresponds to approximately 900 scattering events.

Assuming diffuse light, the transport mean free path between scattering events is 1/15.5 = 0.065 cm. The 2 ns time-of-flight corresponds to an average light path of $3 \times 10^{10}$ cm/s $\times$ $2 \times 10^{-9}$ s = 60 cm. The corresponding number of scattering events is 60/0.065 $\approx$ 900.

## 5.5. MONTE CARLO SIMULATION

Monte Carlo (MC) is a general technique for solving physical problems using probability methods. The modern implementation of MC was first employed by Metropolis and Ulam in 1949 for calculations of neutron diffusion in nuclear reactors and has since been applied to a wide variety of transport processes. The basic concept of MC simulation is that a complex process can be treated as a sequence of microscopic random events. Each event is related to a *random variable* having a definite range of values, each one of which, depending on chance, can be attained with a definite probability. For example, the face of a tossed die is a random variable with six values, each having a probability of 1/6. Radioactive decay exemplifies a natural processes in which the overall rate is determined by random events. The events employed for MC are linked to *random numbers* generated by an algebraic expression. These numbers are referred to "pseudo-random" because a only a finite number of different numbers can be obtained before the sequence repeats.[7] The Monte Carlo method is applied to radiative transfer by recording the history of a large number of photons. The outcome of each microscopic event such as pathlength, absorption and scattering is predicted by a random number with a probability distribution that is based on optical theory. The accuracy of the simulation is proportional to $1/\sqrt{N}$ where $N$ is the total number of photons. Thus, a relative error of 0.1% requires $10^6$ photons. The flow chart of Figure 5.5 summarizes a MC calculation in which $N$ photons are inserted at a single point. The location of photons within the medium and the direction of each photon jump are recorded in an appropriate coordinate system. In spherical coordinates the azimuthal angle is $\theta$ and the longitudinal angle is $\phi$. A photon is "initialized" by locating it at the origin. Each subsequent step is specified by a *probability density function* that relates the physical event to a random variable $x$ between 0 and 1. After initialization the photon enters the layer by jumping a distance $\Delta s$.

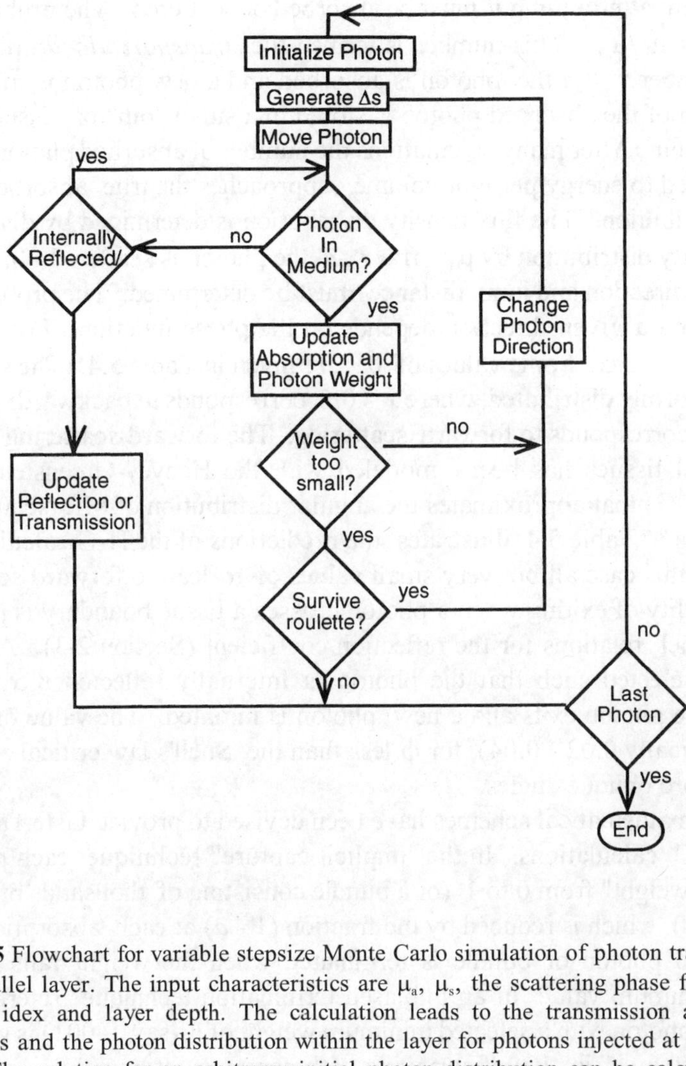


**Figure 5.5** Flowchart for variable stepsize Monte Carlo simulation of photon trajectories in a plane parallel layer. The input characteristics are $\mu_a$, $\mu_s$, the scattering phase function, layer refractive idex and layer depth. The calculation leads to the transmission and reflection coefficients and the photon distribution within the layer for photons injected at a point on the surface. The solution for an arbitrarry initial photon distribution can be calculated from a numerical calculation of the convolution integral. Adapted from Prahl, S. A., Keijzer, M., Jacques, S. I., and Welch, A. J., *Int. Opt. Soc. Eng.,* Vol IS5, Bellingham, 1989. With permission.

A constant step size may be selected or preferably, a variable jump distance is selected such that the photon must be either absorbed or scattered after that step. The probability density function for an exponential step size is: $\Delta s = -\mathrm{Log}_e x$. The calculation is illustrated in Table 5.3. Note that very small and very large jumps are less likely. The dimensionless values of $\Delta s$ in this calculation are scaled to the physical dimensions of the attenuation length: $1/\mu_{att}$. If the photon remains in the medium after a jump it must be absorbed or scattered. The probability of scattering is $\mu_s/\mu_{att}$. This number is termed the *transport albedo* ($a$). If the random number $x > a$ the photon is absorbed and a new photon is initialized. The location of the absorbed photon is stored in a small "bin" of tissue volume at that location. After many calculations the number of absorbed photons in each bin, converted to energy per unit volume, approaches the true absorbed energy density distribution. The flux density distribution is determined by dividing the energy density distribution by $\mu_a$. If $x < a$ the photon is scattered. In this case a new jump direction and jump distance must be determined. The probability of scattering into a given direction depends on the phase function. For isotropic scattering $\theta_{iso} = 2\pi x$. Some values of $\theta_{iso}$ are given in Table 5.4. The scattering angle is uniformly distributed, where $x < 0.5$ corresponds to backward scattering and $x > 0.5$ corresponds to forward scattering. The forward-scattering property of biological tissues has been modeled with the Henyey-Greenstein phase function (H-G) that approximates the angular distribution of Mie scattering by large spheres.[9] Table 5.4 illustrates the predictions of the H-G calculation for $g = 0.5$. In this case all but very small values of $x$ lead to forward scattering. The probability of exiting when a photon crosses a tissue boundary is predicted by the Fresnel relations for the reflection coefficient (Section 2.3). A random number is selected such that the photon is internally reflected if $x < R(\phi)$; otherwise the photon exits and a new photon is initiated. The value of $R(\phi)$ is low for (typically 0.02 - 0.04) for $\phi$ less than the Snell's law critical angle and unity for more oblique angles.

Various mathematical schemes have been devised to provide faster and more accurate MC calculations. In the "implicit capture" technique each photon is assigned a "weight" from 0 to 1 (or a bundle consisting of thousands of photons is initialized) which is reduced by the fraction (1 - $a$) at each absorption event. The depleted photon or bundle is terminated when the weight falls below a selected minimum value. In an unbiased termination technique referred to as "roulette" a photon with a selected minimum weight of $w$ (say, 0.001) is given one chance in $m$ (say, 1 in 10) of surviving with new weight $mw$ (0.01) or else its weight is reduced to zero. An MC calculation for a uniformly illuminated planar layer leads to the transmission coefficient $T$ at the rear interface and the reflection coefficient $R$ at the front interface. The fraction of photons absorbed equals 1 -

*R - T*. The flux density at various points within the medium is calculated by pooling the number of photons absorbed in bins and dividing by $\mu_a$. *Isodose curves* are plotted as the locus of constant absorbed energy (or photon) density. The calculation in which all photons are initialized at the same point also applies for a tissue of infinitely-wide cross-sectional area illuminated by an infinitely-wide and uniform incident beam. The results are of practical accuracy when the beam dimensions are much larger than the mean free path of the photons. An arbitrary initial irradiance profile can be simulated by repeating the calculation for the distribution of point sources. This procedure requires a significantly longer computation. An alternative approach is applicable for certain beam profiles, including "gaussian" and top-hat beams. The MC calculation for a point source is "folded" into the actual beam shape by means of the mathematical operation termed *convolution*.[10] This procedure requires only one additional numerical integration. Figure 5.6 shows the results of a MC calculation for a 20 mm diameter beam of red light incident on a large tissue volume. The region of high flux density is strongly localized near the surface. MC is applicable for time-dependent problems. The photon density distribution at a constant photon track length indicates the flux density distribution at the corresponding transit time. This type of calculation has been used to simulate the spreading of a laser pulse within tissue. A horse-race analogy is the spread of the horses at each quarter-mile post owing to small differences in their performance under the given conditions.

**Table 5.3 Random number calculation of photon step size**

| | | | | | |
|---|---|---|---|---|---|
| Random Number ($x$) | 0.1 | 0.25 | 0.5 | 0.75 | 0.95 |
| Step Size ($\Delta s$) | 2.306 | 1.386 | 0.693 | 0.288 | 0.051 |

**Table 5.4 Random number calculation of scattering angle**

| | | | | | |
|---|---|---|---|---|---|
| Random Number ($x$) | 0.1 | 0.3 | 0.5 | 0.7 | 0.9 |
| $\theta_{iso}$ | 141.13° | 113.58° | 90° | 66.42° | 36.87° |
| $\theta_{HG}$ ($g = 0.5$) | 108.21° | 68.22° | 46.57° | 30.75° | 15.63° |

## 5.6 SOME EXPERIMENTAL RESULTS

Most of the available optical data for tissues is based on *in vitro* measurements. The resultant optical constants do not account for the contributions of blood and, therefore, they are more reliable at wavelengths longer than 600 nm where hemoglobin is weakly absorbing. In a typical approach, the total reflection and total transmission of a thin layer are measured with an integrating sphere or another device. Estimated values of $\mu_a$ and $\mu_s(1 - g)$ are then inserted in a tissue optic theory until the experimental data are reproduced.

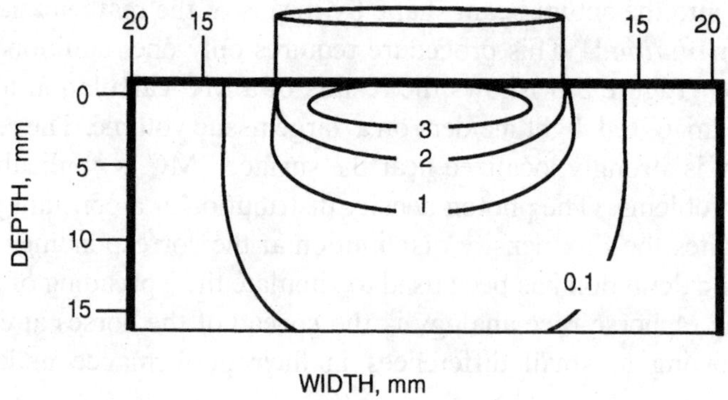


**Figure 5.6** Monte Carlo simulation of light propagation in a thick tissue layer. The incident irradiance is 1 W/cm2 in a 20-mm diameter beam. The profiles are isodose curves for the indicatged flux density in W/cm2. Adapted from Jacques, S.A., *Int. Opt. Soc. Eng.*, Vol. 1645, Bellingham, 1992. With permission.

The DA permits a fast convergence using graphical methods or computer analysis. Approximate values of $g$ can be measured with a goniometer. In this technique, a very thin tissue layer is illuminated with a narrow laser beam and the angular distribution of scattered light $p(\theta)$ is measured with a photodetector. The effective value of $g$ is then calculated by integrating $p(\theta)\cos\theta$ over the range of angles. The goniometer measurement is not identical to the theoretical definition of $g$ for isolated scattering centers. In a novel method for measuring the refractive index of mammalian tissues, the tissue was substituted for the cladding of an optical fiber and the angle at which light exited from the end of the fiber was measured. The surface optical fiber technique is applicable for *in vivo* measurements of penetration depth ($\delta$); see Section 5.3.4. Skin reflection spectra can be measured with an external integrating sphere connected by optical fibers to a recording spectrophotometer. Other non-invasive measurements of tissue optical properties are based on pulsed photothermal radiometry[10], time-resolved reflection and transmission measurements using fast pulsed lasers, and fluorescence emission techniques.

### 5.6.1 Human Skin

Skin is a multilayer tissue incorporating a diversity of light scattering and absorbing entities. Figure 5.7 is a cross-sectional diagram of the *epidermis* and upper (papillary) *dermis*. The Malpighian layer of the epidermis consists of 7-15 layers of tightly packed *keratinocytes* (prickle cells) varying in thickness from

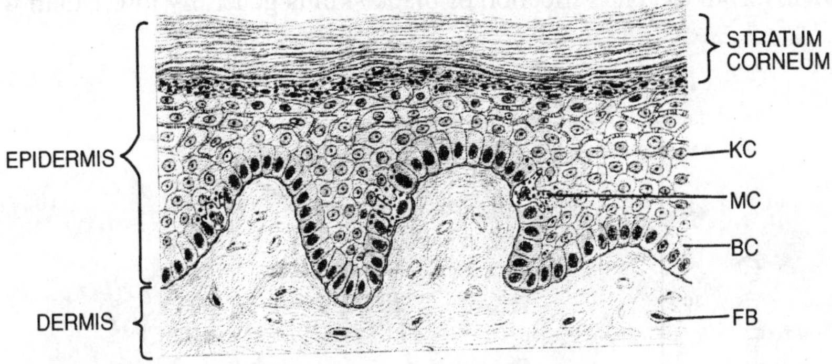


**Figure 5.7** Diagram of skin structure. The indicated cells are keratinocytes (KC), melanocytes (MC), basal cells (BC), and fibroblasts (FB). Adapted from Parrish, J.A., Anderson, R.R., Urbach, F., and Pitts, D., *UV-A*, Plenum Press, New York, 1978. With permission.

60-180 μm. A thin granular cell layer containing *keratohyalin granules* lies between the *stratum corneum* and the prickle cells. Keratinocytes continually migrate outward from a germinative layer of basal cells through the granular cell layer to form the *stratum corneum*, a thin (8-15 μm) surface layer of fused, flattened, dead cells. The pigmentation of skin resides in *melanosomes* which contain dark particles of *melanin*, a high-molecular weight free radical derived from tyrosine. Melanosomes are generated by *melanocytes* at the basal layer with dendritic processes extending between the keratinocytes. The thick dermis (1000-4000 μm) consists of connective tissue incorporating capillaries, lymphatics, and nerves. The sparse *fibroblasts* produce the fibers (collagen, reticulum, elastin), proteins, and viscous materials of the dermis.

Each layer in skin has characteristic optical properties. The stratum corneum is a weakly-scattering layer with a surface reflectivity about 5-8 %. The typical transmission coefficient of the stratum corneum for caucasian skin increases from about 1% at 230 nm, to 20% at 250 nm, followed by a decrease to about 15% at 280 nm from protein absorption, and then a gradual increase to 70% at 700 nm. Light penetration through the stratum corneum into the epidermis is significant from 250 nm through the IR-A region. The epidermis is a highly light-scattering layer incorporating a variable concentration of melanin according to the skin type. Figure 5.8 shows reflection spectra of normal skin on the forearm of a caucasian patient treated for basal cell cancer with PDT. The dips in the reflection coefficient correspond to regions of higher absorption. The overall increase of reflection from 400 nm to 700 nm is largely due to decreasing melanin absorption. The reflection of black skin is generally lower than white

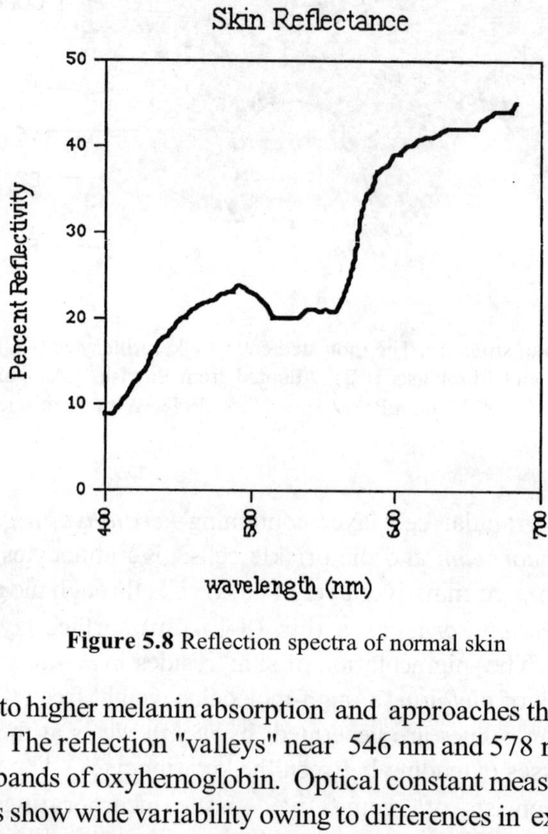


**Figure 5.8** Reflection spectra of normal skin

skin owing to higher melanin absorption and approaches that of white skin in the red region. The reflection "valleys" near 546 nm and 578 nm correspond to the absorption bands of oxyhemoglobin. Optical constant measurements for *in vitro* tissue layers show wide variability owing to differences in experimental methods, tissue preparation, sample geometry, and the tissue optic theory used to extract the results. Typically, $\mu_a$ for the epidermis decreases from about 25 $cm^{-1}$ at 400 nm to 0.2 $cm^{-1}$ at 800 nm and $\mu_s(1 - g)$ decreases from 32 $cm^{-1}$ to 20 $cm^{-1}$ over this spectral range. Table 5.1 summarizes some literature data for red light. Light penetration in whole skin depends on the wavelength. UV-B, UV-A, and blue light are absorbed mostly in the epidermis, while the longer wavelengths of visible light and IR-A penetrate into the dermis.

## 5.6.2 Other Tissues

Light attenuation in tissue is characterized by the optical penetration depth $\delta$. The value of $\delta$ for a highly pigmented tissue can be taken as $1/\mu_{att}$. The optical diffusion depth equals $\delta_d$ for weakly-absorbing, highly-scattering tissues, which can be calculated from the optical constants from Equation (5.12). Wavelength is the principal determinant of light penetration in tissues. Proteins and nucleic acids are responsible for the very strong absorption of UV. Oxyhemoglobin is a principal absorber of visible light. Tissue water is a major chromophore in the near-IR region. Approximate values of $\delta$ in aorta at the wavelengths of surgical lasers are summarized in Table 5.5. This data is useful for the modeling of laser-assisted angioplasty.

**Table 5.5 Optical penetration depth in aorta for different laser wavelengths [a]**

| Laser | Wavelength ( μm) | δ [b] |
|---|---|---|
| ArF | 0.193 | < 1 μm |
| KrF | 0.248 | 15-20 μm |
| XeCl | 0.301 | 30-50 μm |
| Dye | 0.465 | 185-285 μm |
| Argon | 0.515 | 330 l μm |
| Nd:YAG | 1.064 | ≈3 mm [c] |
| Ho:YAG | 2.10 | 300-400 μm |
| Er:YAG | 2.94 | < 1 μ m |
| CO 2 | 10.6 | 15-20 μm |

[a] Adapted from Welch, A. J., Motameti, M., Rastegar, S. L., LeCarpentier, G. ., and Jansen, D., Laser thermal ablation, *Photochem. Photobiol.*, 53, 815, 1991. [b] Calculated as $1/\mu_{att}$.
[c] Estimated as $1/\mu_{eff}$ with Equation (5.15).

NOTES

1. The Kubelka-Munk (K-M ) model is derived for a plane parallel layer with matched boundaries illuminated by diffuse incident flux. The interior flux is divided into a forward stream and a backward stream, each of which is scattered and absorbed according to the linear coefficients $S$ and $K$, respectively. A straightforward solution of two simultaneous differential equations leads the reflection coefficient $R$ and transmission coefficient $T$ in terms of $K$ and $S$. For the special case of a very thick layer where ($T = 0$) the reflection coefficient $R_\infty$ is given by

$$R_\infty = 1 + K/S - \sqrt{[(K/S)(K/S+2)]} \qquad (5.21)$$

$R_\infty$ is high for a weakly absorbing medium and low for a strongly absorbing medium. The K - M model has been extended to collimated incident beams by adding a third stream for directed flux and corresponding forward and backward scattering coefficients. Approximate relationships between the K -M coefficients and the conventional optical constants for isotropic scattering are: $m_a \approx (1/2)K$ and $m_s \approx (4/3)S$. Notwithstanding its limitations, the K - M model remains useful for strongly absorbing layers that do not satisfy the assumptions of the diffusion approximation, especially for diffuse incident illumination such as direct sunlight.

2. The asymmetry parameter is calculated by: $g = \int p(q,f) \cos q \, dW$ where $p(q,f)$ is the phase function and the integration is taken over the entire solid angle.

3. The P-designation originates from a Legendre function expansion of the phase function in powers of cos q. The first terms of the expansion are $P0 = 1$, $P1 = \cos q$, $P2 = (1/3)\cos^2 q$, etc.

4. A delta function designated d(x - x') is the limit of a sharply peaked function at x = x' with the value zero elsewhere. The integral of the delta function is normalized to unity: $\int d(x)\, dx = 1$ for $( -\infty < x < +\infty )$. The integral property of the delta-function is exemplified by: $f(x') = \int f(x) d(x - x')\, dx$ for $( -\infty < x < +\infty )$. This integral returns the value of f(x) and x = x'.

5. The $\nabla$ symbol ("del") is an abbreviation for the three-dimensional gradient (Grad) operator. The gradient of a scalar function is a 3-component vector because it specifies the rate of changes in a specific direction. The gradient of a vector function is a 9-component tensor. The one-dimensional equivalent of $\nabla F$ is the directional derivative **s**(dF/ds) where **s** is a unit vector in the direction of the gradient. $\nabla\cdot$ is an abbreviation for the divergence (Div) operator. The one-dimensional equivalent of Div F is the partial derivative $\partial F/\partial x$. $\nabla^2$ is the Laplacian operator defined as DivGrad. The one-dimensional equivalent of $\nabla^2 F$ is the second derivative $d^2F/dx^2$. The

divergence and Laplacian of scaler functions are scalar quantities.

6. In time-correlated single-photon counting (TCSPC) the delay is measured between a very short excitation light pulse and the time at which the optical signal from the sample arrives at the photomultiplier anode. A very low light intensity is employed to ensure that individual photons are counted. This technique was first employed to measure fluorescence decay lifetimes. Since fluorescence emission is a random process, the probability that a photon is emitted at various times is the same as the time-resolved decay for many fluorophores. A similar relationship exists for transmission and remission of diffusely scattered light. The TCSPC apparatus uses a time-to-amplitude converter (TAC) to convert the time delay signals to voltage amplitude pulses. The voltage pulses are recorded digitally by a multichannel pulse height analyzer (MCPHA). The number of times that each small range of voltages is stored in the digital channels of the MCPHA reproduces the time-resolved fluorescence decay. TCSPC has better time resolution than conventional photoelectric detection techniques because the electronics only need to detect the arrival times of pulses and not the detailed pulse shape.

7. A simple random number generator is the expression: $x_n = C\, x_{n-1} (\mathrm{mod}\ N)$ where C and N are constant integers. $x_0$ should be an odd integer and C can be taken as an odd integer close to $\sqrt{N}$. The notation y = z(mod N) means that if z exceeds the modulus M, then N is subtracted from z as many times as necessary in order to reduce the value of y to the range between 0 and N. For example, 5752 mod 100 = 52. The application of this algorithm for N = 32, C = 5, and $x_0 = 1$ leads to: 5, 25, 29, 17, 9, 13, 1, 5, 25, 29, 17, etc. This sequence repeats after eight numbers which is far too small for practical applications. The range of random numbers is shifted to lie between 0 and 1 by dividing the the random integers by N -1.

8. The H-G phase function is

$$p(\theta) = (1/2)\frac{1-g^2}{(1+g^2-2g\cos\theta)^{3/2}} \tag{5.22}$$

A plot of $p$ *vs* cos θ peaks in the vicinity of θ = 0. For example, for $g = 0.8$, $p(0°)$= 0.45, $p(90°)$= $p(270°)$= 0.17, $p(180°)$= 0.062. The probability density function for the H-G phase function relates cos θ to a random variable x. A straightforward calculation leads to

$$\cos\theta = (1/2g)[1 + 2g^2 - (\frac{1-g^2}{1-g-2gx})^2] \tag{5.23}$$

9. A convolution is a mathematical technique that "folds" one function into another. Assume that the response function of a system to a delta-function input is *I(X)* where *X* is the experimental variable. The convolution integral leads to the output of the system *R(X)* to an arbitrary input signal *S(X)*

$$R(X) = \int S(x) I(X-x) dx$$

The convolution calculation is straightforward although it is not easily visualized. An important relationship exists between convolution and Fourier transformation: The Fourier transform of a convolution of two functions is the product of the Fourier transforms of each function. Thus, a convolution in real space becomes a simple multiplication in Fourier space and *vice versa*. This relationship has applications in x-ray crystallography and image reconstruction.

10. In pulsed thermal radiometry a tissue sample is irradiated with a pulsed laser and the black body radiation emitted by the absorbers in the tissue is measured with a far-IR photodetector.

# CHAPTER 6

# PHOTOCHEMICAL DAMAGE TO BIOLOGICAL SYSTEMS

The effects of light on biological organisms have been investigated at different levels of complexity. Studies on isolated biomolecules provide the most information about the initial photochemical reactions and the least information about their physiologic implications. This approach assumes that the fast early-stage reactions of biological molecules *in vivo* are not strongly perturbed by the cellular environment The most frequently used analytical techniques are optical and magnetic resonance spectroscopy. Short-lived photochemical products can be identified by slowing down their subsequent reactions in rigid media. Alternatively, flash photolysis can be used to identify photochemical products in real time. In this powerful technique, the photochemical reactions are initiated by an intense light pulse and fast optical spectroscopy is used to identify and follow the short-lived reaction products. Nanosecond and picosecond pulsed lasers are used as light sources. Isolated suspensions of cellular organelles and cell cultures are the next higher level of complexity. Assays of biological activity provide a connection between photochemical damage and physiological functions. Cell cultures display important functions under genetic control, including cell division, mutagenesis, and repair. Experimental animal models are used for investigations of complex physiological responses including drug toxicity, pharmakokinetics of drug delivery, and immunological rersponses to light.. Controlled clinical trials provide the most definitive test of a phototherapy However, clinical trials are time-consuming, expensive, and the results are limited to a narrow range of variables owing to the requirement for statistical analysis. Much of the useful clinical information about the established phototherapy procedures has been acquired after the first regulatory approvals.

## 6.1 PHOTOCHEMISTRY OF BIOLOGICAL MOLECULES

The basic structures of proteins, nucleic acids, carbohydrates, and lipids do not absorb visible light. In most cases the visible coloration of biological molecules results from *prosthetic groups* such as heme in hemoglobin. However, visible coloration may not be related to activation by light, for example, light absorption by hemoglobin and cytochromes are not known to be involved in their natural functions. Many chemical and drugs can photosensitize deleterious effects of visible light to biological molecules *in vivo*. Most research attention has been given to enzymes and nucleic acids because a small alteration in the structure can have a large effect on the biological activity. Photochemical studies also are a useful tool for investigation of structure-function relationships owing to the selectivity of photochemical damage.

### 6.1.1 Photochemistry of Proteins

A protein is a long-chain polypeptide consisting of α–amino acids joined by peptide bonds (Figure 6.1).

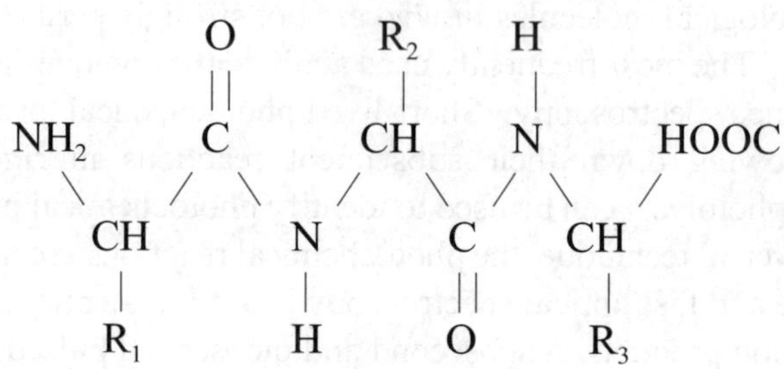


**Figure 6.1** The structure of a protein backbone. $R_1$, $R_2$, etc, are the amino acid *functional groups*.

Organisms utilize proteins for an enormous variety of functions, including the structural framework of cells and tissues, transport and storage of small molecules, antibodies and other functional proteins in blood, and enzymic catalysts of biochemical reactions. All 20 amino acids found in natural proteins are strong absorbers of UV below 230 nm. Of these, the major absorbers at wavelengths longer than 240 nm are the aromatic amino acids, phenylalanine (Phe), tyrosine (Tyr), and tryptophan (Trp), with weaker contributions from histidine (His), cystine (Cys), and the peptide bonds. The "aromatic" absorption band of a "colorless" protein centered near 280 nm originates from the Phe, Tyr, and Trp groups or "residues".

Continuous UV photolysis of an aqueous amino acid generates many stable products. For example, UV irradiation of aqueous Trp gives tryptamine, N-formylkynurenine (NKF), kynurenine, aliphatic amino acids, and ammonia. NFK has been implicated in the aging of the ocular lens. Photochemical products of aqueous Tyr include dihydroxypenylalanine (DOPA), bityrosine, aliphatic amino acids, and ammonia. DOPA is the metabolic precursor of melanin. UV photolysis of aqueous Phe lead to tyrosine and other phenolics, DOPA, aliphatics, and ammonia. Product analysis does not provide definitive information about the primary photochemical reactions because stable products may result from secondary photolysis of intermediates. Pulse irradiation methods provide the best available information about primary mechanisms of protein photochemistry in an aqueous environment under physiological conditions. Flash photolysis studies starting in the 1960s led to the unexpected result that photoionization of aromatic amino acids according to Reaction (4.12) is an important primary reaction.

The transient product from aqueous Trp is the tryptophanyl free radical ($Trp^{\bullet}$) formed by electron ejection at the N-atom of the indole ring followed by dissociation of a proton. Flash photolysis of aqueous Tyr leads to the phenoxyl free radical ($Tyr^{\bullet}$). Aqueous Phe generates a benzyl free radical. In the presence of air, $e^{-}_{aq}$ reacts rapidly with oxygen to form $O_2^{\bullet -}$ that can then generate secondary degradation and oxidation products. The stable UV photolysis products of amino acids are consistent with the initial species identified by flash photolysis augmented by UV photolysis of intermediates.

UV irradiation of a protein in aqueous solution causes extensive changes in virtually all of its physical and chemical properties, including absorption spectra, molecular weight, solubility, heat sensitivity, acid-base titration curves, and catalytic activity. Most attention has been given to inactivation of aqueous enzymes which have been investigated for more than fifty years. The original objective was to relate the UV-induced loss of catalytic activity to photochemical reactions of the constituent chromophoric amino acid residues. Flash photolysis spectra of globular proteins show transient species identified with the constituent chromophoric amino acids. For example, the transient bands obtained by flash photolysis of the bacterial enzyme subtilisin BPN indicate the formation of $e^{-}_{aq}$ (700 nm), $Trp^{\bullet}$ (510 nm), and $Tyr^{\bullet}$ (410 nm). This enzyme contains no cystine (Cys). Some enzymes containing Cys residues show an additional transient absorption band near 420 nm identified with the electron-adduct to a disulfide bond, for example, lysozyme, papain, and trypsin. In addition to direct photolysis of chromophoric groups, *intra*molecular processes can contribute to residue destruction in proteins: (1) Optical excitation energy may be transferred from Phe ⇒Tyr ⇒Trp via "long-range" energy transfer; see Equation (4.18). This process leads to enhanced Trp fluorescence in proteins containing Tyr and Trp. (2) $e^{-}_{aq}$ generated by photoionization of aromatic residues can react at other sites including peptide bonds. (3) Tyr residues can be oxidized indirectly by electron transfer from intact Tyr sites to a photoionized Trp site. (4) Disulfide bridges linking polypeptide chains may be split directly by light absorption and indirectly by intermolecular electron transfer. (5) Long-lived triplet states may initiate secondary *Type 2* reactions in the presence of oxygen.

A simple model relates enzyme inactivation by UV to photolysis of "essential" Trp and Cys residues. An essential residue is defined as a residue involved in binding of a substrate, participation in the catalytic process, or whose disruption alters the normal folding of the active site. For example, three of six Trp residues in lysozyme are part of the active site and two of four Cys residues in ribonuclease maintain the active folding. Quantum yields for photochemical inactivation ($\eta_e$) of enzymes can be estimated with the following semi-empirical expression

$$\eta_e = \Gamma_{trp} f_{trp} \eta_{trp} + \Gamma_{cys} f_{cys} \eta_{cys} \qquad (6.1)$$

where $f_{trp}$ and $f_{cys}$ are the fractional absorptions by all Trp or Cys residues in the enzyme at the irradiation wavelength, $\Gamma_{trp}$ and $\Gamma_{cys}$ are the fraction of essential Trp or Cys residues, and $\eta_{trp}$ and $\eta_{cys}$ are quantum yields for destruction of essential Trp and Cys residues in the enzyme. The approximate agreement between the experimental and calculated values of $\eta_e$ supports the basic assumption that loss of enzymic activity results from photochemical destruction of essential chromophoric residues. Direct enzyme inactivation by UV is less important at the cellular level owing to the more significant effects of UV on the nucleic acids.

### 6.1.2 Photochemistry of Nucleic Acids

This section emphasizes the effects of UV on nucleic acid constituents of cells. The structure and function of the nucleic acids in cells is described in standard biochemistry texts. Briefly, a nucleic acids is a polymeric molecule consisting of 5-member ring furanose derivatives joined by phosphodiester bonds. A purine or pyrimidine base is attached to each sugar. (The term nucleic acid "base" is employed because the isolated compounds are basic in aqueous solution.) The sugar in a *ribonucleic acid* (RNA) is ribose and the sugar in a *deoxyribonucleic acid* (DNA) is deoxyribose.

A small section of a the polynucletotide chain in DNA is shown in Figure 6.4. The 3' OH (tail) of one sugar is joined to the 5' OH (head) of the next sugar by phosphate diester bonds. Two strands of DNA can combine to form an *antiparallel* double helix in which pyrimidine bases form hydrogen bonds to purine bases: Adenine forms two hydrogen bonds thymine and guanine forms three hydrogen bonds with cytosine. The dashed lines in Figure 6.3 depict the hydrogen bonding in the double-helical structure of DNA.

The DNA double helix was discovered by Watson and Crick in 1953, who suggested the unique capability of this structure for transmission of genetic information according to a chemical code. Each type of DNA and RNA in a cell has specific functions. The total amount of chromosomal DNA contained in a cell is the *genome*. A typical human cell contains about 1.8 meters of chromosomal DNA distributed in 46 chromosomes. Chromosomal DNA is packaged by winding around small proteins called *histones* arranged like beads along a string. The *gene* is a section of DNA that is capable of being "read" for

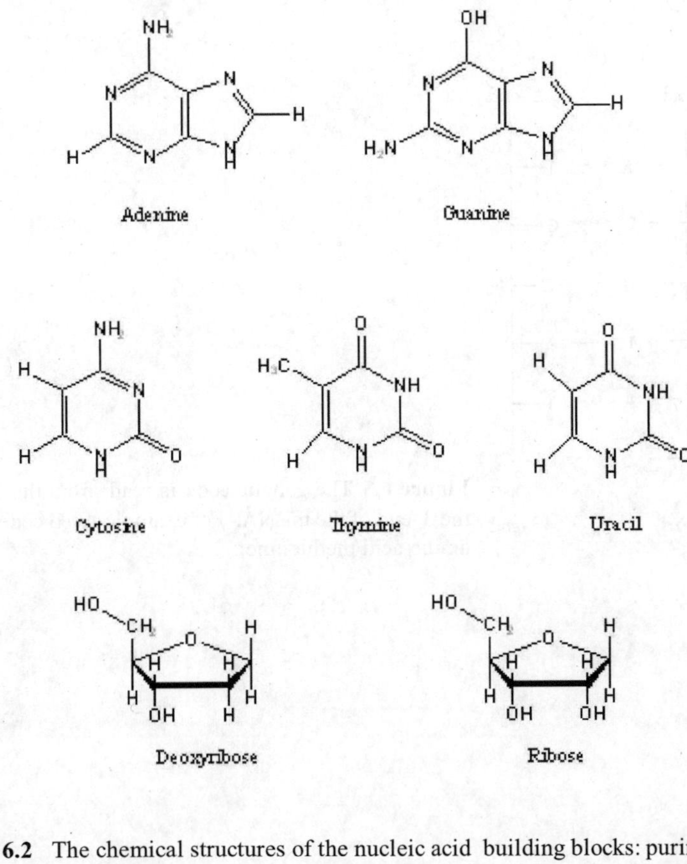


**Figure 6.2** The chemical structures of the nucleic acid building blocks: purines, adenine (A), guanine (G), and the pyrimidines, thymine (T), and cytosine (C). In RNA uracil (U) is present instead of thymine.

synthesis of proteins. The human genome contains about 3 billion base pairs distributed in about 30,000 genes. *Replication* is the information transfer process strand leading to a double-stranded daughter DNA molecule. The other major function of DNA is to to guide the biosynthesis of proteins in the cell. The coding element for each amino acid consists of a sequence of three adjacent bases or *codon*. The first step is the *transcription* of a gene into a complementary single-chain RNA molecule. Transcription generates *messenger RNA* (mRNA) which guides overall protein synthesis or *translation* and a relatively small *transfer RNA* (tRNA) which couples with each specific amino acid. The actual *translation* process takes place on

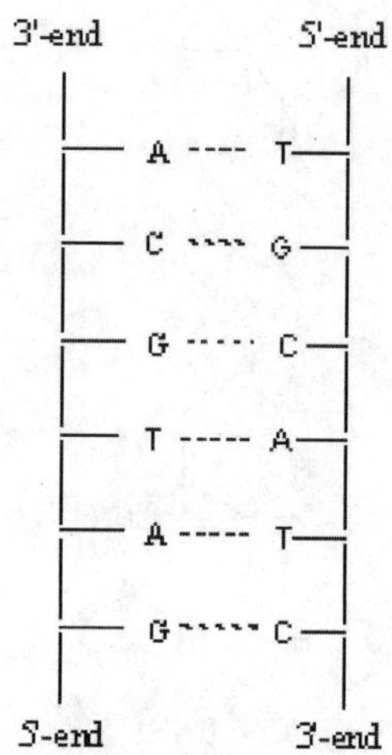


**Figure 6.3** The genetic code is read from the 5'-end to the 3'-end of the mRNA. For example AUG codes for the amino acid methionine.

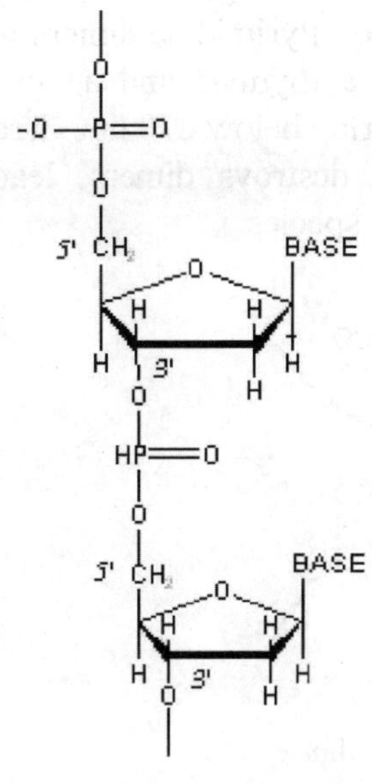


**Figure 6.4** A small section of a the polynucletotide chain in DNA.. The 3' OH (tail) of one sugar is joined to the 5' OH (head) of the next sugar by phosphate diester bonds

*intra*cellular particles, the *ribosomes*. The mRNA bound to the ribosome directs the addition of each tRNA and its coupled amino acid to the protein chain in the order specified by the codons on the mRNA. Replication and transcription are catalyzed by special enzymes known as *polymerases*. Under some circumstances RNA can be *reverse transcribed* to produce DNA. Biochemistry texts describe nucleic acid structures in much more detail and their utilization in the storage and transfer of genetic information. The genetic machinery of the cell is highly sensitive to alterations of the DNA bases by external agents. This process can lead to cell death or mutations in daughter cells if not repaired. A diversity of repair mechanisms have evolved that protect chromosomal DNA from photochemical damage.

The UV absorption of nucleic acids resides almost exclusively in the purine and pyrimidine bases at wavelengths longer than 200 nm. The secondary structure of DNA has a large effect on the intrinsic absorption of the nucleotides.

In pure solutions purines are approximately ten times more resistant to UV photolysis than pyrimidines. Accordingly, most attention has been given to pyrimidine photochemistry. *Pyrimidine dimers* are the most important UV photoproducts. Dimers are formed from the free bases in aqueous solutions and from the nucleotides in nucleic acids. In a thymine dimer, two thymines are linked to each other between their 5 and 6 carbon atoms (Figure 6.5). The resultant four-carbon ring has the cyclobutane structure. Pyrimidine dimers are formed also between two cytosines and between a thymine and a cytosine.The absorption of thymine and thymine dimers overlap below 270 nm. Irradiation in this spectral region both generates and destroys dimers, leading to a *photostationary equilibrium* between the two species.

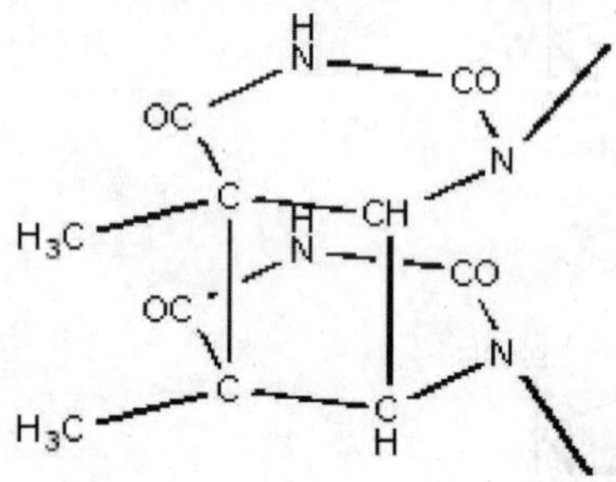


**Figure 6.5** Six isomeric forms of thymine dimer are produced in aqueous solution, but only the *cis-syn* isomer is stable in DNA and RNA.

UV irradiation of DNA generates dimers between adjacent pyrimidines on the same strand. The relative yield of the different types of dimers depends on the base composition of the DNA, the irradiation wavelength, and total light dose.[1] In human tissue cultures, a low dose of 10 $J/m^2$ of UV-C produced approximately one million dimers per cell, while the same dimer yield required 20,000 $J/m^2$ of UV-B at 313 nm and 10,000,000 $J/m^2$ at 365 nm. In one report, 100 minutes of July sunlight in East Lansing, Michigan led to the order of one million dimers per cell. Since the sunlight spectrum at the Earth's surface starts at 290 nm, photochemical reversal of dimers is negligible in natural sunlight. Most attention has been given to pyrimidine dimers owing to their putative role in skin cancer induction. Other types of UV and photosensitized damage have been identified in DNA, including strand breaks, DNA-DNA cross-links, and DNA-protein cross-links. These lesions are less frequent than pyrimidine dimers *in vivo*. The photochemistry of RNA is less studied than DNA owing to the paramount importance of DNA in cells and DNA viruses. RNA viruses are rapidly inactivated by UV-C irradiations.

## 6.2 RADIATION SENSITIVITY AT THE CELLULAR LEVEL

Early experiments showing that UV radiations inactivate bacteria were reported decades before the 1953 Watson and Crick model for DNA structure. The subsequent determination of the *E. coli* gene map made it possible to identify specific repair pathways. A general approach used for studying radiation inactivation is based on *survival curves*. A survival curve is a plot of the surviving fraction of an initial population *vs* the incident dose. Studies of this type led to the identification of cellular repair mechanisms under genetic control.

### 6.2.1 Mechanisms of DNA Repair [2]

UV produces many types of photochemical alterations in DNA, RNA and protein, as well as to structures such as membranes. DNA, however, is the major target for the deleterious effects of UV because it is the largest molecule in the cell, it is present in the fewest copies, it carries the genetic information for a cell, and it absorbs UV radiation very efficiently. In contrast, RNA species and enzymes are present in multiple copies, so that it is much more difficult to inactivate a significant number of these molecules in a cell by UV irradiation. Many different types of damage are produced in DNA by UV irradiation. These include the modification of individual purine and pyrimidine bases, the production of pyrimidine dimers, and the addition of other molecules to the purines and pyrimidines, for example, water molecules and DNA-protein cross-linking. The direct action of UV radiation does not produce strand breaks in DNA but single and double-strand breaks are produced as a by-product of DNA repair. While the most drastic effect of UV on a cell is lethality, UV damage can also result in mutations, which are alterations in the DNA base sequence of a gene such that the capacity of the gene product to perform an essential cellular function is altered. Some mutations can be lethal.

Fortunately, a cell has an enormous capacity to repair all types of damage to its DNA. These DNA repair systems have not evolved just to repair damage produced by UV, they also are required to maintain the integrity of the DNA after damage by the by-products of normal metabolic processes (for example, superoxide), and to repair the heat damage produced at body temperature (for example, deamination). Some of these repair systems are error free (very accurate) and do not produce mutations. Other systems, however, are efficient but error prone (inaccurate) and produce mutations. The first indication that cells could recover from UV damage was the observation that minor modifications in the post-irradiation treatment of cells had a marked effect on the viability of irradiated cells. For example, holding UV- irradiated bacteria under non-growth conditions for various periods of time before plating the cells on growth medium to assay for survival, resulted in increasing levels of survival. This process was called "liquid holding recovery". Thus, the delay of macromolecular synthesis for a time after irradiation led to a higher survival. The ultimate proof that cells could recover from radiation damage was the isolation in 1958 of a radiation sensitive mutant of *E. coli* by Ruth Hill. This discovery inaugurated the era of the genetic approach to the study of DNA repair and mutagenesis.

Genetic studies have been responsible for most of the discoveries about the biochemical basis of DNA repair and mutagenesis. For example, one can measure the initial yield of a particular type of radiation-induced lesion in the DNA of a wild-type (radiation resistant) bacterial strain and in a mutant strain (radiation sensitive), and then follow the kinetics and extent of the repair of that lesion, and thereby determine if the mutation under investigation has an effect on the repair of that type of DNA lesion. Then by constructing strains that carry two mutations that affect DNA repair, one can determine through survival and DNA repair studies whether the two mutations act in the same or in different biochemical pathways for repair. For example, since the *uvrArecA* double mutant of *E. coli* is much more sensitive than either of the singly mutant strains, it suggests that the *uvrA* and *recA* genes function in separate pathways of DNA repair. It was soon discovered that the *uvrA* mutation blocks nucleotide excision repair, a repair system that is largely *constitutive* and error free. The *recA* mutation, however, blocks recombinational repair, which is in part inducible by radiation damage, and therefore requires protein synthesis for it to work. This type of repair has a component that is error prone, that is, it is efficient but makes mistakes, and thus leads to mutations.

The two DNA repair systems are sometimes called "dark repair" systems, since they do not need the presence of light for them to function. However, there is another type of repair called *photoreactivation*, which requires light to function, and is found in a wide range of organisms, including bacteria, yeast, fish, and snakes. The existence of photoreactivation in humans has been the subject of much controversy. Thus, if bacterial cells are exposed to a lethal dose of UV radiation (e.g., 254 nm), lethality can be reduced or eliminated by a second exposure to light in the range of 300-450 nm. This process requires an enzyme, DNA photolyase (a flavoprotein), which binds to pyrimidine dimers. When this enzyme-substrate complex absorbs light in its flavine moiety at the above wavelengths, the pyrimidine dimer is split yielding undamaged pyrimidines, and the enzyme is released. Illumination with blue light splits the dimer and restores the normal DNA structure. Photoreactivation was used in a classic 1977 experiment designed by Robert Setlow to prove a relationship between unrepaired pyrimidine dimers and tumor formation. Thyroid cells of the Amazon molly, a small fish that grows in clones, were irradiated with UV *in vitro* and transplanted without rejection into the abdominal cavity of other mollies. This process led to proliferation of thyroid tumors. The tumor yield was much reduced when the UV irradiations were followed by photoreactivating illumination with blue-light. This elegant experiment demonstrates that pyrimidine dimers can give rise to tumors in this animal model and strongly support the present view that skin cancers in human are initiated by pyrimidine dimers.

Excision repair is responsible for efficient removal of pyrimidine dimers in skin generated by sunlight exposure. In *nucleotide excision repair*, a distortion in the DNA caused by the lesion for example, pyrimidine dimer) is recognized by the uvrABC nuclease. It cuts on both sides of the lesion, producing a gapped structure that is about 12 nucleotides long. Repair replication, using DNA polymerase I, fills in the gap using the undamaged opposite strand of DNA as the template. Finally the break in the repaired DNA strand is sealed by DNA ligase. This repair system is largely constitutive, and is error free. In *base excision repair*, altered purine and pyrimidine bases (for example, thymine glycol, 3-methyladenine, uracil, urea) are recognized by a class of enzymes called DNA glycosylases, which are specific to varying degrees for particular types of altered bases. These enzymes split off the altered base at the N-glycosylic bond that links the base to the sugar in the DNA backbone. This site of the base loss is called an AP site (for apurinic or apyrimidinic). The AP site is degraded to produce a gap in the DNA strand, which is repaired by the same processes used to repair the gaps during nucleotide excision repair. It is these largely constitutive and error free excision repair systems that can function under the conditions of "liquid holding recovery" mentioned above because they do not need protein synthesis to function.

*Recombinational repair* (post-replication repair) is much more complicated, and requires many more gene products. The products of a number of these repair genes are induced by the radiation damage, and therefore this type of repair requires protein synthesis before it can function. Because of its complexity, this type of repair makes mistakes. Recombinational repair requires four strands of DNA ( two sister duplexes), and is of two major types. One type of post-replication repair comes into play after replication proceeds past a lesion in DNA leaving a gap in the replicated strand opposite the lesion. This gap is then filled with material with the correct base sequence from the sister duplex strand. During this process of strand exchange from one sister duplex to another, however, slippage in the base pairing of these DNA strands can occur, leading to mutations, which can be lethal. The second type of recombinational excision repair occurs in the part of the chromosome that was replicated prior to irradiation, where two sister duplexes are present. After the excision of the lesion, the resulting gap is filled by the same recombinational process just described for post-replication repair. Each of these "dark" repair systems involve the cutting of DNA strands, and this can lead to the formation of DNA double-strand breaks, which are lethal if not repaired. Fortunately, cells are very good at repairing DNA double-strand breaks by recombinational processes of the type just described, which require the presence of two sister duplexes. These processes are called homologous recombination, since "proper" base pairing is involved. Cells can also have the capacity for non-homologous recombination, where the ends of a double-strand break are simply sealed together, which usually results in the loss of bases, and therefore results in lethality or mutation. The genes for excision repair and post-replication repair are functional in native DNA. Such processes are referred to as *constitutive*. External stimuli can lead to additional inducible repair mechanisms. The temporary blockage of DNA synthesis after mutagenesis caused by exposure to UV, ionizing radiations, and chemicals induces the genes involved in the *SOS repair* ("last resort") system. Mutations are not only produced by errors in repair, they can also be produced by errors made during normal DNA synthesis. This was first suggested by the finding that some mutations are produced independent of the recombinational repair systems. There are special systems for the prevention, detection, and repair of errors made during DNA replication referred to as mismatch repair.

Most of our early knowledge about the biochemistry and genetic control of the multiple pathways of DNA repair has come from studies on bacteria. In recent years progress has been made in our understanding of DNA repair and mutagenesis in. eucaryotic cells, especially yeast and human cells. Mammalian cells have essentially the same repair systems that were first discovered in bacteria, however, repair is more complicated in mammalian cells because the structure of the genome is more complicated. In bacteria, the DNA is essentially naked, while the DNA in mammalian cells is surrounded by proteins. These proteins must first be removed before DNA repair can proceed. Therefore, more genes are involved in nucleotide excision repair in mammalian cells than in bacteria. DNA repair mutants in yeast can be produced in a manner similar to the production of mutants in bacteria. The cells are treated with a mutagen and then selected for the specific phenotype (characteristic) such as sensitivity to radiation. Fibroblasts from patients with the rare hereditary disease *xeroderma pigmentosum* (XP) are deficient in this "error-free" mode of DNA repair. XP patients must rely on *post-replication repair*, which is "error prone" owing to a high probability of mutations that may originate during this process. The mutants in human cells have largely come from XP patients or patients who have shown a high sensitivity to ionizing radiation detected during radiation therapy for cancer (for example, Ataxia telangectasia). Eukarytotic cells respond to chemical and physical stimuli by cessation of constitutive protein synthesis, coupled with a transient over-expression of *stress proteins*. *Heat-shock proteins* (HSP) are induced by thermal stress, chemical agents, and photosensitization. The expression of HSP is correlated with the development of cellular resistance to subsequent heat shock or *thermotolerance*. A similar mechanism may be involved in resistance of cells to photosensitized oxidative damage and PDT.

## 6.3 PHOTOSENSITIZATION OF BIOLOGICAL SYSTEMS

There is a voluminous literature on biological photosensitization including a number of excellent reviews. The known photosensitizing drugs and pigments number in the thousands, and many of them can act on diverse biological systems ranging from biological molecules *in vitro* to whole organisms. The primary reactions in photosensitization are very different from radiation damage induced by UV. UV radiations are absorbed directly by cellular targets, principally in the 250 nm to300 nm spectral region of nucleic acids and proteins. The colored prosthetic groups in natural macromolecules are usually non-photosensitizing, for example, heme in hemoglobin and 11-*cis* retinal in rhodopsin. However, some colored metabolic intermediates are photosensitizers, for example, serum bilirubin formed by the breakdown of heme in the liver. In early work photosensitization by bilirubin was postulated as an important reaction in the "blue light" treatment of neonatal jaundice, but this has been shown a minor contribution. A photosensitizing agent must be located close to its biological target to be effective. A *Type 1* process is mediated by short-range interactions between the target and photosensitizer molecules. In *Type 2* reactions $^1O_2$ and $O_2^{\bullet-}$ have a short diffusion range in the cellular matrix. The matrix affects the intrinsic photosensitization mechanism because a small shift in electronic energy levels of the photosensitizer can have a large effect on the yield and lifetime of excited states. The *intra*cellular localization of a photosensitizer molecule is determined by its solubility, molecular size, and ionic charge. Typically, anionic dyes are localized in plasma membranes and cytoplasm while cationic dyes bind DNA and mitochondrial membranes. Transport mechanisms are significant in mammals. For example, bilirubin and hypericin, which are insoluble in water, are transported in blood by reversible binding to plasma proteins. Two types of mechanisms have been identified in photosensitization of mammalian cells. One pathway involves generalized necrosis induced by cell membrane fragmentation followed by release of the cellular contents,. This process results in an inflammatory response as lysosomal enzymes interact with adjoining cells. A second mode of cell death involves *apoptosis*, which is a mode of mammalian cell death characterized by a specific pattern of chromatin fragmentation and conversion of cells to apoptotic bodies. These cell fragments are rapidly engulfed by phagocytic cells, minimizing inflammation. Malignant cells often exhibit an impaired ability to undergo apoptosis, an effect associated with reduced response to chemotherapy. Recent evidence suggests that PDT initiates a late step in the apoptotic process.

### 6.3.1 Photosensitization of Viral and Bacterial Contamination in Blood Products

Although the current screening procedures reduce the possibility of viral contamination in blood transfusion to very low levels, there has been continuing interest in decontamination by photosensitization owing to the limits of viral detection, possibility of human error, and emergence of new viruses. Bacterial infection of blood products is a potentially serious problem whose importance has been recognized recently. Viruses are classified according to the type and size of the nucleic acid, the shape and size of the nucleoprotein entity or *capsid*, and the presence or absence of a lipid envelope. Enveloped viruses of current interest include hepatitis B virus (HBV), hepatitus C virus (HVC), and human immunodeficiency virus (HIV-1, HIV-2). HBV is a small DNA virus that contains its own DNA polymerase. HIV is a retrovirus in which the nucleic acid is RNA. After infection, HIV uses the enzyme *reverse transcriptase* to copy its RNA into DNA which is then inserted into chromosomal DNA of the host cell. Other pathogenic enveloped viruses include human T-cell leukemia virus type I, cytomegalovirus, and Epstein-Barr virus. An essential requirement for photosensitized inactivation of blood pathogens is that the function of the cellular components of blood must not be compromised, including red cells, white cells, platelets, and plasma proteins. Blood products that contain red cells are strongly absorbing at wavelengths shorter than about 600 nm. Accordingly, effective photosensitizers for whole blood and packed red cells must have far-red or near-IR absorptions. Results are reported in the literature for various viruses, blood products, and photosensitizers.[3] In early work, herpes simplex virus type 1 in whole blood was inactivated by HPD irradiated at 630 nm.[4] In subsequent experiments, the PDT drug benzoporphyrin derivative monoacid ring A (BPD-MA) irradiated at 690 nm inactivated HIV in whole blood from infected patients without evident damage to blood components.[5] Cationic metallo-phthalocyanine derivatives photosensitized inactivation of HIV-1 in red cell suspensions.[6] In one of the few reports on non-enveloped viruses, metallo-phthalocyanines photosensitized human rhinovirus type 5 in cultured monkey kidney cells.[7] Photosensitization of bacteria has been investigated for many of the same agents employed with viruses. Gram-negative bacteria are more resistant to photosensitization than gram-positive strains, which is attributed to the presence of the outer membrane.

### 6.3.2 Photosensitization of Dental Bacteria

Approximately 10% of all adults suffer from advanced forms of periodontitis leading to extensive loss of tooth-supporting tissues. This disease is characterized by inflammation of the gingiva, formation of periodontal pocket, and tooth loss. The conventional dental treatments involve scaling, root planing, curettage, and most recently topical application of broad-spectrum antibiotics. Gram-positive *Streptococcus mutans and Streptococcus sanguis.* are responsible for dental caries. Recent work showed that oral plaque-forming bacteria harvested as intact biofilms from human patients are inactivated by zinc phthalocyanine excited by visible light.[8] This approach may have potential value for plaque removal in patients who cannot tolerate mechanical techniques. Interest in dye-photosensitized inactivation of oral bacteria developed in the early 1990s. A potential advantage of photosensitization is that the disruption of normal microflora would be localized at the specific treatment sites. Gram-negative *Porphyromonas gingivalis* is a pathogenic black-pigmented bacterium that does not grow in the presence of air. This strain as well as other pathogenic and cariogenic bacteria were inactivated *in vitro* by exposure to 633 nm He-Ne laser light in the presence of the thiazine dye toluidine blue O (TBO). The results for *P. gingivalis* led to evidence for a *Type 2* pathway mediated by $^1O_2$.[9] The clinical applications of photosensitization for inactivation of pathogenic anaerobic bacteria will require photosensityizers which are effective in the absence of air, possibly *via* a *Type 1* pathway or by introducing adjunctive agents.

### 6.3.3 Photosensitization of Red Blood Cell Hemolysis

Photohemolysis refers to the rupture of red blood cell (RBC) membranes and release of hemoglobin induced by the action of light. Although this process has been investigated for more than fifty years, many questions about the mechanism still remain. The current interest in photohemolysis originated from several directions. Many drugs and chemical agents that photosensitize animal and human skin to sunlight also photosensitize RBC hemolysis *in vitro*. The plant pigment hypericin (HY) is a classic example of a skin-photosensitizing agent that photosensitizes hemolysis of RBC (Section 4.7.3). Protoporphyrin PpIX is present in the serum of patients having the hereditary disease *erythropoietic protoporphyria* (EPP). The skin of these individuals is highly light photosensitive and their RBC are hemolyzed by visible light *ex vivo*. More generally, RBC provide well-characterized and reproducible model systems for basic studies on photo-oxidative damage to cell membranes. The *colloid-osmotic model* is the generally accepted mechanism of photohemolysis. According to this theory, photochemical damage to the RBC membrane increases cation permeability leading to swelling and subsequent rupture. The *Type 2* pathway is implicated for many known photodynamic agents including HY, HP, PpIX, BPD-MA, lutetium texaphyrin, and halogenated fluorescein dyes. Photochemical studies indicate that the anion channel membrane protein is the principal target for $^1O_2$. This protein is identified as band 3 on gel-electrophoretic patterns of the proteins extracted from human RBC membranes. Red cells contain other potential targets for $^1O_2$ including cholesterol, lipids, and hemoglobin. Ample evidence exists for photosensitized damage to these targets, but the relationship to hemolysis remains unresolved. Results reported for some agents indicate the occurrence of non-photodynamic mechanisms. Some photosensitizers were active under both oxygenated and anoxic conditions including α-alkylamino-2-arylquinoline-methanol anti-malarial compounds,[10] the non-steroidal anti-inflammatory drug carprofen,[11] and the antifungal drug griseofulvin.[12] Polytriptyline and chlorpromazine generate membrane-bound photoproducts that promote RBC lysis.[13,14]

A typical photohemolysis curve has a *sigmoidal* shape (S-shape). The photohemolysis rate is quantified by the post-irradiation incubation time required for 50% hemolysis ($t_{50}$). A surprising result found for virtually all photodynamic agents is that $1/t_{50}$ increases with the approximate square of the light dose. This dependence implies that initiation of hemolysis requires photochemical damage to two RBC targets. The band 3 protein exists as a dimer, suggesting that both moieties of a given protein entity must be altered. Alternatively, two different band 3 protein entities or a band 3 protein dimer and another membrane target may act in concert. Recent results with PpIX and Lu-Tex show that the photohemolysis rate is faster at higher irradiation temperatures. This effect can be attributed to more efficient reactions of $^1O_2$ with membrane targets. Photohemolysis is accelerated also by higher post-irradiation incubation temperatures. The specific thermally-activated processes responsible for this effect have not been identified.

## 6.4 MODELING OF BIOLOGICAL RADIATION DAMAGE

Analytical modeling of biological radiation damage facilitates the systematic presentation of experimental data and may lead to useful information about the damage mechanisms. A *dose-response curve* shows the probability of achieving a biological endpoint on the dose of an external agent. Dose-response curves have been employed in conjunction with the biological effects of chemicals, ionizing radiation, and light. A *survival curve* is a semi-logarithmic plot showing the fraction of initial entities that remain viable (***S***) after an accumulated dose.

### 6.4.1 Reaction Kinetics

Analytical modeling of survival curves is based on the rate kinetics of physical and chemical processes. The *kinetics order* of a process is characterized by the dependence of the instantaneous rate on the concentrations of the constituents. A starting species may be depleted in the absence of any external stimulus. In a *zero-order* reaction the instantaneous rate of loss does not depend on the concentration of any constituent. Oxidation of alcohol in the liver is a zero-order process. In a *first-order* reaction the instantaneous reaction rate is proportional to the amount of non-reacted material. Radioactive decay is a first-order process. The rate of a first-order process is characterized by the time required for a given fraction of completion, usually the halftime ($t_{1/2}$). It is easily shown that $t_{1/2}$ for a first-order process does not depend on the amount of starting material.

---

**Example 6.1** Half-time of a first-order reaction.

Show that a first-order reaction has a constant half-time that does not depend on the amount of starting material.

Assume that $C_0$ is the initial amount of material and $C(t)$ is the amount of material remaining after a reaction time $t$. In a first-order process the instantaneous loss rate $-dC(t)/dt$ is proportional to the amount of non-reacted material: $-dC/dt = k_1 C$, where $k_1$ is a first-order rate constant. A simple integration leads to: $C(t) = C_0 \text{Exp}(-k_1 t)$. The half-time $t_{1/2}$ corresponds to $C(t_{1/2}) = C_0/2 = \text{Exp}(-k_1 t_{1/2})$. Therefore, $t_{1/2} = \text{Log}_e 2/k_1 = 0.6931/k_1$.

---

In a *second-order* reaction the instantaneous rate of loss depends on the square of the concentration of one constituent or the product of the concentrations of two constituents. Bimolecular reactions in solution exemplify second-order processes. The half-time of a second-order process is inversely proportional to the amount of starting material: $t_{1/2} = C_0/k_2$, where $k_2$ is a second-order rate constant. A second-order reaction may appear to be of first-order if one constituent is present in large excess, for example, the solvent. This type of reaction is termed *pseudo-first order*. A rate process may driven by an external stimulus. The rate of product formation is constant in a zero-order reaction. In higher-order reactions, a maximum yield is attained when a key starting material is used up.

### 6.4.2 Target Theory

Conventional chemical kinetics are not applicable for systems in which the product yield has a sigmoidal dependence on the dose. For example, the "characteristic curve" for blackening of a photographic film has a sigmoidal shape in which the light sensitivity of the emulsion is characterized by the slope of the linear region or "gamma". Photosensitized hemolysis of RBC also have sigmoidal dose-response curves. *Target theory* is a phenomenological approach to biological radiation damage that provides insights into the damage mechanisms. The basic assumption of target theory is that an observable response takes place when a localized region of a biological system or a "target" absorbs one or more units of energy referred to as a *hit*.[15] The relationship between primary hits and a measurable biological endpoint depends on various factors, including the possibility that multiple hits are required on the same or different targets and the operation of repair processes. The mathematical implementation of target theory assumes that radiation damage is a random process. The distribution of hits is predicted by a *Poisson distribution* or the law of small numbers. This statistical distribution applies for a large number of equivalent trials in which the probability of success on a single trial is very low. The application of the Poisson distribution to target theory leads to

$$p(n)=\frac{n_{ave}^{n}}{n!}\exp(-n_{ave}) \qquad (6.2)$$

where *p(n)* is the probability of *n* exactly hits when the average number of hits is $n_{ave}$ after a given dose.

---

**Example 6.2** Poisson distribution

Assume that the number of random typographical errors per chapter in a printed book obeys a Poisson distribution. If the average number of errors per chapter is five calculate the probability of finding: (a) one error; (b) five errors; (a) one or more errors.

(a) The Poisson distribution is: $p(n)$ = Exp(-5) x $(5)^n/n!$. The probability of exactly two errors is: $p(1)$ = Exp (-5) x $5^1$ /1! = 0.03369.

(b) The probability of exactly five errors is: $p(5)$ = Exp(-5) x $5^5$ /5! = 0.1755.

(c) The probability of one or more errors is one minus the number of no errors. The number of no errors is: $p(0) = \text{Exp}(-5) \times 5^0/0! = 0.006738 \times 1/1 = 0.006738$. Therefore, the number of one or more errors is $1 - 0.006738 = 0.9932$.

Note that the probability of exactly one error is much lower than the probability of one or more errors.

The application of the Poisson distribution to hits on a biological target is similar to Example 6.3 where $n_{ave}$ is proportional to the radiation dose. The *one-hit model* assumes that a single hit on one target per entity leads to the observed response. Since *p(0)* is the probability that no hits have occurred, the surviving fraction of undamaged targets equals *p(0)*

$$\boldsymbol{S} = \exp(-n_{ave}) \tag{6.3}$$

The effective photochemical dose can be expressed as

$$n_{ave} = \sigma F \tag{6.4}$$

where σ (cm²) is an *action cross section* corresponding to one hit and *F* (photons/cm²) is the absorbed photon fluence. In terms of incident dose, $n_{ave}$ in Equation (6.4) must be multiplied by the absorption cross section of the relevant chromophore; see Section 5.3. The experimental dependence of σ on wavelength constitutes an *action spectrum*. The action spectrum is a graph relating the probability of achieving a given biological endpoint to the incident wavelength. A comparison of an action spectrum with the absorption spectrum for equivalent irradiation conditions helps identify the species responsible for the observed effect.

#### 6.4.2.1 Detailed Analysis of Survival Curves

A survival curve is a semi-logarithmic plot of the surviving fraction of biological entities *vs* the applied dose. Survival curves are characterized by $\boldsymbol{F}_{37}$, which is defined as the dose at which $\boldsymbol{S}$ decreases from unity to $1/e$ (0.3679).

Therefore, for a system that follows the one-hit model, $\boldsymbol{F}_{37}$ is the dose at where the average number of hits on a target equals unity. These result is easily shown by noting that Equation (6.3) reduces to: exp(-1) = exp($-n_{ave}$) at $\boldsymbol{F}_{37}$. Equation (6.3) can be expressed as

$$\boldsymbol{S}=\exp(-\frac{F}{\boldsymbol{F}_{37}}) \tag{6.5}$$

For a one-hit process, a plot of $\mathrm{Log}_e(\boldsymbol{S})$ *vs* $F$ is a straight line with the slope equal to $1/\boldsymbol{F}_{37}$ This plot leads to the photochemical damage cross section

$$\sigma = 1/\boldsymbol{F}_{37} \tag{6.6}$$

The first survival curve in Figure 6.8 is consistent with the one-hit model.

Target theory can be extended to more complex rate kinetics than the one-hit model. However, $\boldsymbol{F}_{37}$ cannot be taken as the dose at which the average number of hits is unity. Semi-logarithmic survival curves may be less steep or curved at low dose levels followed by a steeper linear decrease at high dose. The initial region is attributed to the production of *sub-lethal damage*. This response can occur if more than one target must be damaged or if more than one hit on a single target is required to initiate the observed effect. Alternatively, the earlier damage may undergo repair by chemical or enzymic processes and be converted to non-lethality. Some survival curves for inactivation of bacteria are concave upwards at high dose levels. Possible explanations for this response include the action of multiple repair mechanisms and non-homogeneity in the radiation response of the original population. The one-hit model is easily extended to the case where one hit is required on each of m equivalent targets. The damaged fraction of targets equals $[1 - \exp(-\sigma F)]^m$ for $m$ targets. The surviving fraction is given by

$$\boldsymbol{S}=1-[1-\exp(-\sigma F)]^{m} \tag{6.7}$$

The low-dose "shoulder" corresponds to *sub-lethal damage* during the initial period while hits are accumulating. Expanding the right side of Equation (6.7) shows that in the limit of high dose: $\boldsymbol{S} \cong m \exp(-\sigma F)$. The survival curve has a linear slope ($1/\boldsymbol{F}_0$) in the high-dose regime. This line intercepts the vertical axis at the *extrapolation number* (m). The "threshold dose" ($\boldsymbol{F}_{th}$) is defined as the initial dose at which the extrapolated high-dose line intercepts $\boldsymbol{S} = 1$. Setting $\boldsymbol{S} = 1$ gives: $\text{Log}_e\, m = \boldsymbol{F}_{th}/\boldsymbol{F}_0$. Examination of Fig. 6.6 shows that: $\boldsymbol{F}_{37} = \boldsymbol{F}_0 + \boldsymbol{F}_{th}$.

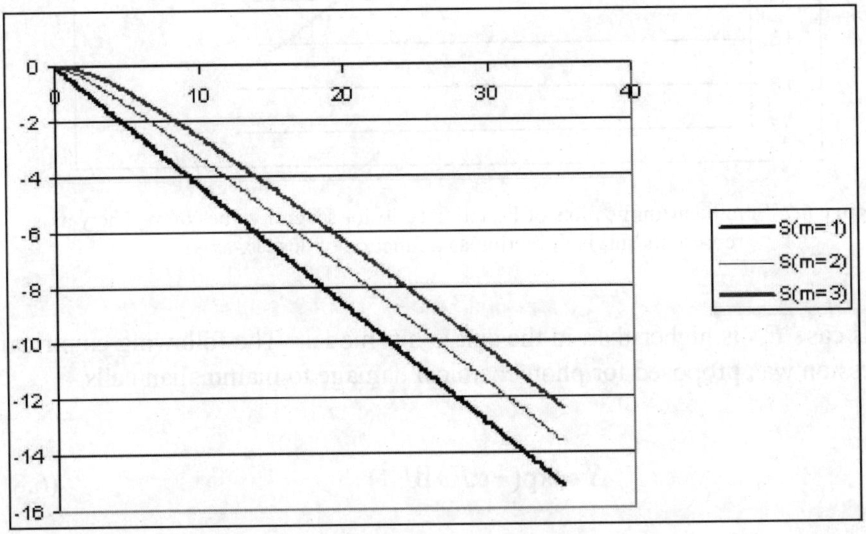


**Figure 6.6** Semi-logarithmic plot of Equation 6.7 for several values of *m*. The y-axis represents survival fraction as a function of dose (x-axis).

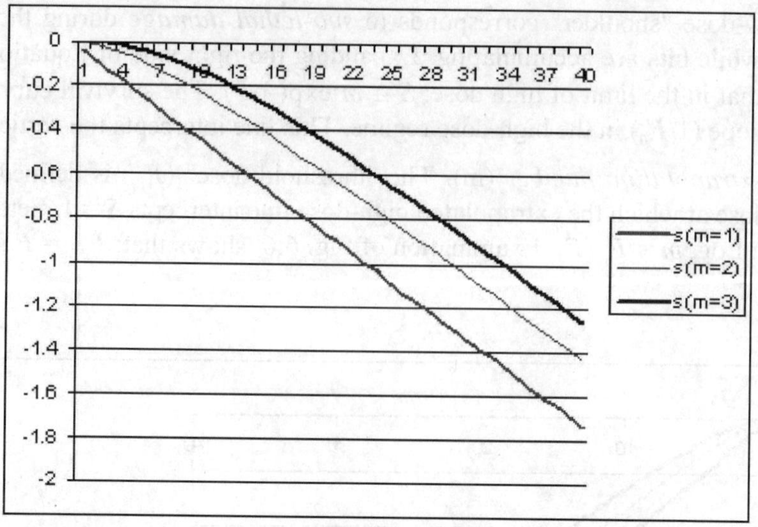


**Figure 6.7** Semilogarithmic plots of Equation (6.9) for several values of m. The y-axis represents survival fraction as a function of dose (x-axis).

In this case $F_{37}$ is higher than in the single-hit model. The following empirical expression was proposed for photochemical damage to mammalian cells

$$S=\exp(-\alpha F+\beta F^2) \quad (6.8)$$

where α is the probability of one-hit damage and β is the probability of two-hit damage. An unrepaired pyrimidine dimer exemplifies one-hit damage. Two-hit damage may be an unrepaired break in both strands of a DNA chain.

Multi-hit kinetics are applicable when at least one target must acquire more than one hit to initiate a response. The semi-logarithmic survival curves do not have a linear region in this case. A plot of the damaged fraction $S^+$ *vs* F ($S^+$ = 1 - $S$) has a sigmoidal shape. Recall that $p(n)$ in Equation (6.2) is the probability of exactly $n$ hits. If $m$ hits are required for the observable event, the fraction of undamaged targets after an absorbed dose $\sigma F$ equals the sum of the terms for n = 0 to n = m - 1. The damaged fraction is 1 minus this summation

$$S^+=\exp(-\sigma F)\sum_{k=0}^{k=m-1}\frac{(\sigma F)^k}{k!} \quad (6.9)$$

Examination of Equation (6.9) shows that the normalized shape of the sigmoidal curve depends only on the hit number *m*. The hit number can be calculated from the relative dose required to attain any two values of $S^+$. For example, if $S_{90}$ is the dose at 90% survival and $S_{50}$ is the dose at 50% survival, then numerical solutions of Equation (6.9) show that the ratio $S_{50}/S_{90} = 3.322$ for $m = 1$, 2.320 for $m = 2$, 1.711 for $m = 5$, 1.472 for $m = 10$, 1.372 for $m = 15$, 1.253 for $m = 30$, etc. Therefore the relative steepness of a sigmoidal damage curve that obeys Equation (6.9) leads directly to the hit number m.

---

**Example 6.3** UV inactivation of a virus

A virus culture is exposed to UV radiation for 5 minutes at a photon fluence rate of $1 \times 10^{10}$ photons/cm$^2$-s. Assume a target cross section for inactivation of $1 \times 10^{-12}$ cm$^2$, which comparable to the geometrical cross section of a virus particle. (a) Calculate the average number of photons "hits" on a virus particle. (b) What fraction of the virus particles receive exactly 0, 1, 2, 3, 4, and 5 hits? (c) Calculate the fraction of inactive virus particles assuming that one hit is required for inactivation. (d) Repeat the calculation for three hits required for inactivation.

(a) $\sigma F = 1 \times 10^{-12}$ cm$^2$ x 5 min x 60 s x $1 \times 10^{10}$ photons/cm$^2$-s= 3 hits per virus particle.

(b) The probability of different numbers of hits are: $p(0) = \text{Exp}(-3) \times 3^0/0! = 0.0498$; $p(1) = \text{Exp}(-3) \times 3^1/1! = 0.149$; $p(2) = \text{Exp}(-3) \times 3^2/2! = 0.224$; $p(3) = \text{Exp}(-3) \times 3^3/3! = 0.224$; $p(4) = \text{Exp}(-3) \times 3^4/4! = 0.168$; $p(5) = \text{Exp}(-5) \times 3^5/5! = 0.101$.

(c) If one hit inactivates a virus particle, the damaged fraction equals one minus the fraction of particles that have received no hits: $p = 1 - p(0) = 0.950$.

(d) If three hits are required for inactivation, the damaged fraction equals one minus the fraction of particles that have received 0, 1, and 2 hits: $p = 1 - (0.050 + 0.149 + 0.224) = 0.557$.

---

NOTES

1. James D. Regan and John A. Parrish *The Science of Photomedicine* Plenum Pub. 1982 p.96

2. This section is based on information provided by Dr. Kendric C. Smith.

3. Ben-Hur E, Horowitz B. Advances in photochemical approaches for blood sterilization. *Photochem Photobiol*. 1995 Sep;62(3):383-8.

4. Matthews JL, Newman JT, Sogandares-Bernal F, Judy MM, Skiles H, Leveson JE, Marengo-Rowe AJ, Chanh TC. Photodynamic therapy of viral contaminants with potential for blood banking applications. *Transfusion.* 1988 Jan-Feb;28(1):81-3.

5. North J, Coombs R, Levy J. Photodynamic inactivation of free and cell-associated HIV-1 using the photosensitizer, benzoporphyrin derivative. *J Acquir Immune Defic Syndr.* 1994 Sep;7(9):891-8.

6. Rywkin S, Ben-Hur E, Malik Z, Prince AM, Li YS, Kenney ME, Oleinick NL, Horowitz B. New phthalocyanines for photodynamic virus inactivation in red blood cell concentrates. *Photochem Photobiol*. 1994 Aug;60(2):165-70.

7. Gaspard S, Tempete C, Werner GH. Studies on photoinactivation by various phthalocyanines of a free or replicating non-enveloped virus. *J Photochem Photobiol B*. 1995 Dec;31(3):159-62.

8. Wood S, Nattress B, Kirkham J, Shore R, Brookes S, Griffiths J, Robinson C. An in vitro study of the use of photodynamic therapy for the treatment of natural oral plaque biofilms formed in vivo. J *Photochem Photobiol B.* 1999 May;50(1):1-7.

9. Bhatti M, MacRobert A, Meghji S, Henderson B, Wilson M. A study of the uptake of toluidine blue O by Porphyromonas gingivalis and the mechanism of lethal photosensitization. *Photochem Photobiol.* 1998 Sep;68(3):370-6.

10. Epling GA, Sibley MT. Photosensitized lysis of red blood cells by phototoxic antimalarial compounds. *Photochem Photobiol.* 1987 Jul;46(1):39-43.

11. De Guidi G, Chillemi R, Costanzo LL, Giuffrida S, Condorelli G. Molecular mechanism of drug photosensitization. 4. Photohemolysis sensitized by carprofen. *J Photochem Photobiol B*. 1993 Mar;17(3):239-46.

12. Fujita H, Inukai N, Matsuo I. In vitro photosensitized lysis of red blood cells by an antifungal drug griseofulvin. *J Photochem Photobiol B*. 1993 Jan;17(1):77-80.

13. Gasparro FP, Kochevar IE. Investigation of protriptyline photoproducts which cause cell membrane disruption. *Photochem Photobiol.* 1982 Mar;35(3):351-8.

14. Kochevar IE, Hom J. Photoproducts of chlorpromazine which cause red blood cell lysis. *Photochem Photobiol.* 1983 Feb;37(2):163-8.

15. The hit concept implies that radiation damage is a discrete process. This picture is appropriate for photochemical damage which is initiated by the absorption of photons. The application of target theory to ionizing radiation damage requires an arbitrary definition as to how much absorbed energy constitutes a hit. Typical assumed values are the order of 50-60 eV.

# CHAPTER 7

# OPTICAL METHODS OF IMAGING AND DIAGNOSIS

During the past decade there has been a major expansion in the development of electro-optical instrumentation for clinical diagnosis and imaging. This field is an important component of biomedical optics, which encompasses optical applications in biology research and medicine. Notwithstanding the high level of university and commercial activity, the clinical applications of new optical diagnostic devices are limited owing to requirement for acquiring regulatory approvals. Instruments for measuring the properties of blood date to the 1970s. The IL 282 CO-Oximeter® (Instrumentation Laboratories, Lexington, MA) is an automated *in vitro* device for determination of total blood hemoglobin (Hb) and the percentages of oxy-Hb, deoxy-Hb, and carboxy-Hb. This instrument measures the optical absorption of diluted whole blood at four wavelengths. An early type of built-in computer calculates the relative concentrations of each component by solving four linear simultaneous equations. Non-invasive measurements of oxygen saturation in blood and tissues or *oximetry* have many clinical applications. Red light and near-infrared (NIR) are used because light penetration is deeper compared to shorter wavelengths. It turns out that oxy-Hb and deoxy-HB have the same extinction coefficient at several wavelengths referred to the *isobestic points*. An isobestic point is used for baseline measurements to correct for perturbing factors such as blood flow and pH.

A diffuse reflection technique for measuring hemoglobin saturation in flowing blood was developed by Reynolds in 1976. This type of oximeter consists of an arterial catheter incorporating two sets of light delivery optical fibers and multiple receiving fibers. The measurements at 685 nm distinguish oxy-Hb from deoxy-Hb and 925 nm is close to an isobestic wavelength. The ratio of the diffuse reflection at 685 nm to 925 nm is scaled to the oxy-Hb concentration based on calculations with the diffusion approximation (DA). A pulse oximeter is a *trans-illumination* device that measures light absorption by oxy-Hb in the thumb or ear lobe. Interference by steadily-flowing venous blood is canceled by locking the measurements to arterial pulses.

*Autofluorescence* is emitted by all tissues excited with blue and near-UV (NUV). The diagnostic applications are based on different autofluorescence spectra and fluorescent lifetimes for normal and diseased tissues. *Extrinsic fluorescence* methods utilize differential uptake and emission properties by an exogenous fluorophore in normal and abnormal tissues. This approach requires administration of a drug and has been applied only in conjunction with PDT. Fluorescence measurements are intrinsically more sensitive than absorption, although not necessarily more accurate, because the fluorescence baseline is a null signal and not the strong transmission signal of a non-absorbing control. *In vitro* fluorescence diagnostic techniques have potential applications for rapid screening, especially when the available tests are not definitive or are not part of routine physicals. A fast, reliable optical screening for cervical cancer to augment the Pap smear technique would have immediate widespread applications.

Optical imaging is an emerging application of high recent activity. In the most direct procedure of trans-illumination an image is obtained by passing CW light directly through the organ onto a light-sensitive medium. Scanning techniques are now available in which the transmission through small regions of the organ are measured using coupled optical fibers for light delivery and detection. The image reconstruction algorithms are borrowed from *x-ray computed tomography* (CT) and *magnetic resonance imaging* (MRI). In *ultrasound-modulated optical tomography* an optical image is derived from a tissue region exposed to pulsed or CW excitation with ultrasound. The differential optical-mechanical properties of normal and abnormal tissues may lead to better resolution of abnormalities than optical imaging alone.

## 7.1 STATISTICS OF DIAGNOSTIC TESTS

Diagnostic tests are never 100% accurate. A number of different criteria are used to quantify the reliability of a given test. Test results can be characterized as true positive (TP), false positive (FP), true negative (TN), and false negative (FN). The standard statistical criteria are as follows:[1]

*Sensitivity* = TP/(TP + FN)

High sensitivity is important when the disease to be diagnosed is serious, should not be missed, is treatable, and false positive results do not have serious adverse consequences for the patient.

*Positive Predictive Value* = TP/(TP + FP)

High positive predictive value is important when the disease to be diagnosed is serious, should not be missed, is treatable, and false positive results may have serious adverse consequences for the patient.

*Test Efficiency* = (TP + TN)/(TP + FP + TN + FN)

High test efficiency is important when the disease to be diagnosed is serious, should not be missed, is treatable, and false positive and negative results are equally serious and potentially injurious to the patient.

*Specificity* = TN/(TN + FP)

High specificity is important when the disease to be diagnosed is serious, but not treatable, and false positive results may have serious adverse consequences for the patient.

*Negative Predictive Value* = TN/(TN + FN)

High negative predictive value is important when the disease to be diagnosed is serious, but not treatable, and false negative results will not have serious adverse consequences for the patient.

Test efficiency is the least well known statistical measure of a diagnostic test performance and is most often the best measure of the clinical utility. For example, a real-time *in vivo* optical diagnosis for dysplasia in Barrett's esophagus might be used to direct laser ablation during the same endoscopic procedure. In this case, the risk of laser ablation of a patient without dysplasia is roughly comparable to the risk of not treating a patient with dysplasia. However, too many tests are disadvantageous. If the probability of abnormal diagnosis on one test is 5%, then 10 independent tests (or parameters measured) have a 40% chance of an abnormal diagnosis, and 50 independent tests have a 92% chance of an abnormal diagnosis. The choice of a study population also affects the statistical criteria. Consider a test with 99% sensitivity and 99% specificity. If a study group has roughly the same number of patients with and without the disease, then the positive predictive value of the diagnostic test is 99%. (This results follows directly from Bayes' theorem in statistics.) However, if the prevalence of the disease in the general population is 0.1%, then for the same 99% sensitivity and 99% specificity, the positive predictive value falls to 9%. Accordingly, a test that has a high positive predictive value in a small scale study may be less reliable when applied to a general population.

## 7.2 NON-INVASIVE DIAGNOSTIC TECHNIQUES BASED ON OPTICAL CONSTANTS

The objective in this approach is to measure a tissue optical property that is related to a parameter of clinical relevance. "Added absorber" measurements are used to determine the concentration of a PDT agent in normal and malignant tissues. CW techniques are feasible for strongly absorbing drugs. Starting with Equation (5.13) it is found that

$$\frac{\mu_{eff}^2}{(\mu_{eff}^2 - \mu_{eff}'^2)} = \frac{\mu_a}{\mu_{as}} \tag{7.1}$$

where the "primed" value refers to the tissue with the additive present and $\mu_{as}$ is the linear absorption coefficient of the additive. The derivation of Equation (7.1) assumes that the additive affects only the absorption, but not the scattering, properties of the tissue and that $\mu_s\,(1\text{-}g) >> \mu_a$. The values of $\mu_{eff}$ can be determined from experimental measurements of the optical penetration depth based on $\mu_{eff} = 1/\delta_d$. The tissue concentration of the additive $C_s$ is related to $\mu_{as}$: according to: $C_s\,(\mu g/g) = \mu_{as}(cm^{-1}) / \gamma_s\ [cm^{-1}/(\mu/g)]$, where $\gamma_s$ is the specific absorption coefficient of the added absorber at the measurement wavelength. The two-fiber geometry of Figure 5.3 can be used to follow the uptake of a PDT drug based on the following semi-empirical expression

$$R_d(t) = R_d(t=0)\frac{\exp[-\mu_{eff}(t)r']}{r'} \tag{7.2}$$

where $t$ is the time interval after drug administration. A constant value of the inter-fiber spacing $r'$ was used and experimental values of $\mu_{eff}(t)$ were substituted in Equation (7.1) to calculate $\mu_{as}$.

Transient measurements are an intrinsically more sensitive procedure compared to CW irradiations. For a strongly absorbing PDT agent, the absorption coefficient calculated with Equation (7.1) can be resolved into the absorption spectrum of the PDT agent and a featureless baseline absorption of the unsensitized tissue. The two-fiber geometry of Figure 5.3 leads to Equation (5.20) for the time-dependent diffuse reflectance $R_d(r',t)$. A typical time-domain measurement of $R_d(r',t)$ on a tissue phantom is shown in Figure 5.4. The absorption coefficient is calculated from the long-time slope of a semi-logarithmic graph of $R_d(r',t)$ *vs* $t$

$$\frac{d}{dt} Log_e R_d(r',t) \Rightarrow c'\mu_a \tag{7.3}$$

where $c'$ is the speed of light in the tissue. The reduced scattering coefficient of the normal tissue is calculated from the time at which $R_d(r',t)$ attains its maximum value ($t_{\max}$) according to

$$\mu_s(1-g) = (1/3r'^2)(4\mu_a c'^2 t_{\max}^2 + 10c' t_{\max}) - \mu_a \tag{7.4}$$

Fast pulsed laser techniques were used to detect very small changes in oxygen saturation of blood in the human brain[2]. The absorption coefficient of Hb at two wavelengths is related to the extinction coefficients of deoxy-Hb ($\alpha$) and oxy-Hb ($\beta$), respectively, by the linear expressions

$$\mu_a^{(1)} = \alpha^{(1)}[deoxy\text{-}Hb] + \beta^{(1)}[oxy\text{-}HB];\ \mu_a^{(2)} = \alpha^{(2)}[deoxy\text{-}Hb] + \beta^{(2)}[oxy\text{-}Hb]$$

where the quantities in [ ] are the concentrations of the Hb derivatives and superscripts (1) and (2) indicate the measurement wavelength. The fractional oxygen saturation $Y$ equals [oxy-Hb]/$\{$[oxy-Hb]+ [deoxy-Hb]$\}$. Solving for $Y$ gives

$$Y = -\frac{\alpha^{(1)} - \alpha^{(2)}[\mu_a^{(1)}/\mu_a^{(2)}]}{-\Delta\alpha^{(1)} + \Delta\alpha^{(2)}[\mu_a^{(1)}/\mu_a^{(2)}]} \quad (7.5)$$

where $\Delta\alpha^{(1)}$ and $\Delta\alpha^{(2)}$ are the absorption coefficient differences at each wavelength. The value of $Y$ does not depend on the total Hb concentration. Measurements of transient diffuse reflectance bracketing the 800 nm isobestic point were made with pulsed lasers emitting at 754 nm and 816 nm. The value of $\mu_a^{(1)}/(\mu_a^{(2)}$ at these wavelengths varies from 2.1 at Y = 0 % to 0.6 at Y = 100 %. Equation (7.5) was used with time-domain technique for calculations of $\mu_a$. The corresponding frequency-domain measurements utilized the changes in phase shift between the incident and reflected sinusoidal signals.

## 7.3 FLUORESCENCE-BASED DIAGNOSIS

Optical fluorescence of tissues follows the same photophysical principles as indicated by the Jablonski diagram for an organic dye-like molecule (Figure 4.6). In the primary act of light absorption one molecule of a chromophore is excited to a more energetic state by absorption of one photon.[3] The first product is a "hot" molecule corresponding to an electronically- and vibrationally-excited state of the chromophore. After a sequence of steps requiring about $10^{-11}$ s, the excited molecule relaxes to its lowest-energy electronically-excited state in which it is in thermal equilibrium with the surroundings. A typical lifetime of a *fluorescent state* ranges from 0.1 - 10 ns depending on the molecular structure and the environment. Several possible processes may then take place, one of which is fluorescence. Radiationless decay and internal conversion to the triplet state are competing processes. After photon emission the molecule in its ground electronic state may still possess excess vibrational energy which is dissipated by a fast, radiationless decay process. It is evident from the Jablonski diagram that:

(1) The shortest wavelength fluorescence band of a chromophore is centered at longer wavelengths than the longest wavelength absorption band. This relationship is the Stokes' shift.

(2)The fluorescence emission spectrum is the same for any excitation wavelength.

Several experimental techniques are used to measure fluorescence. A *fluorescence emission spectrum* displays the emission spectrum at a fixed excitation wavelength. This measurement provides spectral information about the structure and environment of the *fluorophore*.[4] In a *fluorescence excitation spectrum*, the emission intensity at a fixed wavelength is measured over a range of excitation wavelengths. An excitation spectrum provides information about the specific chromophores that are responsible for the observed fluorescence.[5] For example, the excitation spectrum of a self-aggregated species may show that the fluorescence originates mostly from the monomer component. The kinetics of fluorescence decay can be measured with time-domain and frequency-domain techniques. The decay function is usually resolved into a summation of exponential terms having different lifetimes and amplitudes. The emission spectrum may change during the short lifetime of the decay process owing to dynamic interactions of the fluorophore with the environment. A *time-resolved emission spectrum* measures fluorescence spectra at various times during the decay process.

### 7.3.1 Autofluorescence

Essentially all tissues emit autofluorescence when illuminated with blue light and NUV. Some diagnostic techniques are based on local differences in autofluorescence induced by tissue abnormalities. The approximate emission wavelengths of common tissue fluorophores are: proteins (340-350 nm), collagen (405 nm), reduced nicotinamide-adenine dinucleotide (440-460 nm), β-carotene (520 nm), and melanin ( > 540 nm). Early applications of autofluorescence for diagnosis were based on spectral differences between normal and abnormal tissues. Autofluorescence is usually less intense for malignant tissues compared to the corresponding normal and benign tissues, and there may be a weak peak near 630 nm from the accumulation of endogenous porphyrins. Laser-induced autofluorescence is being investigated for diagnosis of atherosclerosis. The emission spectrum depends on the chemical composition of the plaque. Attempts have been made to distinguish tumors from normal tissue by measuring the intensity ratio of autofluorescence at several wavelengths. An early study on human breast tissues led to a fluorescence intensity ratio at 340 nm: to 440 nm of about 16 for cancerous sections and 4-5 for benign and normal tissues.[6] Although fluorescence is not likely to replace histological examination, it may serve as a screening procedure and assist in eliminating ambiguities. Fluorescence decay provides an additional signature of cancerous tissues.[7] A typical tissue emission excited by a short pulse of visible light has two decay components, a "fast" decay of about 0.2 ns lifetime and a "slow" decay of about 2 ns lifetime. The amplitude ratio of the fast-to-slow decays is about two-fold higher for malignant tissues.

### 7.3.2 Sensitized Fluorescence

Fluorescence diagnosis exploits the ability of some fluorescent agents to concentrate in malignant tissues. This technique is referred to as *photodynamic diagnosis* (PDD) because PDT drugs have been used for most applications. However, the resolution of malignancies is relatively poor because PDT drugs are usually weak emitters. This approach is exemplified by studies since 1978 at Tokyo Medical College in conjunction with PDT treatments of early stage bronchogenic cancer.[8] A recent implementation uses a pulsed excimer laser-pumped dye laser emitting at 405 nm as the light source. The image of the bronchoscope is amplified by means of *CCD* camera and a *microchannel plate* and viewed on a fluorescent screen.[9,10] Emission measurements at 530-570 nm and 670 nm distinguish between the green autofluorescence and the red fluorescence of the PDT drug. Measurements on lung cancer patients showed that tumors are best visualized as the ratio of the red to green fluorescence. PDD is also being used in conjunction with PDT of bladder cancer with ALA.[11] The superposition of three color images led to the best resolution of tumors, consisting of the red PpIX fluorescence image, the green autofluorescence image, and a blue image of remitted (diffusely reflected) light. Since PDT drugs have adverse effects, the awarding of regulatory approvals for diagnostic PDD without PDT remains an open question.

## 7.4 OPTICAL IMAGING OF TISSUES

In conventional optical imaging, light reflected or transmitted by each element of an object is collected by an optical system and projected on the retina or a light-sensitive device. Optical imaging has the major potential advantage over x-rays and other ionizing radiation techniques of being non-carcinogenic. In addition, the intrinsic resolution is higher than is possible with ultrasound imaging and apparatus can be smaller and less expensive than MRI devices. Trans-illumination imaging of tissues with CW light has limited applications because light scattering destroys the spatial and phase relationships between the rays originating from different points of the object. Recent work shows that useful optical images of tissues can be achieved with picoscond laser techniques. Alternatively, frequency-domain methods can be employed in conjunction with tomographic image reconstruction techniques. The second approach can be best described by reviewing the general aspects of x-ray *computed tomography* (CT). Optical imaging has been a subject of much research interest since the 1990s. Most of the published work has been reported for tissue phantoms incorporating buried objects. The most general approach to optical imaging is based on radiative transfer. The time-dependent diffusion approximation has been employed for highly scattering and weakly absorbing objects. Measurements based on CT geometry can utilize the image reconstruction techniques developed for x-ray CT.

The objective in CT is to reconstruct a two-dimensional image based on measurements confined to a thin slice of a three-dimensional object. Similar image reconstruction techniques are used for various methods of generating sectional images, including x-ray transmission CT, single photon emission CT (SPECT), nuclear magnetic resonance CT (MRI), and ultrasonic imaging. Each of these methods is based on a different type of interaction of the external stimulus with tissue and, consequently, the apparatus and measurement techniques are not at all similar. However, all CT techniques provide physical signals at multiple locations in the object plane that are probed along linear "projections" from a source to a receiver.

Recent approaches to optical imaging utilize short laser pulses based on either frequency-domain or time-domain techniques. The experimental and theoretical objective is to measure only the first few scattered photons emerging from the object which are less affected by light scattering within the object. This section summarizes the results of several recent studies that exemplify various aspects of the problem. The voluminous recent literature in this field describes other techniques. However, most of these embody the same basic stages:

(1) multiple transient measurements of transmitted light in CT geometry;

(2) calculation of photon densities within the object from intensity data based on a theoretical model;

(3) image reconstruction and refinements.

### 7.4.1 Frequency-Domain Imaging: Optical Mammography

The potential of optical mammography in breast cancer lies in the sensitivity of red/NIR light to the hemoglobin concentration and saturation, and to modifications to the tissue architecture. Diagnosis of breast cancer by *diaphanography* was developed in the 1970s. The empirical approach was highly operator-dependent and yielded a large number of false positives.

A more recent technique utilizes frequency-domain measurements at 70 MHz and four red and NIR wavelengths (690 nm, 750 nm, 788 nm, 856 nm)[12]. Light from diode lasers delivered by optical fibers is scanned over the breast between glass plates and measured by coupled receiving optical fibers. The amplitude and phase of the transmitted photon density wave produces a two-dimensional projection image at each wavelength. The relative optical attenuation is calculated as a dimensionless parameter termed the "N-image". Results are reported for a clinical study on 63 patients (106 breasts), 45 of whom had malignant breast tumors and the others had benign breast lesions. The sensitivity for detecting breast cancer in this study was 73% and the specificity was 49%. The results were taken as encouraging because the analysis employed did not discriminate absorption from scattering inhomogeneities and only the 788 nm data were utilized.

### 7.4.2 Time-Domain Imaging

One approach to optical imaging is based on the effect of the tissue optical properties on the decay of ultra-short laser pulses. This technique is exemplified by measurements on a tissue phantom simulating the optical constants of a human breast with a more highly absorbing and less scattering region in the center.[13] The light source was a mode-locked Ti-sapphire laser generating 2 ps pulses at 800 nm operated at a repetition rate of 76 MHz. Tomographic geometry was employed with the receiving optical fiber collinear with the laser beam at 1369 positions on the back face of the phantom. The number of early-arriving photons at each location was measured by means of time-correlated single photon counting (see Chapter 5, footnote-6). A typical measurement shows the distribution of arrival times for a light beam passing through the simulated lesion and the reference beam through the bulk material.

The normalized fractional transmittance was defined as

$$T^f(x,y) = \frac{\int_0^{\tau_f} N(x,y,\tau)\,d\tau}{f\int_0^{\infty} N(x_{ref}, y_{ref}, \tau)\,d\tau} \tag{7.18}$$

where $(x_{ref}, y_{ref})$ are locations far from the insert. and $f$ is the integrated fraction of photons detected up to $\tau_f$ at the reference locations relative to the total number of photons detected at the reference locations. Higher values of $f$ led to poorer images owing to the effects of scattering on the later-arriving photons. The amplitude and phase of the corresponding photon density waves were calculated by Fourier transformation of the time-of-flight. The normalized Fourier amplitudes and the phase shifts at a selected modulation frequency also led to trans-illumination images of the buried object.

### 7.4.3 Image Reconstruction Based on the Time-Dependent Diffusion Equation

The objective of the trans-illumination imaging techniques is to minimize the effects of light scattering. In a potentially more accurate approach the absorption and light-scattering properties of the tissue are simulated by the time-dependent diffusion approximation. A theoretical feasibility study applicable for CW imaging with NIR follows the analysis employed in x-ray CT[14]. A solution to Equation (5.19) for an infinite diffusive medium and a constant light source source of strength $S_0$ ($s^{-1}$) located at $\boldsymbol{r}_S$ gives the photon density at a detector position $\boldsymbol{r}_D$

$$\Phi(\boldsymbol{r}_D)=\frac{S_0}{4\pi c'D}\frac{\exp[-\mu_{eff}|\boldsymbol{r}_D-\boldsymbol{r}_S|]}{|\boldsymbol{r}_D-\boldsymbol{r}_S|} \tag{7.19}$$

Equation (7.19) shows that the photon flux density reaching the detector is proportional to the source strength and decreases with increasing distance between the source and detector. The photons traveling from the source to the detector travel on different paths. The light path probability density for rays originating from a source and reaching a detector in an infinite diffusive medium has the form

$$P(\boldsymbol{r})=\frac{\exp[-\mu_{eff}|\boldsymbol{r}-\boldsymbol{r}_S|]}{|\boldsymbol{r}-\boldsymbol{r}_S|}\cdot\frac{\exp[-\mu_{eff}|\boldsymbol{r}-\boldsymbol{r}_D|]}{|\boldsymbol{r}-\boldsymbol{r}_D|} \tag{7.20}$$

Software developed by Philips Medical Systems for CT is used to calculate the projections of $\Phi(\boldsymbol{r}_D)$. The photon density distributions is then converted into attenuation coefficients $\mu_{eff}$ using Equation (7.19). The photon density projection data is then treated as with x-ray CT projections using the filtered back-projection technique.

## 7.4.4 Image Reconstruction Based on the Equation of Transfer

The diffusion approximation to the equation of transfer is least accurate for the early-arriving photons that can provide the best optical images. A more rigorous approach is based on perturbative solutions of the radiative transfer equation, assuming that the inhomogeneity has slightly different absorption and scattering coefficients than the bulk tissue[15]. The transfer equation is solved for an arbitrary point source at a specific instant of time. The complete solution is then obtained by integrating the point-source solution over the relevant range of positions and signal decay times. In mathematical physics this technique is referred to as the method of *Green's functions*. This procedure leads to the "point spread function" (PSF), which is the probability that a photon injected at a point $\boldsymbol{r}_0$ at time $t_0$ passes through the field point at $\boldsymbol{r}'$ before time $t$ and is collected by the detector at point $\boldsymbol{r}$ at time $t$. Equation (7.19) is a simpler Green's function for the time-dependent DA and a constant light source. In x-ray CT a small volume element of the object or *voxel* located on the trajectory contributes 100% to the PSF and all other voxels contribute 0%. The situation is different in a turbid medium because off-axis scattered photons can reach the receiver. Each voxel in the object can be assigned a weighting factor determined by the photon path distributions. In an extension of the x-ray CT technique, the contribution of each voxel to a source-detector measurement was weighted by the density of photon paths across it. The direct x-ray CT algorithm leads a considerably enlarged image of a spherical object, while the DA leads to a distorted image of approximately the correct size. The more accurate imaging was achieved with the Green's function calculation incorporating a correction for "causality". In simple terms, causality takes into account that diffusion starts only after the light pulse traverses the medium. This study demonstrates that refined image reconstruction algorithms based on photon migration measurements, have the potential to provide high quality optical CT images of the breast, the brain, and perhaps elsewhere in the body.

NOTES

1. Fitzmaurice M. Principles and pitfalls of diagnostic test development: implications for spectroscopic tissue diagnosis. *Journal of Biomedical Optics* 2000 April 5(2) 119-130.

2. Levy WJ, Serle L, Chance B. Near-infrared measurement of cerebral oxygenation: correlation with electroencephalographic ischemia during ventricular fibrillation. *Anesthesiology* 1995;83:738-746

3. At high laser intensities fluorescence can be excited by the simultaneous absorption of photons. This technique has been employed to improve light penetration in tissues by exciting fluorescence with weakly-absorbed long wavelength light.

4. An experimental fluorescence emission spectrum requires corrections for the spectral response of photodetector. Otherwise the data is referred to as "uncorrected". One method of correction is based on comparisonsnton the absolute emission spectrum of a standard substance.

5. An experimental fluorescence excitation spectrum requires corrections for the emission spectrum of the excitation lamp and the spectral efficiency of the light delivery system. This can be done by using a "quantum counter" consisting of a strongly absorbing dye solution. Since the fluorescence emission spectrum should be independent of the excitation wavelength, the output of the quantum counter wavelength is proportional to the spectral intensity distribution of the incident photons.

6. Gupta PK, Majumder SK, Uppal A.Breast cancer diagnosis using N2 laser excited autofluorescence spectroscopy. *Lasers Surg Med* 1997;21(5)417-22.

7. Fluorescence lifetimes can be measured by pulsed techniques in real time and by repetitive pulse methods based on the phase shift between the incident and emitted light. Complex fluorescence decays are not infrequent which usually are resolved into the superpositionn of two or more exponential decay lifetimes.

8. Lamn S, Palcic B, McLean D, Hung J, Korbelik M, Profio AE. Detection of early lung cancer using low dose Photofrin II. *Chest* 1990 Feb: 97(2):333-7.

9. A charge-coupled device (CCD) is a self-scanning semicoductor unit in which the light sensing element consists of a closely spaced array of metal oxide semiconductor (MOS) capacitors. The stored charge in each unit is proportional to the number of incident photons. The charge distribution on the CCD is read out sequentially and synchronized with the scanning frequency of a CRT monitor.

10. An image intensifier boosts the intensity of an optical image to the poit when it becomes useful. The primary image on a photocathode surface can be intensified by increasing either energy or number of the photoelectrons. In on type of device the photoelectrons are accelerated by an applied voltage to a photoanode coated with a phosphor screen. The microchannel plate exemplifies another type of image intensifier consisting of a thin slab of insulating material with a high density of channels cut through it with slightly conducting inner walls. Primary electron entering a channel at a slight angle strikes the walls and generates secondary electrons which are accelerated down the channel by an applied voltage and strike a phosphor screen.

11. Landry JL, Gelet A, Bouvier R, Dubernard JM, Martin X, Colombel M. Detection of bladder dysplasia using 5-aminolaevulinic acid-induced porphyrin fluorescence. *BJU Int.* 2003 May:91(7):623-6.

12. Franceschini MA, Moesta KT, Fantini S, Gaida G, Gratton E, Jess H, Mantullin WW, Seeber M, Schlag PM, Kaschke M. Frequency-domain techniques enhance optical mammography: Initial clinical results. *Proc. Natl Acad Sci USA* June 1997 Vol 94 p. 6468-6473.

13. Grosenick D, Wabnitz H, Rinneberg HH. Contrast and spatial resolution of time-resolved transillumination images. *Proceedings of SPIE* Volume 2626 Dec 1995 p. 206-217.

14. Colak SB, Paioannou DG, t'Hooft GW, van der Mark MB. Optical image reconstuction with deconvolution in light diffusing media. *Proceedings of SPIE* Volume 2626 December 1995. P. 306-315.

15. Chen K, Perelman LT, Zhang Q, Dasari RR and Feld MS. Optical computed tomography in a turbid medium using early arriving photons. *Journal of Biomedical Optics* 2000 5(2) p.144-154.

# CHAPTER 8

# LIGHT DOSIMETRY MODELING FOR PHOTOTHERAPY

Controlled clinical trials are the final stage of experimentation prior to regulatory approval of a phototherapy. The essential problem with clinical trials, as necessary as they are, is that their high cost and long duration limits the range of variables that can be evaluated. The treatment parameters employed in clinical trials may not be optimal and even inappropriate for some patients. The accumulation and reporting of relevant clinical information is a lengthy process before a new procedure has received regulatory approvals. Analytical modeling attempts to "bridge the gap" between theory and practice by relating the treatment planning parameters to clinical endpoints. Treatment planning for a phototherapy is more complicated than for a conventional drug therapy because the responses depend on interactions between light and a chemical agent. Analytical modeling is especially useful for PDT in view of the rapid increase in the number of PDT drugs and the recent extensions of the original procedure to non-cancer disorders.

## 8.1 ELEMENTARY APPROACH TO LIGHT DOSIMETRY MODELING

The basic tenant of modeling is that a phototherapy is *threshold* type of process. It is assumed that a minimum optical energy must be absorbed by the effective chromophores in tissue in order to activate the desired clinical endpoint. This assumption is supported for PDT by animal model experiments and clinical data. The existence of a light dose threshold is less certain for PUVA and the "blue light" phototherapy of neonatal jaundice. However, thresholds are implicit for these procedures owing to the desirability of achieving clinically useful results within a practical time frame. The role of the threshold dose is depicted in Figure 8.1 for PDT of a large tumor, showing that necrosis band migrates deeper into the tissue as the irradiation proceeds.

According to tissue optic theory, the energy density absorbed by a tissue chromophore of type-*j* at location $\boldsymbol{r}$ after an exposure time $t$, $W_j(\boldsymbol{r},t)$ (J/cm$^3$), is given by

$$W_j(\boldsymbol{r},t)=\mu_{aj}\Phi(\boldsymbol{r})t \tag{8.1}$$

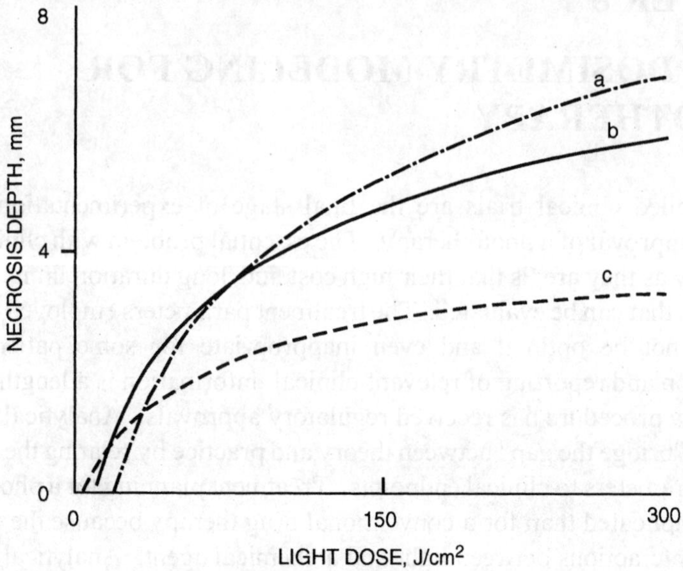


**Figure 8.1** Calculated dependence of necrosis depth on 630-nm light dose for PDT; Photofrin® administered at 2 mg/kg, based on Equation 8.15; $q^* = 40$ J/cm$^2$, and the optical constants in Table 5.1 (a) White brain tissue, (b) Nonpigmented tumor tissue, (c) White epidermis on a thick tissue "bed".

where $\Phi(\boldsymbol{r})$ (W/cm$^2$) is the local fluence rate and $\mu_{a\,j}$ (cm$^{-1}$) is the linear absorption coefficient of constituent-*j* in the tissue. Equation (8.1) assumes that $\mu_{a\,j}$ and $\Phi(\boldsymbol{r})$ have the same values throughout the tissue and do not change in time. For a threshold type of process $W$ must exceed a certain value $W^*$ in order to initiate the physiological response. If $\boldsymbol{r}^*$ denotes a tissue location in which the threshold dose is just attained, the light dosimetry condition can be expressed as

$$W(\boldsymbol{r}*) > W* \tag{8.2}$$

Combining Equation (8.1) and Equation (8.2) leads to an expression that relates the threshold dose to tissue optic quantities

$$W^* = \mu_{aj}\Phi(\boldsymbol{r}^*)t \tag{8.3}$$

$W^*$ (J/cm$^3$) is the energy absorbed by the photosensitizer per unit volume of tissue required for necrosis or another physiological endpoint. Equation (8.3) is the basic light dosimetry relation. The first step in modeling requires a calculation of $\Phi(\boldsymbol{r})$ in terms of the optical constants and the incident fluence rate. This function is substituted in Equation (8.3) and solved for $\boldsymbol{r}^*$ which is the effective treatment depth. It is convenient to define an overall photosensitivity parameter $q^*$ (J/cm$^2$) as the threshold energy fluence within the tissue required for a given clinical endpoint. The expression for $q^*$ from Equation (8.3) is

$$q^* = \Phi(\boldsymbol{r}^*)t \qquad (8.4)$$

where $t$ is the treatment time. The numerical value of $q^*$ depends on the treatment parameters, including the wavelength, local concentration of the photosensitizer, optical properties of the tissue and the intrinsic photosensitivity of the tissue. Changes in the treatment parameters can be modeled in terms of the effect on $q^*$. For example, the dependence of $q^*$ on temperature may be useful for modeling of PDT in conjunction with tumor hyperthermia. The parameters $q^*$ and $W^*$ are related by

$$W^* = \gamma_s C_{s0} q^* \qquad (8.5)$$

where $\gamma_s$ [cm$^{-1}$/(μg/g)] is the linear absorption coefficient of the photosensitizer per unit concentration and $C_{s0}$ (μg/g) is the initial concentration of the photosensitizer in the tissue.

---

**Example 8.1.** PDT threshold dose

Assume that the average Photofrin® concentration in a tumor is 3 μg/g at an injection dose of 2 mg/kg and $\gamma_s$ = 0.005 cm$^{-1}$/(μg/g) at 630 nm For $q^*$ equal to 40 J/cm$^2$, calculate the absorbed energy per unit tissue volume required for the initiation of tumor necrosis.

The linear absorption coefficient of Photofrin® in the tumor is: 3 μg/g x 0.005 cm$^{-1}$/(μg/g) = 0.015 cm$^{-1}$. The absorbed energy density required for tumor necrosis is: $W^*$ = 40 J/cm$^3$ x 0.015 cm$^{-1}$ = 0.60 J/cm$^3$. On a molecular basis, this threshold dose corresponds to approximately 530 absorbed photons for each Photofrin® molecule in the tumor.

---

## 8.2 TISSUE OPTIC CALCULATIONS FOR PHOTOTHERAPY APPLICATIONS

The specific implementations of a model depend on the tissue geometry and the mode of light delivery. Idealized modes of light delivery to tissue are depicted in Figure 8.2. Light delivery by a collimated external source is referred to as

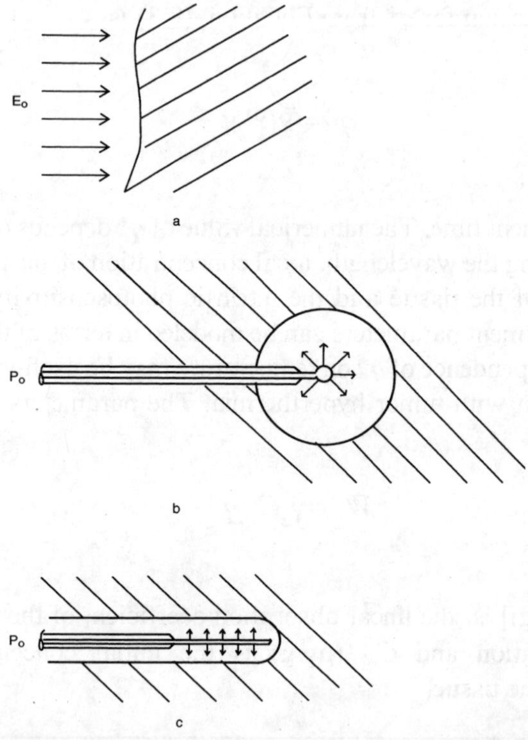


**Figure 8.2** Idealized modes of light delivery to tissues. (a) Front surface (FS) illumination of an exposed tissue surface at uniform incident irradiance. (b) Point surface (PS) illumination of a spherical cavity by an isotropic tip optical fiber. (c) Cylindrical surface (CS) illumination of a lumen with a cylindrical tip optical fiber. (Continued on next page)

*front surface* (**FS**) delivery. This mode is used for light delivery to a tissue layer by an external source. **FS** delivery is appropriate for PUVA, phototherapy of skin diseases, and PDT of superficial tumors. In some PDT applications a small region of a cavity wall is irradiated by an inserted optical fiber. This procedure is referred to as either **FS** delivery or *focal* delivery. Uniform irradiation of a spherical cavity in a large tissue mass by a small isotropic source is termed *point surface* (**PS**) delivery. **PS** delivery is applicable for PDT of the urinary bladder

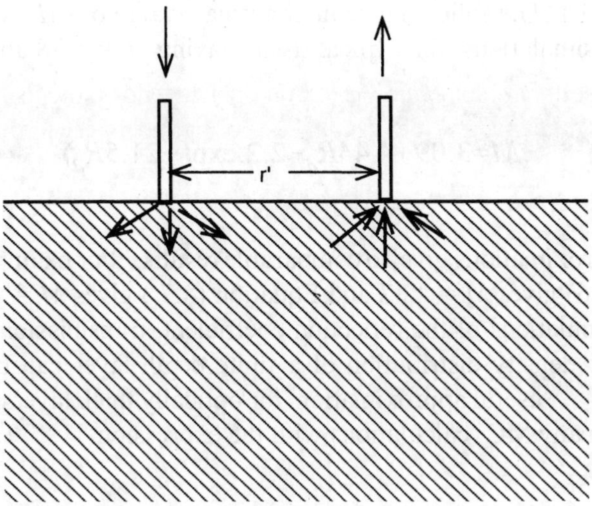


**Figure 8.2 (continued)** (d) Surface illumination of a large tissue volume with a surface optical fiber. The optical fiber surface probe is used for reflection measurements.

with a centered optical fiber having a spherical applicator tip. Irradiation of a cylindrical cavity in a large tissue mass by a linear source is termed *cylindrical surface* (**CS**) delivery. **CS** delivery is applicable for PDT of the esophagus and bronchus. Light delivery by an optical fiber inserted directly in a large tissue mass is described as *cylindrical insertion* (**CI**) delivery.

## 8.2.1 Modeling for a Thick Planar Tissue Layer

In many cases of practical interest it may be assumed that a given tissue is "optically dense" such that all incident light is eventually absorbed. This approximation may be applicable even for a thin tissue layer, such as the epidermis, if it is assumed that the thin layer lies on a bed of thick tissue having approximately the same optical properties. However, more realistic multilayer tissue models provide for different optical constants in each layer. Equation (5.14) is an approximate solution of the diffusion equation for a semi-infinite layer ignoring the source term. The same approximation gives for the coefficient $M$

$$M=3[1+(2/3)bR_d] \tag{8.6}$$

where $R_d$ is the diffuse reflection coefficient and $b \equiv (1 + R_{int, diff})/(1 - R_{int, diff})$ (see Section 5.4.3.). The following semi-empirical value of $M$ was calculated by fitting MC simulations for typical tissue having $n_r = 1.38$ and $g = 0.7\text{-}0.9$

$$M = 3.09 + 4.44R_d - 2.2.\exp(-21.5R_d) \tag{8.7}$$

The two approximations give to comparable results. For example, for $R_d = 0.3$, Equation (8.6) gives $M = 4.87$ and Equation (8.7 ) gives $M = 4.72$. [1] More accurate calculations of the reflection coefficient, transmission coefficient, and interior flux density distribution of a finite depth layer can be found by MC simulations. A useful approximation for optically thick layers was obtained by correlating many MC calculations with the expression

$$\Phi(z) = C_1 \exp(-k_1 z/\delta) - C_2 \exp(-k_2 z/\delta) \tag{8.8}$$

where $z$ is the depth into the layer and $\delta$ is the optical penetration depth (see Section 5.4.2.). The parameters for $n_r = 1.38$ and $g = 0.7\text{-}0.9$ are

$$C_1 = 3.09 + 4.44R_d - 2.12\exp(-21.5R_d)$$

$$C_2 = 2.09 - 1.47R_d + 2.2.\exp(-21.5R_d)$$

$$k_1 = 1 - (1 - 1/\sqrt{3})\exp(-20.1R_d)$$

$$k_2 = 1.63\exp(3.40R_d)$$

Flux density profiles based on Equation (5.14) and Equation (8.8) are compared in Figure 8.3. The MC calculation indicates a buildup of flux near the illuminated surface resulting from internal reflection of diffuse flux at the mismatched boundary. This buildup is shown also in an improved implementation of the diffusion approximation by incorporating refractive index mismatch and using the delta-Eddington phase function.

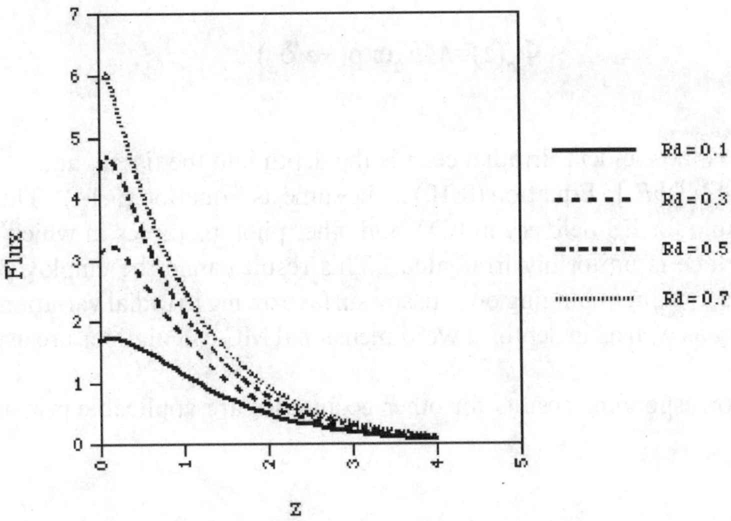


**Figure 8.3** Calculated flux density profiles for uniform incident irradiance.

## 8.2.2 Modeling for Planar, Spherical and Cylindrical Geometries

The diffusion approximation of Equation (5.10) with $S(r) = 0$ leads to approximate analytical expressions for other mode of light delivery. The incident irradiance at the illuminated surface is related to the energy fluence rate by the following approximate boundary condition derived from the diffusion model

$$\Phi_d(a) = 4E(a) - 2F_d(a) \tag{8.9}$$

where $a$ is the coordinate at the interface at which light enters the tissue. The first term on the right side of Equation (8.9) relates the isotropic diffuse flux density just inside the layer to the incident irradiance and the second term is the contribution of the interior diffuse flux flowing towards the interface. An approximate correction can be made for refractive index mismatch by equating the flux density deep within the tissue to the results derived for **FS** delivery. The following expressions are derived with Equation (8.9) for the dependence of the interior fluence rate on distance from the interface.

**FS** Delivery

$$\Phi_d(z) = ME_0 \exp(-z/\delta_d) \tag{8.10}$$

where $E_0$ is the incident irradiance, $z$ is the depth into the tissue, and $M = 3[1 + (2/3)bR_d]$. Equation (8.10) is the same as Equation (5.14). This result is applicable for **FS** delivery in PDT and other phototherapies in which a large planar surface is uniformly irradiated. This result cannot be employed for a focused laser beam impinging on a tissue surface owing to radial variations in the flux density as well as in depth. Two-dimensional MC calculations are useful for this case.

The corresponding results for other geometries are applicable primarily for PDT.

**PS** Delivery

$$\Phi_d(r) = \frac{M'P_0/\pi a^2}{1 + 2D/a + 2D/\delta_d}(a/r)\exp[-(r-a)/\delta_d] \tag{8.11}$$

where $P_0$ is the total power delivered to the fiber, $r$ is the radial distance from the center of the spherical applicator, $a$ is the radius of the spherical cavity, and $M' = (1/4)(1 + 2D/\delta_d)M$. The diffusion constant $D$ is related to the optical constants by Equation (5.11).

**CS** Delivery

$$\Phi_d(r) = \frac{(2M'P_0/\pi aL)}{1 + 2D/\delta_d}\sqrt{(a/r)}\exp[-(r-a)/\delta_d] \tag{8.12}$$

where $P_0$ is the total power delivered to the fiber, $L$ is the length of the emitting part of the cylindrical applicator, $r$ is the radial distance from the fiber axis, $a$ is the radius of the cylindrical cavity, and $M' = (1/4)(1 + 2D/\delta_d)M$.

**CI** Delivery

$$\Phi_d(r)=(P_0/LD)\sqrt{(\delta_d/2\pi r)}\exp(-r/\delta_d) \tag{8.13}$$

where $P_0$ is the total power delivered to the cylindrical fiber, $L$ is the length of the emitting part of the cylindrical applicator and r is the radial distance from the fiber.

### 8.2.2.1 Integrating Sphere Effect

The effective fluence in PS delivery is enhanced by multiple remission of the directly incident light. This integrating sphere effect leads to a significant enhancement of the effective light dose in PDT of the urinary bladder wall. The enhancement factor β calculated with the DA is given by

$$\beta=\frac{3\mu_{tr}R}{1+\mu_{eff}R}-2 \tag{8.14}$$

where $\mu_{tr} = \mu_a + \mu_s (1 - g)$ and $R$ is the diameter of the spherical cavity.

## 8.2.3 Calculation of Treatment Depth for Planar Geometry

Equations (8.10)-(8.13) are approximate expressions for the diffuse flux density $\Phi_d$ within an irradiated tissue in terms of the incident optical power and the optical constants of the tissue. Each equation can be applied to light dosimetry by substituting $\Phi_d$ in Equation (8.3) and solving for $r^*$ which is the depth at which the threshold dose is achieved.. It is convenient to express the dosimetry results in terms of $q^*$ as defined by Equation (8.4). This approach is illustrated for FS delivery. Inserting Equation (8.3) in Equation (8.10) and setting $z$ equal to the effective depth $d_n$ leads to

$$d_n=\delta_d Log_e[ME_0t/q^*) \tag{8.15}$$

This relatively expression relates the necrosis depth for PDT, or the effective depth for other procedures, to the incident light dose ($E_0$ $t$) in terms of the intrinsic photosensitivity parameters $q^*$ and $M$. The value of $q^*$ must be determined from experimental results for the specific treatment conditions. However, $q^*$ should not depend on the geometry and the same values can be used for other light delivery modes. $M$ can be estimated the diffuse reflection coefficient of the tissue using Equation (8.6) or Equation (8.7). It turns out that the numerical values of $d_n$ are insensitive to the numerical values of $q^*$ and $M$ because they occur within the logarithmic term. For example, doubling $q^*$ from 40 J/cm$^2$ to 80 J/cm$^2$ increases $d_n$ by only 20%. The application of Equation (8.14) is illustrated for a hypothetical photosensitization of viral contamination in blood..

---

**Example 8.2** Photochemical irradiator for blood detoxification

Blood detoxification is carried out by adding a photosensitizer to banked blood and irradiating with 630 nm red light. The irradiator provides a uniform incident irradiance of 10 W/cm$^2$ and has a depth of 5 mm. Assume that the TIR coefficient of blood in a glass container is 0.50. (a) Calculate the irradiation time required to achieve a fluence of 20 J/cm$^2$ at the rear of the irradiator. (b) In a flow system, calculate the maximum flow rate if the irradiated volume is 10 cm$^3$.

(a) The optical properties of whole blood at 630 nm from Table 5.1. are: $\mu_a$ = 1.3 cm$^{-1}$, $\mu_s$ (1 - $g$) = 7.1 cm$^{-1}$, $\delta_d$ = 1/$\mu_{eff}$ = 0.196 cm. Therefore, $N'$ = 7.1/1.3 = 4.69 and $b$ =(1 + 0.5)/(1 - 0.5) = 3.00. Substituting in Equations (5.16 ) and Equation (8.6) gives $R_{diff}$ = 0.26 and $M$ = 5.5 Inserting these values in Equation (8.14) with $q^*$ = 20 J/cm$^2$ and $d_n$ = 0.5 cm gives ($E_0$ $t$) = 47 J/cm$^2$. For $E_0$ = 10 W/cm$^2$, the required irradiation time is 47/10 = 4.7 s.

(b) If a flow system is employed, the irradiated volume divided by the flow rate should equal 4.7 s. Therefore the maximum flow rate is 10 cm$^3$ / 4.7 s = 2.1 cm$^3$/s.

---

## 8.3 MODELING FOR PDT LIGHT DOSIMETRY

Accurate light dosimetry is especially important for PDT because the treatment light is applied in a single session. The requirements are less stringent PUVA and phototherapy of neonatal jaundice because the light dose can be adjusted to the clinical responses during the course of treatment. Equation (8.14) is applicable for FS light delivery. The dosimetry relations for other idealized geometries can be derived using the appropriate expressions for $\Phi_d$ from Section 8.2.2. The set of results have the same general form

$$d_n = \delta_d Log_e(FG) \tag{8.16}$$

where $F$ is the incident fluence in appropriate units scaled to $q^*$ and $G$ is a function that depends on the optical constants of the tissue (Table 8.1). The key input parameters are δ and q*. Typical values of $\delta_d$ for non-pigmented tissues range from 0.5-1.5 mm for blue to yellow light, 2 - 4 mm for red light, and 3 -6 mm for near-IR. Some values of $\delta_d$ and $R_d$ calculated from experimental optical constants are given in Table 8.2. Numerical calculations based on Table 8.1 can be performed with a hand calculator. The occurrence of $d_n$ in $G$ requires an iterative calculation that converges rapidly. Alternatively, the calculations can be performed using a computer program such as Mathcad® and Mathematica®. Applications of this formalism to hypothetical PDT procedures are illustrated in Chapter 10.

**Table 8.1 Necrosis depth calculations for PDT light delivery**

| Light delivery mode | $d_n=\delta_d Log_e(FG)$ F | G |
|---|---|---|
| Front surface (**FS**) | $ME_0t/q^*$ | 1 |
| Point surface (**PS**) | $\dfrac{P_0tM'}{\pi a^2q^*}$ | $\dfrac{a}{a+d_n}\,\dfrac{1}{1+2D/a+2D/\delta_d}$ |
| Cylindrical surface (**CS**) | $\dfrac{2P_0tM'}{\pi Laq^*}$ | $\dfrac{1}{1+2D/\delta_d}\sqrt{a(a+d_n)}$ |
| Cylindrical insertion (**CI**) | $\dfrac{P_0t}{LDq^*}$ | $\sqrt{\delta_d/2\pi d_n}$ |

$M = 3[1 + (2/3)bR_d]$; $M' = (1/4)(1 + 2D/\delta_d)M$; $D = (1/3)[\mu_a + \mu_s(1 - g)]$

## 8.4 PHOTOBLEACHING OF PHOTOSENSITIZING AGENTS

Light exposure induces photochemical reactions in virtually all photosensitizing agents. This effect is termed "photobleaching". Fading of a dyed fabric in sunlight is a common example of photobeaching. The least complicated situation obtains when photobleaching is not accompanied by the formation of colored photoproducts. This case can be analyzed by extending the modeling analysis to account for a light-induced decrease in the effective drug concentration.

It is assumed that the instantaneous rate of photobleaching in the tissue is proportional to the local rate of energy absorption. Therefore, the time rate of change of the photosensitizer concentration at position $\boldsymbol{r}$ and time $t$, $C_s(\boldsymbol{r},t)$ μg/g, is proportional to the time rate of change of the local rate energy absorption, $W(\boldsymbol{r},t)$ J/cm³. This relationship is expressed by the differential rate equation

$$-\frac{\partial C_s(\boldsymbol{r},t)}{\partial t}=\frac{K_s[\partial W(\boldsymbol{r},t]}{\partial t} \tag{8.17}$$

where $K_s$ [(μg/cm³)/(J/cm³)] isthe photobleaching rate constant. However, the energy absorption rate in the right side of Equation (8.17) depends on the product of the local energy fluence rate $\Phi(\boldsymbol{r})$ and $C_s(\boldsymbol{r},t)$ according to

$$-\frac{\partial W(\boldsymbol{r},t)}{\partial t}=\gamma_s C_s(\boldsymbol{r},t)\Phi(\boldsymbol{r}) \tag{8.18}$$

where $\gamma_s$ is the linear absorption coefficient of the photosensitizer per unit concentration. Substitution and integration over time leads to the local concentration of photosensitizer

$$C_s(\boldsymbol{r},t)=C_{s0}\exp[-\gamma_s K_s \Phi(\boldsymbol{r})t] \tag{8.19}$$

where the initial photosensitizer concentration $Cs_0$ is assumed to be uniform.

Equation (8.19) indicates that the photosensitizer concentration decreases exponentially with time. The effect of photobleaching on light absorption is calculated by substituting Equation (8.19) in Equation (8.18) and integrating over time

$$W(\mathbf{r},t)=(C_{s0}/K_s)[1-\exp[\gamma_s K_s \Phi(\mathbf{r})t]] \tag{8.20}$$

An important conclusion derived from Equation (8.20) is that the absorbed energy saturates after a prolonged irradiation. Therefore, photobleaching may protect the skin of PDT patients from damage induced by exposure to sunlight if $(Cs_0/K_s)$ is sufficiently large that $W(\mathbf{r},t) < W^*$. Equation (8.20) is related to light dosimetry by setting $W(\mathbf{r},t) = W^*$ and inserting Equation (8.4) and Equation (8.5). The result for FS delivery is obtained by substituting Equation (8.10) for $\Phi(z)$ and solving for the necrosis depth

$$d_n=\delta_d Log_e \frac{1-\gamma_s K_s M E_0 t}{Log_e(1-K_s W^*/C_{s0})} \tag{8.21}$$

Equation (8.21) predicts the effect of photobleaching on the effective depth of the phototherapy. This result is applicable to PDT and other modalities and reduces to Equation (8.14) in the limit of $K_s \Rightarrow 0$. An important conclusion derived from Equation (8.21) is that the PDT drug dose and light dose are not "reciprocal". For example, halving the drug dose is not compensated by doubling the light dose. The extent of "non-reciprocity" can be estimated by calculating the dependence of $(E_0\, t)$ on $C_{s\,0}$ at the same $d_n$. The results for PDT with Photofrin® at 630 nm indicate that halving the drug dose requires an approximately seven times higher light dose to achieve the equivalent clinical result (see Chapter 10).

# CHAPTER 9

# LASER INTERACTIONS WITH TISSUES

The primary effects of light absorption in biological tissues are heat generation, light emission, and initiation of photochemical reactions. Heat generation is an undesirable effect in phototherapy and optical diagnosis because tissue damage occurs at a relatively low temperatures. Clinical applications of laser radiation were investigated with the earliest types of lasers. The first applications to surgery had to rely on the then available industrial and research lasers. Medical lasers are now designed to accomplish specific clinical objectives. Current applications include general surgery, fragmenting of kidney and gall stones, arterial angioplasty, ablation of vascular lesions, laser reshaping of the cornea,[1] laser dentistry, and laser-assisted fertilization.[2] The most important characteristics of medical lasers are the emission wavelength, power level, and temporal mode of operation. These properties jointly determine the nature of the laser-tissue interactions. Table 3.2 summarizes the properties of some important types of biomedical lasers. Recent developments in tissue optics have been extended to modeling of laser-tissue interactions. This problem involves the microscopic interactions between high-power laser radiations and tissues which are not as well understood as the optical interactions. The effects of heating on the optical and thermal properties of tissues are a major complication. The interactions of laser radiation with tissues are categorized as "thermal" and "non-thermal". Thermal processes involve tissue heating and vaporization. High-power continuous wave (CW) and pulsed lasers induce rapid ejection of material referred to as *ablation*. Pressure waves and visible plasmas are generated by very high-power pulsed lasers. The UV excimer laser is described as a "cold laser" in which the ablation results from photomechanical pressure waves initiated by photoionization. This chapter describes the basic principles of photothermal tissue interactions with illustrative examples taken from some recent clinical applications.

## 9.1 PHOTOTHERMAL PROPERTIES OF TISSUES

Typical responses of tissue to increasing temperature are summarized in Table 9.1. Cell killing induced by mild heating is used in hyperthermia treatments of tumors. Laser surgery applications involving annealing of soft tissues utilize tissue denaturation and coagulation. Tissue cutting is based on vaporization and thermal ablation. Tissues are optically and thermally inhomogeneous substances. The conventional thermal parameters employed for other materials are applicable for those phenomena that extend over relatively large tissue regions. *Heat capacity* (*C*) is the temperature rise per unit mass generated by one unit of absorbed energy. A typical value of *C* for a tissue having 70% water content is about 3.5 kJ/kg-°C. In comparison, *C* for liquid water near room temperature is 4.2 kJ/kg-°C. Conductive heat flow is characterized by the thermal conductivity (*K*). Typical values of *K* in units of W/m-°C are 0.2 for bone, 0.5 for muscle, and 0.6 for blood plasma. By comparison, *K* is about 400 for copper and 0.2 for wood. Fick's first law is the basic equation of steady-state heat conduction (see Section 5.4.1).

**Table 9.1 Tissue responses to photothermal processes**

| Temperature (°C) | Biological Effect |
|---|---|
| 42-45 | cell killing<br>tissue retraction |
| 45-50 | enzyme inactivation<br>lipid gelation |
| 50-60 | protein and DNA denaturation<br>closing of vessel lumens |
| 80 | collagen denaturation<br>membrane permeabilization<br>carbonization |
| 100 | water vaporization |
| 140 | elastin denaturation |
| >100 | explosive tissue ablation<br>rupture of vacuoles |

**Example 9.1** Conductive heat flow

(a) Calculate the rate at which body heat flows through clothing of a skier if the skin surface temperature is 33 °C and the surface of the clothing is at 1°C. Assume that the body surface area is 1.8 $m^2$, the clothing is 1.0 cm thick, and the thermal conductivity of the clothing is 0.040 W/m-°C. (b) What is the rate of heat loss if the clothing is soaked with water assuming $K = 0.6$ W/m-°C? (c) Compare the heat loss with the metabolic heat generation rate.

(a) From Fick's first law: $F_q = K(\Delta T/\Delta x)$ where $\Delta T/\Delta x$ is the temperature gradient and $F_q$ is the heat flow rate. Substituting gives: $F_q = 0.040$ W/m-°C x (32°C/0.01 m) = 128 W/$m^2$. Multiplying by the surface area 1.8 m gives 230 W.

(b) The heat loss with wet clothing is 128 x (0.6/0.04) = 1920 W.

(c) Typical values of the metabolic heat rate are 70 W/$m^2$ for a standing person and 200 W/$m^2$ during hard exercise. The skier would have no problem maintaining body temperature with dry clothing and major problems with wet clothing.

The generalized form of Fick's first law relates a flow rate to the gradient of the driving force. The proportionality parameter is the *thermal diffusivity* which is defined: $\alpha = (K/\rho C)$ where $\rho$ is the mean density. The numerical values of $\alpha$ for tissues are less variable than $K$ and range from about (0.5-2.5) x $10^{-7}$ $m^2$/s. The time required for diffusion of a heat pulse is related $\alpha$. The *thermal relaxation time constant* $\tau_{th}$ characterizes the average time required for a heat pulse to flow over a given distance $d$. The value of $\tau_{th}$ for two-dimensional heat flow is given by

$$\tau_{th} = d^2/4\alpha \qquad (9.1)$$

Equation (9.1) shows that tissue damage by heating can be confined to the irradiated region by delivering the energy in pulses the duration of $\tau_{th}$.

**Example 9.2** Flow of a heat pulse

What is the laser pulse duration required to confine the heating effect to a 0.1 mm diameter arteriole? Assume $\alpha = 1.0 \times 10^{-7}$ $m^2$/s.

According to Equation (9.1): $\tau_{th}$. $=(0.0001)^2/4 \times 10^{-7} = 0.025$ s. A 10 ms laser pulse would be confined to the target.

CW laser heating leads to deeper and more uniform temperature elevations than repetitive pulses because energy is continually being pumped into the tissue. However, if the interval between laser pulses is much shorter than $\tau_{th}$ the heating effect is equivalent to CW.

The changes in tissue optical and thermal constants with temperature are an important factor in high-power photothermal phenomena. Familiar examples are increased light scattering by egg white during cooking and browning of bread by toasting. Experimental data indicate that $\mu_a$ and $\mu_s$ of soft tissue increase several-fold from 45°C to 60°C which has a modest effect on the rate and spatial distribution of energy absorption. Vaporization leads to a large change in optical properties because $\mu_a$ of water vapor is significantly lower than that of liquid water. The changes in optical properties of tissue are accompanied by changes in thermal constants. *K* of liquid water increases by about 30% from 30°C to 80°C. Water vapor has an approximately 30 times lower thermal conductivity than liquid water which strongly limits conductive heat flow above 100°C. Rapid uptake of heat occurs during a vaporization process even though the temperature remains approximately constant. The vaporization latent heat of liquid water at 100°C is 2.26 kJ/g compared to 0.25 kJ/g required to heat water from 37°C to 100°C. Heat uptake during the vaporization process leads to sharp discontinuities in heating curves at 100°C.

Two additional features are involved in biological heat flow *in vivo*. Transfer of heat from tissue to flowing blood provides an important heat loss mechanism for localized heating. The rate depends on the blood perfusion rate (*P*) of the tissue. Typical values of *P* in units of ml/min-g are 0.15 for skin and 0.5 for brain. The product of *P* and $\rho$ corresponds to a first-order rate constant: $Q = \rho P$, where $1/Q$ characterizes the time required to perfuse a given quantity of tissue by the same volume of blood. Blood flow limits the maximum depth of heat flow in tissues for CW heating. The *thermal penetration depth* is defined by

$$\delta_{th} = \sqrt{(\alpha/Q)} \tag{9.2}$$

Typical values of $\delta_{th}$ are given in Table 9.2. The rate of metabolic heat generation is an order of magnitude lower than the power delivered by a 100 mW/cm$^2$ laser beam and can be neglected for localized heat injection.

**Table 9.2 Thermal penetration depth in human tissues**

| Tissue | $\delta_{th}$ (mm) |
|---|---|
| Fatty | 22-25 |
| Muscle | 10-20 |
| Skin | 4-7 |
| Brain | 3-4 |
| Kidney | 1.5 |
| Thyroid | 1.3 |
| Choroid | 1.0 |

**Example 9.3** Thermal penetration depth

The blood perfusion rate in human brain is about 0.5 ml/min-g. Estimate the thermal penetration depth. Assume $\rho_t = 1.2$ g/ml and $\alpha = 1.2 \times 10^{-7}$ m²/s.

$Q = (0.5/60) \times 1.2 = 0.01$ s$^{-1}$. From Equation (9.2), $\delta_{th} = \sqrt{(1.2 \times 10^{-3}/0.010)}$ = 0.35 cm = 3.5 mm.

## 9.2. THE LASER BIOHEAT EQUATION

The thermal energy balance between the various heat generation and loss contributions in a small tissue region can be summarized as

| Heat Accumulation | = | Heat Input | - | Heat Flow | - | Heat Removal by Blood Flow |
|---|---|---|---|---|---|---|

The equivalent mathematical expression is the *laser bioheat equation*

$$(1/\alpha)(\partial T/\partial t)_r = \nabla^2 T - (1/\delta_{th}^2)(T - T_0) + (\mu_a/K)\Phi(\mathbf{r},t) \qquad (9.3)$$

Equation (9.3) assumes that tissue and blood have the same heat capacity. Additional heat losses by radiation, convection, and water evaporation take place at air-tissue boundaries. The net effect is approximated by Newton's law of cooling, which assumes that the rate of heat flow across a boundary layer is proportional to the temperature difference between the interior and exterior regions. The mathematical statement of this boundary condition is

$$dQ/dt = hA(T - T_0) \tag{9.4}$$

where $T_0$ is the ambient temperature, $A$ is the surface area, and $h$ is a surface heat transfer coefficient. A typical value of $h$ for convective heat loss by superficial tissues at body temperature is about 0.005 W/cm$^2$ -°C. The laser bioheat equation resembles the time-dependent photon diffusion equation, Equation (5.19), with tissue temperature equivalent to the photon fluence rate. However, the solutions of the heat flow equation are more complicated because both heat flow and photon flow are involved. The particular solutions of Equation (9.3) depend on the geometry, the initial temperature distribution, and boundary conditions, which may be constant or variable in time. Heat flow can be described as a wave motion when the heat source varies in time.[3]

## 9.3 TISSUE HEATING BY LASER RADIATIONS

Laser radiations induce biological damage in tissues *via* photochemical, photothermal, and photomechanical interactions. The type of response depends on the optical and thermal properties of the tissue, wavelength, laser power density, and pulse duration. The optical absorption spectra of tissue chromophores are shown in Figure 9.1.

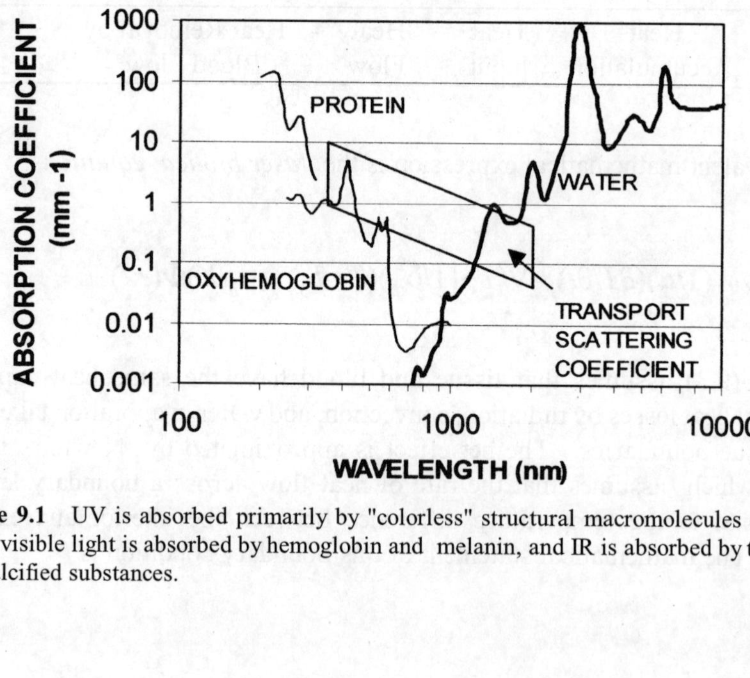


**Figure 9.1** UV is absorbed primarily by "colorless" structural macromolecules and nucleic acids, visible light is absorbed by hemoglobin and melanin, and IR is absorbed by tissue water and calcified substances.

The average power density within the irradiated region of the tissue is inversely proportional to the optical penetration depth. The latter parameter is designated as $\delta_{op}$ in this chapter to distinguish it from the thermal penetration depth $\delta t_h$. The spectra in Figure 9.1 shows that both UV and IR laser radiations are strongly absorbed by tissues. However, there is a significant difference in the physical mechanisms. IR lasers heat and ablate tissue by macroscopic thermal and photomechanical effects, while the basic interactions for UV lasers are initiated by photoionization and photodissociation of the constituent molecules. High values of $\delta_{op}$ are desirable for applications requiring deep light penetration such as hyperthermia and coagulation. The Nd:YAG laser is especially useful for coagulation because the 1064 nm emission lies in a "window" between the strong absorptions of hemoglobin and tissue water. Small values of $\delta_{op}$ are useful for controlled vaporization and ablation as provided by carbon dioxide and excimer surgical lasers. The pulsed erbium-YAG laser is approved in the U.S. for dental applications. The 2.94 μm IR emission is strongly absorbed by hydroxyapatite, the predominant crystal structure in dentin and enamel. In recent work the pulsed 9.6 μm carbon dioxide laser is being tested for dental applications.

### 9.3.1 Pulsed Photothermal Radiometry

Pulsed photothermal radiometry (PTTR) is useful for non-destructive measurements on opaque materials including biological tissues. The tissue is exposed to a short pulse of laser light which induces instantaneous heating. The initial temperature distribution in depth parallels the flux density distribution. An external detector measures the intensity of IR thermal radiation at the surface. The decay of the heat pulse owing to thermal diffusion is controlled by the optical and thermal properties of the tissue. For a strongly absorbing tissue, the initial IR signal $S(t=0)$ is given by

$$S(t=0)=(1/\rho C)\frac{\mu_a\mu_{IR}}{\mu_a+\mu_{IR}} \tag{9.5}$$

where $\mu_a$ is the absorption coefficient at the excitation wavelength and $\mu_{IR}$ is the attenuation coefficient at the measurement wavelength[4].

## 9.3.2 Tumor Hyperthermia

After many decades of study, cancer treatments by the application of full-body or localized hyperthermia (HT) remains an experimental therapy. Various methods of heat delivery have been tested including implantation of a hot object, electrical heated wires, ultrasound, and diathermy with microwaves and radio frequency waves. Laser HT is a recent implementation. Dosimetry for HT depends on the combination of temperature and treatment time. Experimental observations have shown that the time required for cell killing approximately doubles for each centigrade degree of higher temperature. Typically, tissue necrosis is initiated after about 15 min at 45 °C, 30 min at 44 °C and 60 min at 43 °C. An empirical relation defining the HT dose ($D_{HT}$) is

$$D_{HT}=a\exp(bT)t_{HT} \qquad (9.6)$$

where $t_{HT}$ is the treatment time and $T$ is the absolute temperature. The empirical time and temperature results lead to: $a = 1.862 \times 10{-99}\ s^{-1}$ and $b = 0.693\ K^{-1}$. Equation (9.6 ) applies for constant temperature heating and must be integrated over the temperature-time profile during the heating and cooling period. A relation approximating Equation (9.5 ) can be derived from the Arrhenius equation for the rate of a thermally-activated chemical reaction; Equation (3.7). The ratio of the hyperthermic rates at two temperatures equals

$$k_1/k_2=\exp[-\Delta H(T_1-T_2)/T_1T_2] \qquad (9.7)$$

Taking $T_1$ as a standard temperature at which HT requires $t_1 = 1/k_1$ leads to an expression equivalent to Equation (9.6) with $1/a = t_1 \exp(\Delta H/RT_1)$ and $b = \Delta H/RT_1{}^2$. Experimental values of $\Delta H$ for tissue coagulation range from 400-600 kJ/mol, which are comparable to the activation energy for the heat denaturation of proteins. Accurate temperature measurement is a major problem for HT. Tissue surface temperature can be measured with an IR thermometer. This type of device utilizes a semiconductor photodiode to measure the near-IR black-body radiation emitted by the tissue, usually InAs (3 μm - 5 μm) and cooled HgCdTe (8 μm - 15μm). A thermograph is a scanning IR detector that generates a visual image. Thermal measurements are highly sensitive to the ambient conditions and must be calibrated for humidity and air flow. Invasive probes are required for accurate measurements of interior temperatures, such as micro-thermocouples and thermisters. Optical fiber temperature sensors are a recent development. A practical fiber device utilizes the temperature dependence of the optical band gap of a GaAs crystal located in the tissue. The sensor crystal is illuminated *via* an optical fiber with light from an external GaAs LED and the fractional transmission is calibrated as local temperature. Another device uses the temperature variation of the luminescence emitted by a phosphor excited by UV radiation. An advantage of fiber optic thermometers compared to thermo-couples and thermisters is the insensitivity to strong electromagnetic fields as generated by microwave and radiowave heating.

### 9.3.3 CW Laser Heating of Tissue

Photothermal damage to isolated cells has been analyzed with survival curves in which the surviving fraction of active cells after an exposure time $t$ is related to the temperature. Assuming that the Arrhenius equation is applicable to the thermal damage leads to

$$-Log_e \frac{C(t)}{C(0)} = \int A \exp[-\frac{\Delta H}{R_g T(t)}]dt \qquad (9.8)$$

where $\Delta H$ is the activation energy for initiating thermal damage. Equation (9.8) is applicable only to optically "thin" or well-stirred systems. The right side of Equation (9.8) is termed the "damage integral".

More general steady-state solutions to the laser bioheat equation are obtained by setting the left side of Equation (9.3) equal to zero. An exact solution can be found for the case of a semi-infinite layer uniformly irradiated with a collimated laser beam. The energy fluence rate within the layer $\Phi(z)$ is given by Equation (5.19). The following result is obtained with the convection boundary condition of Equation (9.4)

$$T(z)-T_0 = \frac{\mu_d M E_0 / K}{(1/\delta_{op})^2-(1/\delta_{th})^2}[\frac{(K/\delta_{th})+h}{K/\delta_{op})+h}\exp(-z/\delta_{th})-\exp(-z/\delta_{op})] \qquad (9.9)$$

Higher values of $\delta_{th}$ lead to higher initial temperatures and deeper penetration. The peak temperature occurs within the tissue owing to surface cooling. The profile for $\delta_{th}$ = 10 mm is comparable to the expectation for a moderately vascularized superficial tumor with a calculated peak temperature rise of 12 °C at 1 mm depth.

### 9.3.4 Tissue Heating by Pulsed Lasers

An elementary approach to transient tissue heating ignores heat flow while the laser energy is applied. This approximation is satisfactory for short- duration single pulses and repetitive pulses with a low duty cycle. The energy balance between heat generation within the tissue and local heating leads to

$$\rho_t C \Delta T(\boldsymbol{r}) = \mu_a \Phi_d(\boldsymbol{r}) t \tag{9.10}$$

The changes of the optical and thermal parameters with temperature are not included in Equation (9.10). Equation (5.14) can be employed for a simple case of a uniform irradiance laser beam, ignoring beam spread within the tissue. The depth average energy fluence rate $\Phi_{ave}$ equals $ME_0/e$ leading to

$$\frac{\Delta T_{ave}}{t} = \mu_a E_0 \frac{M}{2.72 \rho_t C} \tag{9.11}$$

where $\Delta T_{ave}$ is the average temperature rise. Equation (9.11) is an approximate estimate of the tissue heating rate for "short" optical pulses relative to heat flow. In this limit the pulse duration $t_p$ must satisfy

$$t_p \ll \frac{\delta_{op}^2}{\alpha} \tag{9.12}$$

**Example 9.4** Tissue heating by an Nd:YAG laser

Consider heating of liver tissue by an Nd-YAG laser. Assume that the optical constants of liver at 1064 nm are $\mu_a = 0.7\ \text{cm}^{-1}$ and $M = 4.0$ and the thermal parameters are $C = 3.5$ J/g-°C, $\alpha = 0.001\ \text{cm}^2/\text{s}$, and $\rho_t = 1.1\ \text{g/cm}^3$.

(a) Estimate the optical pulse duration required to confine the heat pulse to the the optical penetration depth. (b) Assume that the laser delivers a power density of 100 W/cm$^2$. What pulse duration is required to heat the tissue from 37°C to an average temperature of 60°C?

(a) From Equation (9.12), $t_p << 0.3^2/0.001 = 90$ s.

(b) From Equation (9.11), $t = (60\text{-}37) \times 2.72 \times 3.5/(0.7 \times 100 \times 4) = 0.8$ s. The heat pulse would be confined to the 3 mm optical penetration depth with a corresponding temperature gradient.

### 9.3.5 IR Laser Ablation of Soft Tissues

Soft tissues consist of 70% - 80% water, which is the principal chromophore in the IR region. The absorption coefficient of water maximizes near 3 μm with a approximate values of 13,000 cm$^{-1}$ at 30°C and 15,000 cm$^{-1}$ at 100°C. The photoablation mechanism has been modeled as a one-dimensional problem in which an ablation front propagates into the tissue. Absorbed energy is deposited as latent heat in the tissue region above the ablation front. The temperature is approximately constant at 100°C up to the ablation front. Heat flow leads to lower temperatures below the ablation front where energy is not deposited This region has been modeled by the one-dimensional laser bioheat equation

$$\rho_t C \frac{\partial T}{\partial t} = \frac{\partial}{\partial t} K \frac{\partial T}{\partial z} + \mu_a \Phi(z) \tag{9.13}$$

where $T(z,t)$ is the temperature at depth $z$ and time $t$. From left to right, the terms in Equation (9.13) describe accumulation of heat in a differential layer at depth $z$, heat flow out of the layer allowing for a temperature-dependent thermal conductivity, and the rate of energy absorption at depth $z$. Starting at 37° C, the surface temperature would attain 72°C after 0.5 s, 87 °C after 1 s, and 124 °C after 3 s (The temperature-depth-time profiles were calculated for a semi- infinite tissue layer with a concentrated heat source of strength $E_0 = 10$ W/cm$^2$ at an insulating air-tissue interface with $K = 0.005$ W/cm-°C and $\alpha = 0.002\ \text{cm}^2/\text{s}$).

The ablation front is located at the depth in which the rate of forward heat flow provides the latent heat required for ablation ($H_{abl}$)

$$K\frac{\partial T(z',t)}{\partial z}=\rho_t H_{abl}\frac{dz'}{dt} \tag{9.14}$$

where $z'$ is the location of the ablation front at time $t$. The left side of Equation (9.13) describes the rate of heat flow across the ablation front and the right side describes the heat utilized for vaporization at the ablation front, where $dz'/dt$ is the ablation velocity. The calculations show a small positive temperature gradient moving at 50 mm/s which could initiate ablation if the conditions of Equation (9.13) were satisfied. Experimental measurements of CW ablation of collagen fibers with an argon-ion laser delivering 60 W/cm$^2$ led to an ablation front velocity of 150 mm/s. More detailed calculations including the temperature dependence of the water absorption coefficient show that the depth of the crater increases linearly with the duration of the laser pulse.[4] For the 2.94 μm emission from an erbium-YAG laser the model predicts an approximately 5 μm depth crater at 2 J/cm$^2$ after 200 μs and a 35 μm depth at 10 J/cm$^2$.

### 9.3.6 Fast-Pulsed Laser Ablation

Tissue ablation by fast pulsed lasers is a highly complex process. Two phenomenological models have been proposed. The thermal model treats ablation as a fast surface evaporation that propagates into the tissue. The depth of ablation ($D$) is given by

$$D=\frac{\Phi-\Phi_{thr}}{H_{abl}} \tag{9.15}$$

where $\Phi_{thr}$ is the threshold fluence for ablation. Equation (14) predicts a linear dependence of $D$ on pulse energy. "The "blow-off" model considers ablation as an explosive event in which a mass of tissue extending to depth $z$ is removed entirely when the energy density exceeds a threshold value $H_{abl}$. This limit should be applicable when the pulse duration ($t_p$) is shorter than the thermal relaxation time as given by Equation (9.12). For a strongly absorbed radiation the fluence at depth $z$ is given by

$$\Phi(z)=(1-R)(E_0 t)\exp(-z/\delta) \quad (9.16)$$

where $(E_0 t)$ is the pulse energy, $R$ is the reflection coefficient, and $\delta = 1/\mu_a$. The fluence at $z = D$ must equal the threshold fluence for ablation $\delta H_{abl}$. Solving for $D$ gives

$$D=\delta Log_e[(1-R)(E_0 t)/\delta H_{abl}] \quad (9.17)$$

The blow-off model predicts a logarithmic dependence of ablation depth on pulse energy. The experimental data of Walsh for weight loss per pulse are not in strict agreement with either of these models. The threshold fluence for ablation of different tissues irradiated with 2μs pulses from a "transverse excitation atmospheric" (TEA) carbon dioxide laser was about 1 J/cm². The ablation rate for all tissues was approximately the same below 5 J/cm². This result is consistent with the fluence requirement for water evaporation of 4.5 J/cm².[5] At higher fluence levels water vaporization was accompanied by explosive tissue removal, which requires more energy for a mechanically strong tissue such as skin and less energy for a weak tissue such as liver.

Ablation of hard tissue such as bone and calcified plaque by a pulsed IR laser involves plasma formation at high fluence levels. Irradiation of beef shank bone by 400-1000 ns pulses from an HF laser (2.7-3.0 μm) was accompanied by an audible crack above 10 J/cm². The sound was augmented by a bright blue and yellow spark at high pulse energies indicative of plasma formation. The craters had smooth walls with an approximately gaussian shape and minimal charring. Plasma production has been associated with the very strong E-fields at high irradiance which induce dielectric breakdown of the target material. The associated shock wave generates a localized mechanical rupture where the pressure rise exceeds the mechanical strength of the tissue. The photoablative interaction of the UV excimer laser is attributed to direct rupture of chemical bonds by high-energy photons. There has been considerable interest in the 193 nm ArF excimer laser for corneal reshaping. The 0.5-1.0 μm penetration depth produces an extremely sharp cut with less than 1 μm of thermal damage. The use of this laser is limited by the lack of a convenient delivery fiber and potential mutagenicity. Excimer laser angioplasty systems have been developed around the 308 nm XeCl laser. This wavelength is not mutagenic although a higher fluence per pulse is required to ablate tissue.

### 9.3.7 Dermatological Laser Applications

The objective in dermatological applications of lasers is to deliver an absorbed energy density that accomplishes the surgical while confining the damage specific targets. The laser wavelength should match the specific targets for laser surgery of specific structures, as in tattoo removal and treatment of vascular lesions. Pulsed dye lasers are generally used for these applications. The IR carbon dioxide and erbium-YAG lasers have been employed for laser skin resurfacing (LSR in which case tissue water is the principal target). The pulsed dye laser has been used for laser treatments of port wine stain (PWS), which is a vascular condition resulting from an overabundance and/or enlargement of dermal capillaries. Typical procedures use 300 μs - 500 μs pulses at 577 nm or 585 nm and power levels on the order of 3 J/cm$^2$ - 10 J/cm$^2$. This wavelength region optimizes low absorption in the epidermis and dermis and moderately high absorption by oxy-Hb and deoxy-Hb. The laser treatment of PWS has been a major application of Monte Carlo modeling. In an early approach the optics of human skin were simulated by a hypothetical 5-layer model, consisting of the upper epidermis (50 μM), lower epidermis (10 μM), dermis (300 μM), and deep dermis (540 μM), where the quantities in ( ) are the assumed layer thickness.[6] A 100 μM layer of blood was located at a depth of 300 μM below the below the epidermis. Optical constants at 577 nm were assumed for each layer. The strongest absorbers are the highly-scattering lower epidermis and weakly-scattering blood layer. The fluence distribution was determined by subdividing the tissue into a large number of absorbing bins and the average fluence in each box was calculated as the local absorbed energy per unit volume divided by the local value of $\mu_a$. The results for practical irradiation parameters shows a sharp spike in the epidermis and some light penetration through the upper region of the dermis into the hypothetical blood layer. The local instantaneous temperature rise was calculated by the formula

$$\Delta T = \frac{E}{C\,m} \tag{9.18}$$

where $E$ (J) is the energy absorption in a bin, $C$ (J/kg-°C) is the specific heat of the tissue, and *m* (kg) is the tissue mass of the bin. The calculated results indicate the formation of heat "spikes" in the epidermis and blood layer. In a more realistic MC calculation human skin was modeled by embedding cylindrical blood vessels in the dermis.[7] The absorption maximum for the assumed 60 μm blood vessel is deeper than the 120 μm blood vessel. Additional calculations for vertical, curved, and multiple blood vessels led to the general conclusion that 585 nm should be preferable to 577 nm for treatment of PWS because the energy deposition is more uniform. In a recent MC simulation the tissue geometry is based on a computer-based tomographic reconstruction of 6 μm sections from a PWS biopsy specimen.[8] The "damage integral" of Equation (9.8) is used to predict the distribution of thermal damage. A typical calculation for 1 W/cm$^2$ at 585 nm and a wide laser beam shows that peaks in the energy deposition rate occur in the epidermis and in those blood vessels located along a line perpendicular to the skin surface. The predictions of this model are that blood-specific damage occurs at $t_p = 0.5$ ms, vascular and perivascular damage occurs at $t_p = 5$ ms, and widespread damage in superficial regions occurs at $t_p = 10$ ms.

## 9.4 LASER SURGERY HAZARDS

Specific atandards for the safe use of surgical lasers are proposed in the report *America National Standard for Safe Use of Laser in Health Care Facilities* (ANSI Z136.3-1996, American National Standards Institute, New York, 1996) Clinical personnel must be protected from the plume generated during laser surgery. Hundreds of volatile organic compounds have been detected in laser plume some of which are known to be harmful to humans. The biological hazards of laser plume are mainly contaminated particles with human pathogenic viruses and bacteria. The potential risks are reduced to low levels with adequate plume evacuation and room ventilation. Additional personal protection is provided by the surgical mask, goggles, gloves, and gown. The extent of ocular damage depends on the laser wavelength and power level. UV and IR radiations are absorbed by the cornea, while visible and near-IR light are absorbed by the retina. The increase in irradiance from the cornea to the retina is given approximately by the ratio of pupil area to that of the retinal image. Some typical values of retinal irradiance are: TV screen ($10^{-6}$ W/cm$^2$), incandescent lamp ($10^{-4}$ W/cm$^2$), sunlight (5 W/cm$^2$), 1 mW laser (100 W/cm$^2$), 20 kW xenon arc (500 W/cm$^2$), 1 W laser ($10^6$ W/cm$^2$). The threshold for burning the retina after a 10 s exposure decreases from about 500 W/cm$^2$ for a small laser source to 10 W/cm$^2$ for direct sunlight. Skin is less susceptible to laser radiations than the eye, although blistering, ulceration, and scarring can result from exposure to high irradiance levels. A hazard in laser surgery results from reflection of light from skin to the eye. Protective goggles are required for all clinical applications of lasers. Care must be exercised to use the goggles appropriate for each type of laser because goggles are designed to optimize visibility at other wavelengths. This requirement is not a limitation for UV- and IR-emitting lasers. However, goggles designed for visible light-emitting lasers also restrict normal vision which can be a problem for laser surgery.

NOTES

1. Excimer laser surgery of the cornea is used to correct vision errors. In photoreactive keratectomy (PRK) the top layer of the cornea, the epithelium, is scraped away to expose the stromal layer underneath. In laser assisted *in situ* keratomileusis (LASIK) a flap is cut in the stromal layer and folded back.

2. It was shown in 1985 by Ashkin and coworkers that particles of cellular dimensions can be trapped and manipulated by a focused laser beam ("optical tweezers"). A single beam gradient force trap consists of a laser beam with a gaussian intensity profile focused to a spot smaller than the particle. The particle is confined to a location just below the focal point of the laser beam in the axial direction and centered in the beam in the transverse direction. In conjunction with lasers cutting ("laser scalpel") this technique has been applied to a wide variety of biological and medical problems, including i*ntra*cellular microsurgery, gene transfer, force measurements on single cell organisms, sperm manipulation, and *in vitro* laser assisted embryo hatching.

3. Solutions of Equation (9.3) for periodic temperatures leads to highly damped thermal waves. In the absence of blood flow, the thermal wave velocity is given by: $V = \sqrt{(4\pi f\alpha)}$ where $f$ is the frequency of the temperature oscillations. Note that thermal waves are always dispersive because the wave velocity $V \neq 1f$. The thermal penetration depth is given by: $d_{th} = \sqrt{(a/p\,f)}$. Diurnal heating of the ground by alternate sunlight and night cooling can be estimated by thermal waves. Taking $a = 5 \times 10^{-7}\ m^2/s$ for earth gives $d_{th} = (5 \times 10^{-7} x24 \times 3600/p = 0.12$ m. Thus, a 12 cm layer of earth will average out day and night temperatures. A similar calculation for the summer-winter temperature cycle leads to $d_{th} = 2$ m.

4. Olmes A, Franke H-G, Raible M, Lubatschowski H, Ertmer W, Baensch E, Dzuik G. Numerical simulation of infrared photoablation. *Proc SPIE* Vol 2923 p. 144-154, Laser-Tissue Interaction and Tissue Optics II; Dec. 1996.

5. The absorption coefficient of water at 10.6 mm is 794 cm-1. For a tissue of 70% water, ma = 0.7 x 794 = 560 cm-1. Taking 2.5 kJ/cm3 for the heat of vaporization leads to: Ethr = 2,500/560 = 4.5 J/cm2.

6. Miller ID and Veitch AR. Optical modeling of light distributions in skin tissue. *Lasers Surg Med.* 1993;13(5):565-71

7. Lucassen GW, Verkruysse W, Keijzer M, van Gemert MJ. Light distriutions in a port wine stain model containing multiple cylindrical and curved blook vessels. *Lasers Surg Med.* 1996;18(4):345-57.

8. Pfefer TJ, Barton JK, Smithies DJ, Milner TE, Nelson JS, van Gemert MJ, Welch AJ. Modeling laser treatment of port wine stains with a computer-reconstructed biopsy. *Laser Surg Med.* 1999:24(2):151-66.

# CHAPTER 10

# PHOTODYNAMIC THERAPY: SCIENCE AND TECHNOLOGY

## 10.1 HISTORY OF PHOTODYNAMIC THERAPY

The term "photodynamic action" was coined in the early 20th century to describe photosensitization of biological effects. In current usage, a photodynamic action is a light-activated biological process that requires the presence of molecular oxygen. Photodynamic action takes place *via* the *Type 2* photosensitization pathway, in which the triplet state of an optically-excited photosensitizer molecule transfers its energy to molecular oxygen. This process generates singlet molecular oxygen ($^1O_2$) which is responsible for the biological endpoints. Some authors extend the definition of photodynamic action to include generation of superoxide and other active oxygen intermediates. This extension is reasonable because some photosensitizing agents generate singlet oxygen or superoxide depending on the reaction conditions. The exact role of singlet oxygen in PDT remain unclear. Peroxidation of membrane lipids has been proposed as a major process. In 1903 von Tappenier and Jesionek attempted unsuccessfully to treat skin cancers with topical eosin dye followed by exposure of the lesions to sunlight. After a hiatus of more than fifty years, in 1960 R. L. Lipson synthesized the first practical PDT drug which he termed hematoporphyrin derivative (HPD). Subsequent analysis showed HPD consists of a mixture of monomeric porphyrin derivatives plus a higher molecular weight constituent that is responsible for most of the anti-tumor activity. The uptake of HPD in tumors is identified by a red fluorescence excited by blue or near-UV light. This test was used occasionally for tumor diagnosis and visualization in the 1960s and 1970s and is currently receiving new impetus as *photodynamic diagnosis* (PDD). In 1976 Kelley and Snell achieved a partial-tumor response in one patient with small, multicentric bladder tumors. HPD was given intravenously and after 24 hours the bladder wall was irradiated for 30 minutes with white light from an arc lamp through the urethra *via* a quartz rod.. Worldwide interest in PDT developed in 1978 when the group of T. J. Dougherty at Roswell Park Cancer Institute reported promising preliminary results for 25 patients with recurrent skin cancers. The procedure was first referred to as "photoradiation therapy" which was then changed to "photodynamic therapy" to avoid confusion with cancer treatments by ionizing radiations. Several HPD preparations were labeled "Photofrin I" and "Photofrin

II" until the present lyophilized powder became available as Photofrin® porfimer sodium. Clinical trials of Photofrin® are reported for localized malignancies, including tumors of the skin, head and neck, eye, brain, urinary bladder, lung, gastrointestinal tract and female genital tract. The annual number of patients treated with the HPD and Photofrin® increased from about 100 in 1985 to approximately 4000 patients in 2000. In 1993 Photofrin® received full regulatory approval in Canada for PDT of superficial bladder cancer. This action was followed by approvals for PDT of specified malignancies in the U.S. and other countries. Virtually all patients receiving Photofrin® acquire photosensitivity of their skin to direct sunlight which may persist for one to three months. This adverse reaction probably contributed to the slow acceptance of PDT by the medical community, although severe reactions in compliant patients have been uncommon. Numerous "second-generation" PDT drugs have been proposed and some of these are in clinical trials. The designation "PDT" is being used also to describe non-cancer disorders photosensitized by an anti-cancer photosensitizing drug, including "wet" age-related macular degeneration (AMD), actinic keratosis (AK), and atherosclerotic vascular disease. This usage is not only confusing but it may be incorrect, because the involvement of singlet oxygen has not been demonstrated for those applications.

## 10.2 PDT DRUGS

The essential criteria for a useful PDT drug are low dark toxicity, uptake and retention by tumors, and effective tumor photosensitizing properties. Emphasis has been given to new compounds with stronger far-red and near-IR absorption bands than Photofrin® and more rapid clearance from skin. Longer wavelength is advantageous because the useful depth of light penetration in non-pigmented tissues increases from about 5 mm at 630 nm to 1 cm - 2 cm at 700 nm - 800 nm. Each new PDT drug requires an appropriate treatment protocol, and in many cases a special light source and drug delivery vehicle. The properties and structures of some currently important PDT drugs are given in Table 10.1 and Table 10.2.

### 10.2.1 Hematoporphyrin Derivative

Drugs derived from HPD are the most widely-used PDT photosensitizers at present. HPD is synthesized by treating commericial hematoporphyrin (HP) with a mixture of glacial acetic and sulfuric acids, followed by filtration, precipitation

**Table 10.1 Photophysical Properties of PDT Drugs**

| *Photosensitizer* | *Abs* (nm) [a,b] | $^1O_2$ *Yield* |
|---|---|---|
| Photofrin® | 630 (1) | 0.9 |
| *meso*-tetra(sulfonatopheny) porphine (TPPS) | 630 (x) | 0.6 |
| mono-aspartyl chlorin $e_6$ (NP$e_6$) | 653 (12) | 0.5 |
| benzoporphyrin derivative monoacid ring A (BPD-MA) | 690 (10) | 0.5 |
| tetra(*meta*-hydroxyphenyl) chlorin (mTHPC) | 650 (12) | 0.5 |
| tin ethyl purpurin (SnET$_2$) | 650 (15) | 0.7 |
| hexyl ether derivative of pyropheophorbide-*a* (HPPH) | 665 (20) | 0.6 |
| aluminum phthalocyanine sulfonate (AlPcS) | 675 (40) | 0.4 |
| silicon naphthalocyanine (Pc4) | | 0.2 |
| lutetium (III) texaphyrin (Lu-Tex) | 730 (12) | 0.3 |
| δ–Aminolevulinic acid (ALA) | 630 (2) [c] | 0.6 |
| Victoria Blue BO (VB-BO) | 532 (110) | ≈0 |
| merocyanine 540 (MC 540) | | 0.004 |
| hypericin (HY) | 592 (14) | 0.4 |

[a] Approximate values based on data from many different sources.

[b] Based on measurements in diverse solubilizing media. The intensity of the longest wavelength absorption band relative to Photofrin ® at 630 nm is given in braces.

[c] For protoporphyrin IX

**Table 10.2 Chemical Structures of PDT Drugs**

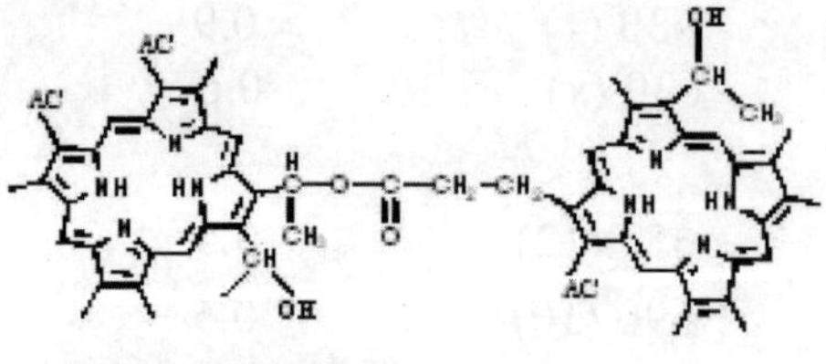

AC= $CH_3CH_2COOH$

**di-hematoporphyrin ether**

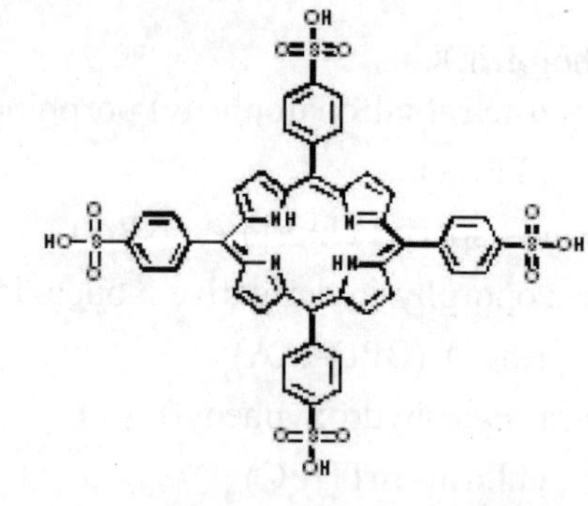

***meso*-tetra(sulfonatophenyl)porphine**

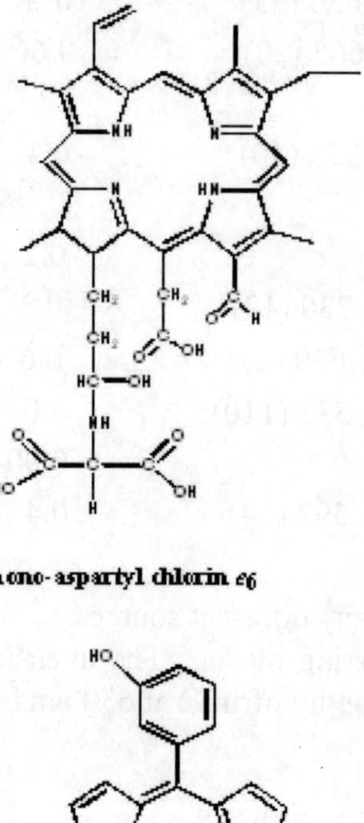

**mono-aspartyl chlorin *e6***

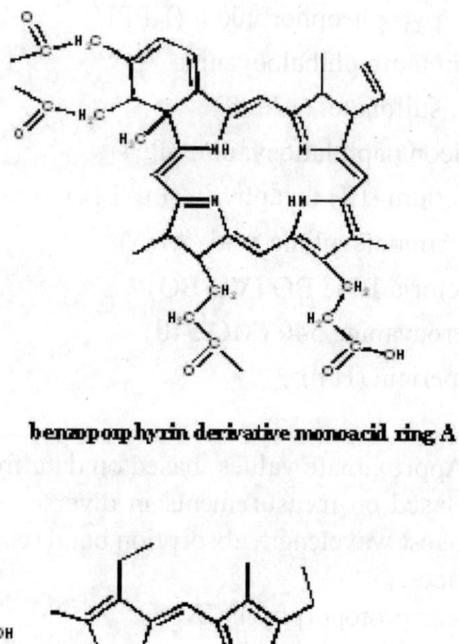

**benzoporphyrin derivative monoacid ring A**

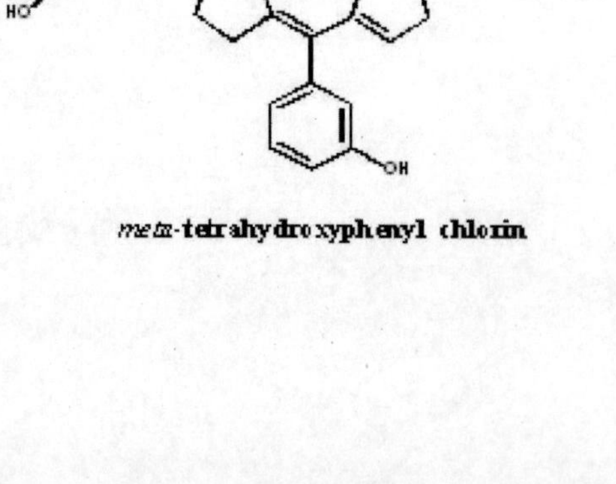

***meta*-tetrahydroxyphenyl chlorin**

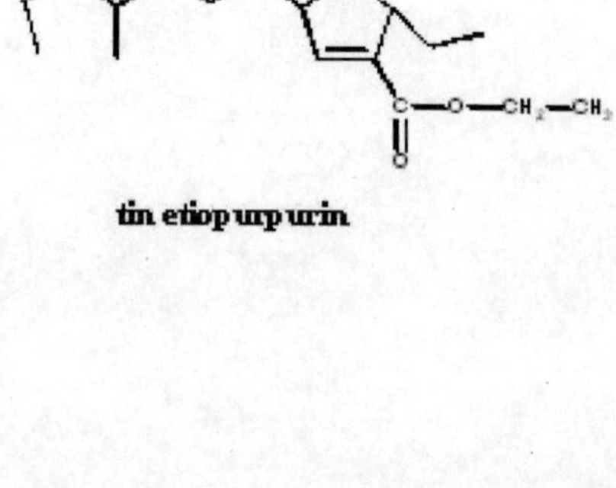

**tin etiopurpurin**

**Table 10.2 Chemical Structures of PDT Drugs (cont.)**

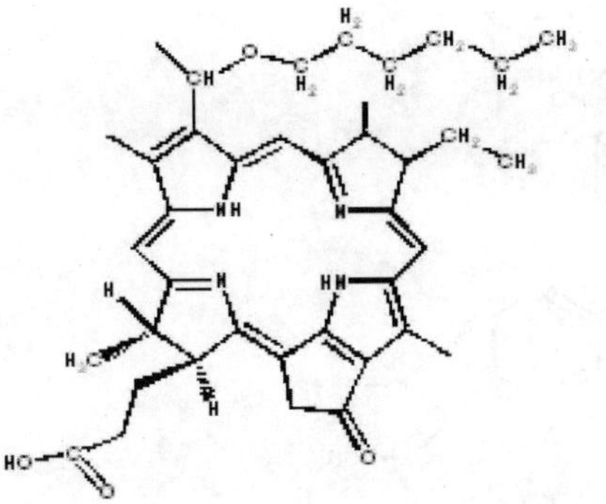

**Hexyl ether derivative of pyropheophorbide-*a***

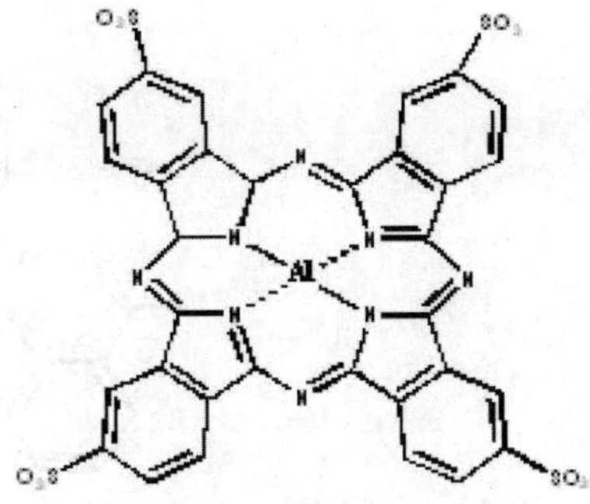

**aluminum phthalocyanine tetrasulfonate**

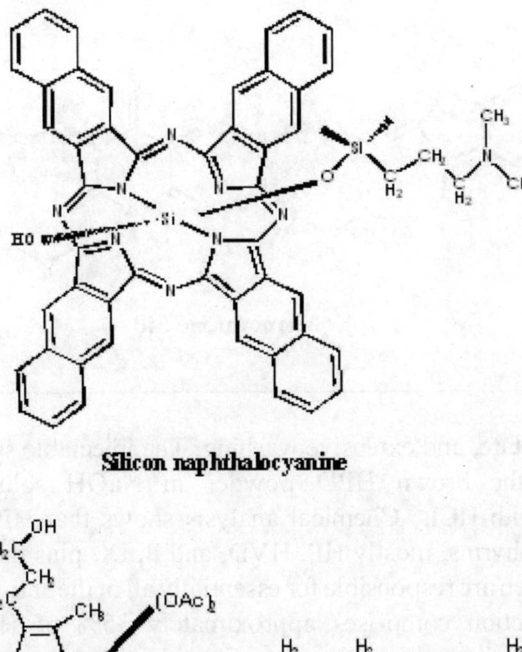

**Silicon naphthalocyanine**

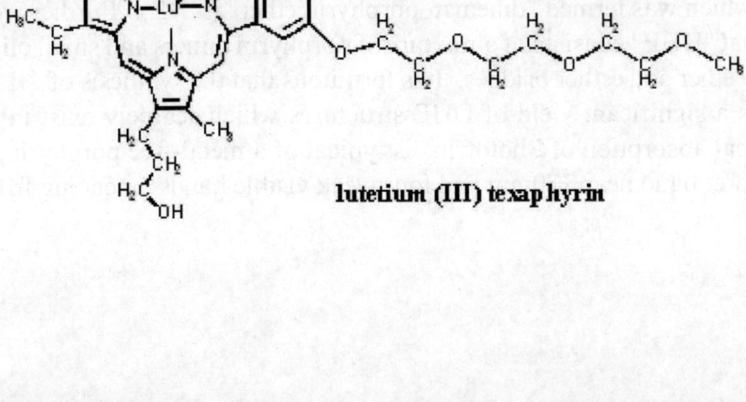

**lutetium (III) texaphyrin**

**Table 10.2 Chemical Structures of PDT Drugs (cont.)**

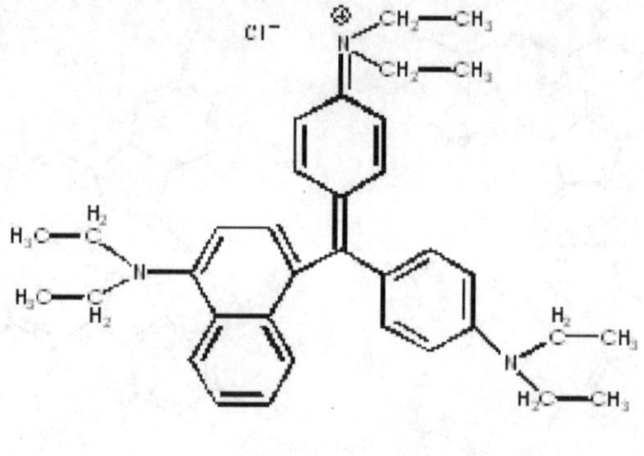

Victoria Blue BO

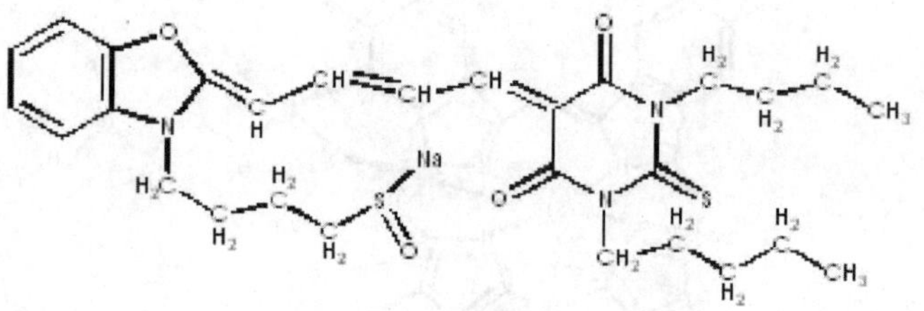

Merocyanine 540

with sodium acetate, and extensive washing. The injectable solution is prepared by dissolving the brown HPD powder in NaOH solution followed by neutralization with HCl. Chemical analysis shows that HPD is a mixture of monomeric porphyrins, mostly HP, HVD, and PpIX, plus the higher-molecular constituents which are responsible for essentially all of the anti-neoplastic activity. This active fraction comprises approximately 35% of HPD and 85% of Photofrin®. Dougherty and coworkers first identified the active component in HPD as a dimeric molecule consisting of two porphyrin units joined by an ether linkage which was termed "dihematoporphyrin ether" (DHE). Further work has shown that "DHE" consists of a mixture of porphyrin dimers and small oligomers joined by ether and esther bridges. It is fortuitous that the synthesis of HPD from HP gives a significant yield of DHE structures which scarcely exist in nature. The optical absorption of Photofrin® is typical of a metal-free porphyrin, with a strong Soret band near 370 nm and four weak visible bands extending from 510 nm to 630 nm ( Figure 10.1a). The fluorescence emission bands are located near 630 nm and 680 nm (Figure 10.1b). Significant spectral shifts take place in the presence of agents that reduce self-aggregation, including detergents, emulsions, and serum proteins. The slow clearance of Photofrin® from serum and skin has been a major motivation in the search for new PDT drugs. However, a long serum lifetime is not necessarily disadvantageous because it permits effective follow-up light treatments many weeks after the initial administration of Photofrin.®

## 10.2.2 Second-Generation PDT Drugs

Virtually all PDT drugs were shown to act as photodynamic agents in photochemical systems and animal tumor models. The new drugs are typically porphyrin analogs synthesized to achieve strong long-wavelength absorptions and other desirable properties. Some cationic dyes have been tested as PDT drugs based on their ability to photosensitize cell cultures. The aqueous solubility of a PDT drug and its interactions with serum proteins and biological membranes are important factors in the pharmacokinetics of drug delivery. *Hydrophilic* compounds are freely soluble in water at physiological conditions. Compounds with three or more charged peripheral substituents are usually hydrophilic, including DHE, TPPS, THPC, AlPcS, NP$e_6$, and Lu-Tex. *Hydrophobic* compounds are virtually insoluble in water and alcohol. Molecules without charged substituents are usually hydrophobic, including HY and Pc4. *Amphiphilic* compounds have present in their structures both a hydrophobic and hydrophilic region. They are usually soluble in alcohol and lipids. Examples include BPD-MA and HPPH. The spectral data in Table 10.1 refer to the longest wavelength absorption band as measured in a solubilizing medium. The numbers in ( ) are the approximate extinction coefficients at that wavelength scaled to 3000 liters/mol-cm for Photofrin® in PBS at 630 nm. The quantum yield for $^1O_2$ generation ($\Phi_\Delta$) is highly dependent on the measurement conditions and the experimental method. The values in Table 10.1 are representative of widely variable published data with a probable accuracy of ±25%. With the exception of VB-BO and MC-540, all of the listed agents have relatively high values of $\Phi_\Delta$ *in vitro*. The effective $^1O_2$ yield *in vivo* may be modified by the local pH, availability of molecular oxygen, the polarity of the tissue matrix, and binding interactions of the drug with membranes and macromolecules. Metabolic alterations of the drug structure also may affect the photosensitizing properties *in vivo*.

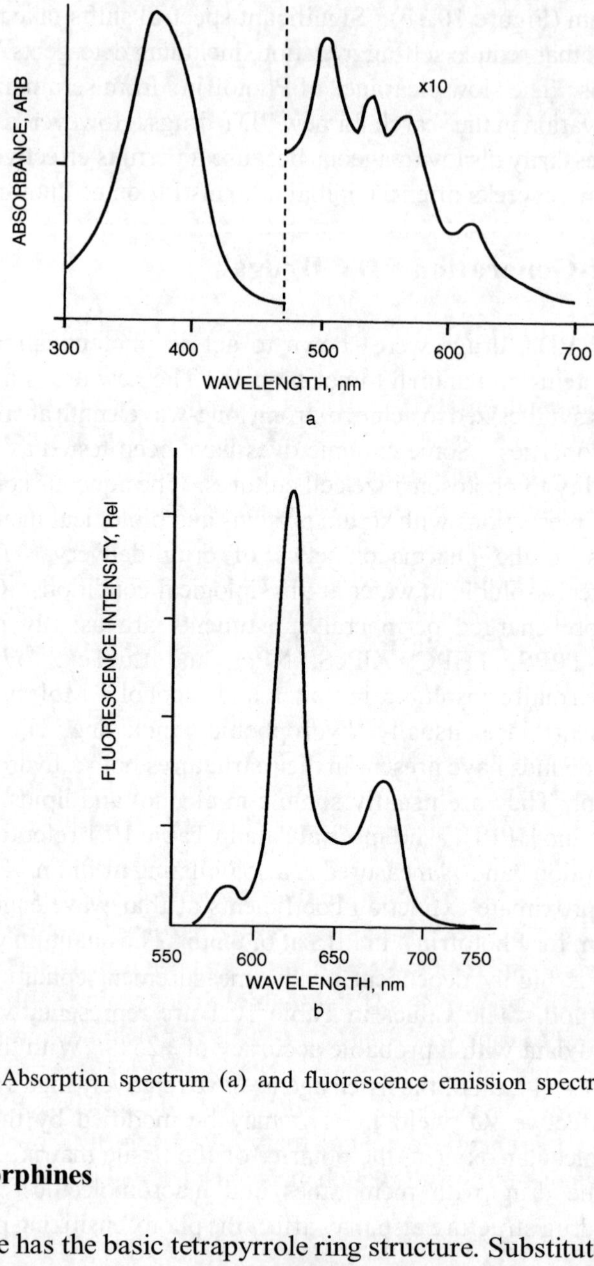


**Figure 10.1** Absorption spectrum (a) and fluorescence emission spectrum (b) of aqueous Photofrin®

### 10.2.2.1 Porphines

Porphine has the basic tetrapyrrole ring structure. Substituted porphines are among earliest second-generation PDT drugs. *Meso*-tetra (sulfonatophenyl) porphine (TPPS) is a hydrophilic porphine which has been tested for PDT of skin cancers. The major limitation on the use of TPPS is a risk of neurotoxicity.

#### 10.2.2.2 Chlorins and Related Compounds

Chlorins are reduced porphyrins characterized by a strong enhancement of the far-red absorption band compared to metal-free porphyrins. Mono-aspartyl chlorin $e_6$ (NP$e_6$) was one of the first new PDT photosensitizers. The aspartic acid group attached to the propionic group leads to solubility in water. NP$e_6$ has a high affinity for serum albumin (ALB) and high density lipoprotein (HDL) compared to more hydrophobic agents with significant binding to low density (LDL) and very low density lipoproteins (VLDL). NP$e_6$ requires only a 4 - 8 hour time delay after systemic administration. Benzoporphyrin derivative monoacid ring A (BPD-MA, Vertaporphyrin®) is an amphiphilic agent with a strong far-red absorption. Uptake of BPD-MA requires 30-150 minutes after intravenous application in liposome carriers. Skin photosensitization by BPD-MA lasts about five days. Inhibition of "wet"AMD and PDT of psoriatic plaque are non-cancer phototherapy applications of BPD-MA. Tetra(*meta*-hydroxyphenyl) chlorin (THPC, Foscan®) is a hydrophobic chlorin derivative. PDT with Foscan® requires much lower drug dose and light dose levels than Photofrin®. Skin photosensitization lasts only 1-2 weeks, but it may be severe owing to the high photosensitizing activity of the drug[1]. Purpurins are benzochlorin derivatives with additional conjugation through substitution. Tin ethyl purpurin (SnET$_2$, Purlytin®) is a hydrophobic chlorin analog that has been tested for PDT of skin cancers and inhibition of AMD. Pheophorbides are metal-free derivatives of chlorophyll without the long phytol side chain. The hexyl ether derivative of pyropheophorbide-*a* (HPPH, Photochlor®) is a designed drug that has been tested for PDT of esophageal cancer[2]. HPPH is cleared rapidly from normal skin.

#### 10.2.2.3 Phthalocyanines and Naphthalocyanines

Phthalocyanines and naphthalocyanines have been tested as alternatives for porphine and chlorin derivatives. Aluminum phthalocyanine sulfonate (AlPcS, Photosens®) is used routinely in Russia for PDT. The clinical preparation is a mixture of differently sulfonated water-soluble derivatives with a strong absorption band near 670 nm. Silicon naphthalocyanine (Pc4) is a hydrophobic agent with higher therapeutic potential than AlPcS for PDT of skin cancers. Indirect evidence indicates that the action mechanism of Pc4 involves direct killing of cancer cells.

#### 10.2.2.4 Texaphyrins

Texaphyrins are synthetic "pseudo-porphyrins", first synthesized at the University of Texas in Austin. Texaphyrin molecules contain five nitrogen donor atoms, not four as in porphyrins. The molecular dimensions are roughly 20% larger than porphyrins and, therefore, texaphyrins can coordinate larger metallic cations than porphyrins. The 22 *pi*-electrons in the macrocycle, compared to 18 in porphyrins, shift the longest wavelength absorption band from the far-red to the near-IR near 730 nm. Lutetium (III) texaphyrin (Lu-Tex, Lutrin®) permits PDT irradiations at 2-4 hours after intravenous administration. Other Lu-Tex preparations are being tested for inhibition of AMD and arterial photoangioplasty. Gadolinium (III) texaphyrin (Gd-Tex) is being tested as an adjuvant agent for radiotherapy of brain tumors. The putative mechanism of radiosensitization is based on the extreme electron affinity of texaphyrin molecules. They can act as redox modulators by repeated electron captures from cytoprotective reducing agents, such as glutathione and ascorbate, and generate reactive oxygen species *via* electron transfer to molecular oxygen.

#### 10.2.2.5 δ–Aminolevulinic Acid

δ–Aminolevulinic acid (ALA, Levulan®) is the biosynthetic precursor of all porphyrins in nature. In the animal pathway, two molecules of ALA condense to form porphobilinogen (PBG), followed by the reaction of four PBG molecules and several decarboxylation steps to form protoporphyrin IX (PpIX) (Figure 10.2). In the final step in the biosynthesis of heme, an iron atom is inserted into PpIX catalyzed by ferrochelatase. Increased PBG in rapidly proliferating cells and lower ferrochelatase activity promote accumulation of metal-free PpIX in cancer cells. Skin cancers have been treated by applying topical ALA in an oil-in-water emulsion and irradiating the lesions at 630- 635 nm after 3-6 hours. The fluorescence of ALA-induced PpIX is used for tumor visualization. ALA phototherapy of AK is approved in the U.S.

#### 10.2.2.6 Cationic Photosensitizing Agents

The possible use of cationic photosensitizing dyes (CPS) as PDT agents is based on the electrostatic potential difference that exists across all biological membranes, which is the order of -60 mV at the plasma membrane of cells and -180 mV at mitochondria. Owing to this potential difference, a cationic molecule is concentrated in the cytoplasm by a factor of 10 compared to the outside and by an additional 1000 times within mitochondria. *In vitro* cell culture studies have confirmed the uptake of CPS by malignant cells. However, high light dose levels

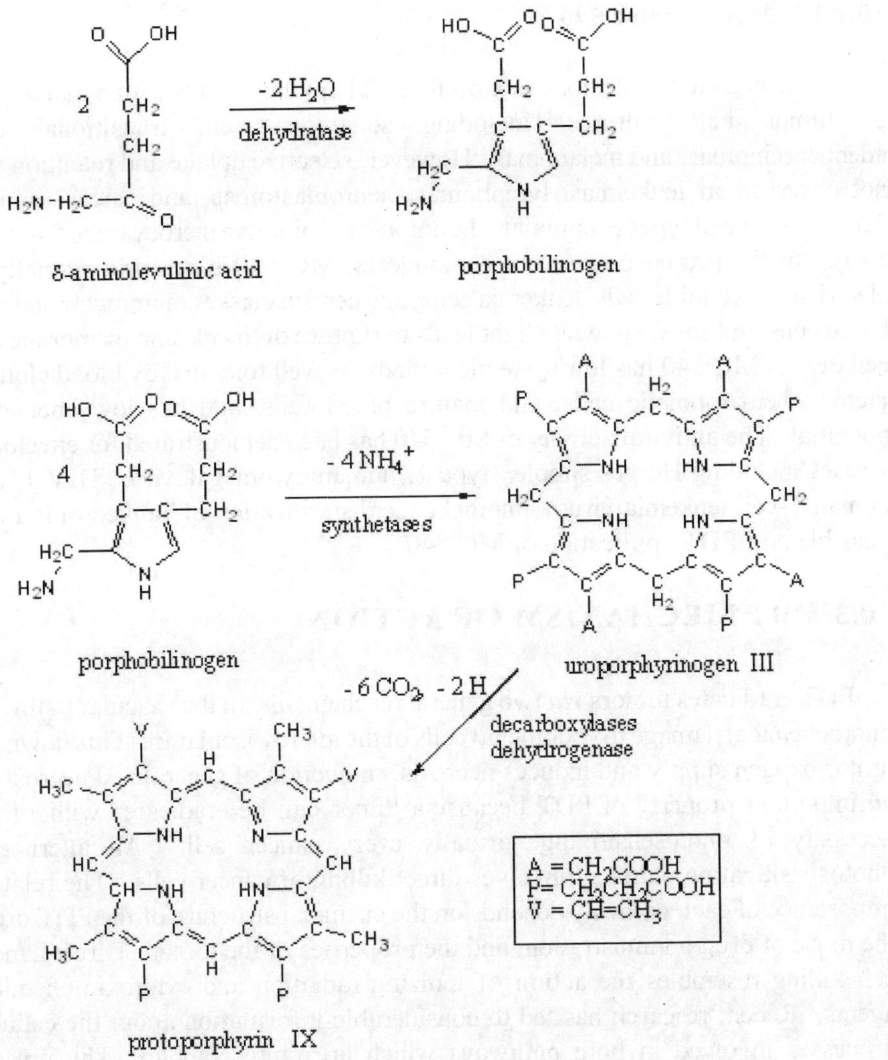


**Figure 10.2** Biosynthesis of protoporphyrin

were required for phototoxicity compared to porphyrins. CPS dyes of interest for PDT include triarylmethanes, thiazines, oxazines, and kryptocyanines. The yields of $^1O_2$ are very low compared to anionic porphyrins. Spin-label EPR studies on the triarylmethane dye Victoria Blue BO (VB-BO) have shown that illumination in aerated aqueous solutions generates superoxide ($O_2^-$) and possibly hydroxyl radicals (OH·) but not $^1O_2$.

#### 10.2.2.7 Merocyanine 540

Screening studies demonstrate preferential uptake of CPS within many types of tumor cell cultures, including squamous cell, transitional cell, adenocarcinomas, and melanomas. However, selective uptake and retention was not observed in leukemias, lymphomas, neuroblastomas, and osteosarcomas. These cancer cell types accumulate the anionic cyanine dye merocyanine 540 (MC 540). In the presence of serum components, MC 540 binds preferentially to electrically excitable cells, leukemia cells, and certain classes of immature normal blood cells. Exposure to white light leads to rupture of the plasma membrane and cell death. MC 540 has low systemic toxicity, is well tolerated by blood clotting factors, hematopoietic cells, and mature blood cells, and has low oncogenic potential. The antiviral activity of MC 540 has been demonstrated for enveloped viruses including Herpes simplex type 1, human cytomegalovirus, HIV-1, and human T-cell leukemia virus. Photochemical sterilization of banked blood is a possible non-PDT application of MC 540.

## 10.3 PDT MECHANISM OF ACTION

PDT eradicates tumors *via* two general mechanisms. In the vascular pathway, photochemical damage to endothelial cells of the microvasculature shuts down the tumor oxygen supply and induces necrosis. Induction of generalized necrosis is an important property of PDT because a tumor can be eradicated without the necessity of photosensitizing virtually every cancer cell. An alternative photosensitization pathway involves direct killing of cancer cells. The relative importance of each pathway depends on the chemical structure of the PDT drug, the mode of drug administration, and the properties of the tumor. Direct cancer cell killing resembles the action of ionizing radiation and cytotoxic chemical agents. Recent research has led to considerable information about the cellular processes involved in both pathways which are quite complex. The present discussion describes only the elementary aspects of the action mechanism.

### 10.3.1 Pharmocokinetics of PDT Drugs

Administration of the drug is the first step in PDT. For systemic administration the drug dose is specified as milligrams per kilogram of body weight. Hydrophilic drugs (for example, Photofrin® and $Npe_6$) are administered by intravenous injection in an aqueous preparation. Aqueous solutions of porphyrins are subject to concentration-dependent self-aggregation. Reversible aggregates are usually dimers while irreversible aggregates may consist of larger entities. Dilution and binding to serum proteins acts to break up porphyrin aggregates after injection. Measurements of Photofrin® binding to human serum proteins led to the following distribution: 40% ALB, 37% HDL, 16% LDL, 7% VLDL. In general, hydrophilic compounds are more strongly bound to ALB and less to LDL. Np$e_6$ exemplifies this effect with 80% binding to the ALB fraction and 1-2% binding to LDL. A vehicle is required for administration of a hydrophobic agent, for example, a liposome suspension or an oil-in-water emulsion. Amphiphilic drugs (for example, BPD-MA) may be administered either in an aqueous medium or *via* a carrier. Liposomes are usually employed for PDT applications in order to facilitate cellular uptake. Aqueous vehicles such as Hyskon® (dextran 70 in dextrose solution) are employed for topical PDT with ALA.

Accurate assays of PDT drug concentrations in tumors and serum are difficult and time consuming. Radioactive tracer techniques were used in early animal model studies. In a more recent fluorimetric assay, the detergent Solvable® fully dissolves tissue and converts the weakly fluorescent active constituent in Photofrin® to more strongly fluorescent porphyrins. Only a small fraction of an injected PDT drug is localized and retained by solid tumors. A rough estimate for Photofrin® is 2 μg/g - 3 μg/g. In one study, the serum concentration of Photofrin® in human PDT patients did not show a definitive dependence on the injection dose from 0.875-2.0 mg/kg. The mean serum concentration at 48 hours after administration was 3 μg/ml - 4 μg/ml[3]. Photofrin® elimination from serum followed a three-compartment model with mean half-lives equal to ≈ 16 hours, 7.5 days, and 155 days.[4] The loss of Photofrin® in serum is much faster than the decrease of cutaneous photosensitivity. Typically, skin photosensitivity lasts about 4 - 6 weeks while plasma Photofrin® decreases by about 50-fold after one week. There is some evidence that elimination of Photofrin® from skin can be accelerated by daily exposures to low levels of sunlight. Less information is available about the pharmocokinetics of the new PDT drugs. Intravenous administration of BPD-MA in liposomes led to serum concentrations ≈ 1 μg/g - 2 μg/ml at dose levels from 0.2 mg/kg - 0.5 mg/kg[5]. The serum elimination half-life of BPD-MA was 5 - 6 hours with negligible skin photosensitization after 5 days.

### 10.3.2 Tissue Distribution of PDT Drugs

PDT drugs accumulate in tumor cells, infiltrating host cells, in components of the tumor vasculature, and in the tumor bed. Studies on *in vitro* cell cultures suggest a relationship between the number of negatively-charged groups attached to the photosensitizing "core" of the molecule and the mode of entry into cells. Those agents with a net cationic charge are localized in mitrochondria (for example, VB-BO) and those with a net anionic charge of -2 or more accumulate in lysosomes (for example, $Npe_6$ and $AlPcS_4$). TPPS localizes in lysosomes with lesser accumulations in the nucleus and cytoplasm. The tissue distribution of PDT drugs is structure dependent. Photofrin® accumulates preferentially in stroma compared to tumor paryenchyma, and especially in endothelial cells. The uptake of endogeneous porphyrins in tumors is well known. However, the reason that DHE is retained for much longer times than the monomeric porphyrins in HPD and Photofrin® remains unclear. Other tissue binding sites for Photofrin® include free and fixed macrophages of connective tissue, fibrous tissue matrix, mast cells, and Kupffer cells of the liver. Early research with mice showed that highest tissue concentrations of Photofrin® occur in tissues high in reticuloendothelial component in the following order: liver, adrenal gland, urinary bladder > pancreas, kidney, spleen > stomach, bone, lung, heart > skin > muscle > brain. The *tumor-peritumoral tissue ratio* (TTR) is relevant for PDT because photochemical damage to normal tissues which are exposed to the PDT light sets an upper limit on the light dose. Approximate values of TTR reported for experimental animals after administration of the drug in PBS are the order of 20 for $AlPcS_4$, 3 for $Npe_6$, and 7 for chlorin $e_6$[6]. Hydrophobic agents were administered in saturated liposomes, and Cremophore emulsions®. Some reported values of TTR are approximately 6 for zinc phthalocyanine and 10 for silicon naphthalocyanine. Amphiphilic agents were administered in PBS, Cremophore EL, and surfactant micelles. The TTR was approximately 3 for BPD-MA in different vehicles, while the TTR for pheophorbide *a* was strongly dependent on the vehicle.

## 10.3.3 PDT Photophysics and Phototochemistry

Photosensitization of tumors by Photofrin® and most other PDT drugs is attributed to the *Type 2* pathway mediated by $^1O_2$. The photochemical mechanism is depicted in Figure 10.3 where $S_0$, $S_1$, and $T_1$ are the ground state, first-excited singlet state, and lowest triplet state of the photosensitizer. According to this mechanism, the short-lived $S_1$ state populates $T_1$ *via* intersystem crossing (ISC) in competition with emission of fluorescence (hf') and radiationless decay. The ISC efficiency is quantified by the triplet yield $\Phi_T$. Published values of $\Phi_T$ show a strong dependence on the solvent. The metastable $T_1$ state may transfer its excitation energy to unexcited oxygen, thereby generating $^1O_2$ in competition with quenching reactions of $T_1$ with substrate molecules. Values of $\Phi_\Delta$ *in vitro* for good photosensitizers are typically 80%-90% of $\Phi_T$. Little information is

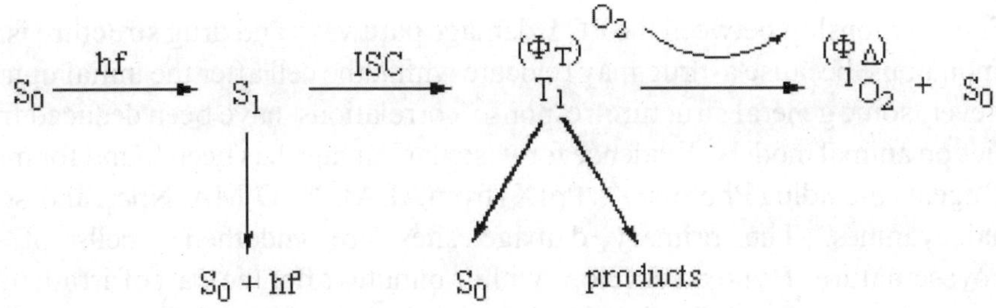


**Figure 10.3** Type 2 photosensitization pathway

available about *in vivo* values of $\Phi_{\Delta}$. Molecules of $^1O_2$ can diffuse < 15 μm during the decay lifetime in a cellular matrix.[7] Accordingly, the initial reactions must take place close to the localization sites.

### 10.3.4 Cellular and Tissue Responses to PDT

The initiating photochemical reactions in PDT may lead to damage to the tumor microvasculature, killing of cancer cells, and modification of the immune system[8]. Necrosis and apoptosis are distinct forms of cell death. Necrosis is a passive process characterized by gross damage and spillage of the intracellular contents. The intracellular targets are the plasma membrane and lysosomes. Apoptosis is a highly regulated and controlled process that avoids inflammation and damage to the surrounding tissue. Apoptosis is an indispensible process during normal development, tissue hemostasis, development of the nervous system, and regulation of the immune system. Disregulation of programmed cell death has been implicated in the pathogenesis of many diseases, including immune deficiency syndrome, neurodegenerative disorders, and ischemic injury. Apoptosis is characterized by a recognizable pattern of DNA followed by the conversion of cells to "apoptotic bodies". Mitochondria and DNA are the likely targets for the apoptotic response. Both the vascular and apoptotic pathways are implicated in the immunological responses to PDT. The details are very complex because alterations in the immune system can both promote and inhibit different stages of the tumor eradication process.

#### 10.3.4.1 Vascular Effects

The relationship between the PDT damage pathways and drug structure is not unambiguous because a drug may relocate within the cell after the initial uptake. However, some general structure-response correlations have been deduced from studies on animal models. Evidence for vascular damage has been found for many PDT agents including Photofrin®, PpIX (from ALA), BPD-MA, Np$e_6$, and some phthalocyanines. The primary damage sites are endothelial cells of the microvasculature. Hypoxia develops within minutes after the start of irradiation. Gross edema and erythema usually precede any detectable tumor cell or endothelial cell damage, as evidenced by blanching and hemorrhaging. This response involves three stages: (1) initial oxygen consumption; (2) reduction in regional blood supply; (3) total vascular exclusion (ischemia). Since the formation of $^1O_2$ requires oxygen, it may be possible to optimize the efficacy of the vascular pathway by intermittent applications of the PDT light. The maximum vascular damage for hydrophilic agents occurs at later times, well after the peak serum levels are attained. Hydrophobic and amphiphilic agents show a different pattern in which the vascular damage parallels the serum concentration at the time of irradiation. The reported cellular effects of PDT on epithelial cells include release of clotting factors, calcium influx, and alterations of the cytoskeleton. These effects may take place at light dose levels well below those required to kill or inactivate cancer cells. The loss of tight junctions between cells and extent of cell rounding are sufficient to activate circulating platelets and polymorphonuclear leukocytes and induce thrombogenic responses. Leakage of lactate dehydrogenase (LDH) and release of eicosanoids (prostaglandins and thromboxanes) and histamine are implicated in the vascular damage component. This action is accompanied by the induction of heat stress proteins. Release of fast acting vasoacative compounds can induce arteriole constriction and increase in vessel permeability to macromolecules. The combined effects of acute vessel constriction and aggregation of platelets accounts for blood flow stasis. The mobilization of the host is a unique feature of the vascular mechanism. The inflammatory metabolites of membrane lipids are a potent signal for attracting and activating non-specific immune effector cells which may inactivate large numbers of cancer cells[9]. At the same time, PDT may promote inherent physiological control elements that have developed to prevent excessive damage to normal tissue by various insults. In the early stage of PDT the induction of stress proteins may diminish the generation of phototoxic lesions by $^1O_2$ and $O_2^{\cdot -}$. In the later stages, PDT- induced immunosuppression may hinder the activity of host immune cells for killing cancer cells. This aspect of the PDT action mechanism is an active area of current research and many questions are unresolved.

### 10.3.4.2 Apoptosis

Programmed cell death by apoptosis is characterized by distinct morphological alterations including nuclear condensation, cell shrinkage, bleb formation, and the absence of inflammatory responses of the affected tissue[10]. The available evidence associates oxidative stress and mitochondrial damage with apoptosis. Accordingly, lipophilic CPS are likely to be effective PDT agents. A postulated mechanism involves alterations in the normal uptake and release of $Ca^{2+}$ across the inner mitochondrial membrane leading to general leakiness and inhibition of ATP synthesis. Leakage of the mitochondrial proteins, cytochrome *c* and apoptosis-inducing factor, has been implicated. *In vivo* studies with animal tumor models indicated a correlation between a rapid apoptotic response and PDT efficacy for Photofrin®, ALA, $SnET_2$, and Lu-Tex[11]. A given photosensitizer may act *via* both vascular and apoptotic pathways. The contribution of direct tumor cell kill has been estimated in animal models as 20-30% for Photofrin® and close to 100% for AlPcS.

## 10.4 PDT LIGHT DOSIMETRY

There is no practical way to measure the actual optical flux density within a solid tumor. Empirical light dosimetry for PDT is based on the results of clinical trials and the medical literature. The major function of analytical modeling based on tissue optics is to assist in the development of improved clinical procedures and instrumentation. The key assumption made in PDT light dosimetry is that tumor eradication requires a *threshold* of absorbed energy by the PDT drug localized in tumor tissue. An approximate calculation in Example 8.1 indicates that the energy threshold is the order of $0.6\ J/cm^3$ for Photofrin® administered by intravenous injection at 2 mg/kg and irradiated 24 - 48 hours at 630 nm. It is convenient to define these parameters as *standard conditions* for Photofrin®. One objective of analytical modeling is to relate the energy density of the effective light to the administered drug dose and delivered light dose. The existence of a dose threshold provides a margin of safety because a higher light dose can be employed to compensate for uncertainties in the localized drug concentration. Clinical results with Photofrin® do not indicate a significant difference in tumor responses when the PDT light was applied at different times ranging from 24-48 hours after injection. The timing between drug administration and light treatment may be a crucial factor with the more rapidly eliminated drugs.

## 10.4.1 Empirical PDT Light Delivery

The dosimetry units employed for PDT depend on the mode of light delivery. It is convenient to review the relevant aspects of optical radiometry from Chapter 2 and Chapter 5. The generic term *flux density* ($W/cm^2$) refers to the optical *power* per unit surface area. Flux density varies with the measurement geometry. The *irradiance* ($E$) is defined as the flux incident from all forward directions on small planar surface divided by the area of the surface. The *fluence rate* ($\Phi$) is defined as the flux density incident from all directions on a small sphere divided by the surface area of the sphere. $\Phi$ is termed the "spherical irradiance" in the older literature. $E$ and $\Phi$ have the same units but they measure different quantities. In practical cases, $E$ is used to specify the flux density in a collimated light beam and $\Phi$ is used to specify the average flux density at a point within a medium. For a collimated light beam $\Phi = E$ and for ideally diffuse light $\Phi = 4E$. The incident light dose in PDT is always specified in *energy* units. Energy equals the integrated product of power and time. Since flux density is defined as optical power per unit area, it is convenient to define a corresponding energy unit, *fluence*, as optical energy per unit area. Accordingly, the corresponding power unit is *fluence rate*. For a constant rate of illumination the fluence within tissue is $\Phi t$ ($J/cm^2$). Note that *energy density* has units of energy per unit volume which differs from the units of fluence which are energy per unit area. The PDT threshold is calculated as the absorbed energy density ($J/cm^3$), which equals the product of the local fluence ($J/cm^2$) and the linear absorption coefficient of the PDT drug ($cm^{-1}$); see Equation (8.1). It is virtually impossible to measure $\Phi$ directly by a non-invasive method and the PDT light dose is specified by the delivered energy in units that are appropriate to the geometry of light delivery. The four most common modes of PDT light delivery are described in Section 8.2. Table 10.3 summarizes the treatment time calculations for these modes.

### 10.4.1.1 Front Surface Light Delivery

Front surface (**FS**) light delivery is employed for PDT of superficial lesions. The appropriate units are joules incident on a unit area of the treatment field. A "microlens" optical fiber delivers a circular spot having an approximately uniform irradiance. In some applications an optical fiber inserted within an endoscope directs light to a small surface area of an interior wall. This mode of **FS** light delivery has been described as "focal". One way of determining irradiance is to measure the total power ($P$) delivered by the fiber tip using an integrating sphere type of power meter (see Section 2.6.4.). $E$ is then calculated by dividing $P$ by the area of the circular field. The **FS** light dose is the product_ of $E$ (W/cm$^2$) and the treatment time $t$ (s). If a mask is used to block part of the beam, the full circular area of the unblocked field must be employed for this calculation. Alternatively the irradiance of a light beam can be measured directly with an optical power meter. Most instruments measure the total

**Table 10.3 Light dose calculations for PDT**

| *Mode* | *Applicator* | *Dose Unit (D)* | *Dose Calculation* |
|---|---|---|---|
| **FS** | microlens | J/cm$^2$ | $t$(min) = 13.09 $D\ d^2$ (cm$^2$)/$P$(mW) [a] |
| **PS** | spherical | J | $t$(min) = 16.67 $D$ /P(mW) |
| | | J/cm$^2$ | $t$(min) = -80.60 $D\ V^{0..67}$ (cm$^3$)/$P$(mW) [b] |
| **CS** | cylindrical | J/cm | $t$(min) = 16.67 $D\ L$(cm)/$P$(mw) [c] |
| | | J/cm$^2$ | $t$(min) = 52.36 $D\ L$(cm) $d$(cm)/$P$(mw)[d] |
| **CI** | cylindrical | J/cm | $t$(min) = 16.67 $D\ L$(cm)/P(mw) [e] |

[a] $d$ is the diameter of the circular field. $P$ is the laser power.
[b] $V$ is the volume of the spherical cavity; $P$ is the laser power.
[c] $L$ is the length of the fiber applicator; $P$ is the laser power.
[d] $d$ is the diameter of the cylindrical lumen; $L$ is the length of the fiber applicator; $P$ is the laser power.
e $L$ is the length of the fiber applicator; $P$ is the laser power.

optical power incident on the photosensor, in which case $E$ is calculated by dividing the power indicated by the meter (typically in mW or W) by the area of the receiving element in the power meter "head". Some types of power meters measure over a wide acceptance angle while others block all but the forward-directed beam. Laser-fiber optic systems were employed for virtually all early PDT procedures because other light sources did not deliver adequate power levels of red light. New non-laser light sources are being developed which are appropriate for PDT. For this type of application the PDT protocol must specify both the light dose and the device used for the measurements. The spectral band width of a non-laser source is broader than that of a laser line and the spectral response of the power meter must be considered. In general, the photosensitivity of a solid state photodiode detector is highly dependent on the wavelength while a thermal detector is essentially wavelength independent (see Section 2.6.6.). Table 10.3 summarizes the dosimetry calculations for **FS** and other "standard" modes of PDT light delivery. Multi-fiber and other more complicated arrangements require *ab initio* calculations.

**Example 10.1** PDT treatment of skin cancer

PDT of basal cell skin cancer is performed using Photofrin® at standard conditions and a light dose of 30 $J/cm^2$. The light is delivered with a microlens optical fiber providing a 2 cm diameter treatment field. In order to minimize tumor heating the maximum power density is taken as 150 $mW/cm^2$. (a) Calculate the maximum power delivered by the fiber. (b) Calculate the treatment time for this power level.

(a) The area of the treatment field is $(\pi/4)2^2 = 3.14\ cm^2$. The maximum output power is given by: $150\ mW/cm^2 = P/3.14\ cm^2$ ; $P = 471$ mW.

(b) The treatment time is given by: $30\ J/cm^2 = (471/1000)\ W \times t/3.14\ cm^2$ ; $t = 200$ seconds.

#### 10.4.1.2 Point Surface Light Delivery

Point surface (**PS**) light delivery is employed for interstitial PDT. For PDT of the urinary bladder wall, an optical fiber with an approximately isotropic applicator tip is inserted *via* a cystoscope and centered in the bladder. The nominal **PS** light dose is specified as the total power delivered by the fiber in units of joules. Alternatively, the average surface dose can be calculated for an assumed total surface area. A balloon containing saline may be located in the bladder to improve uniformity of light delivery. An integrating sphere type of power meter is used to measure the total power output of an isotropic fiber. The actual surface dose depends on the location of the applicator within the bladder and the distance from the fiber tip to each section of the bladder wall. In an improved technique developed by Star and coworkers, the applicator was a 3 mm diameter spherical diffuser located within a special cystoscope that provides for the insertion of three movable catheters incorporating isotropic light probes used to measure the irradiance. It was found that the actual surface dose was approximately 5 times higher than the average non-scattered fluence delivered by the fiber owing to light trapping in the *integrating sphere effect* (see Section 8.2.2.1.).

**Example 10.2** Integrating sphere effect

Using Equation (8.14) estimate the integrating sphere effect for bladder wall PDT. Assume the following optical constants at 630 nm: $\mu_a = 1.4$ cm$^{-1}$, $\mu_s = 140$ cm$^{-1}$, g = 0.93. Assume the bladder radius is 3.3 cm (150 ml volume).

For these parameters, $\mu_{tr} = 1.4 + 140\ (1 - 0.92) = 12.6$ cm$^{-1}$, $\mu_{eff} = \sqrt{[3 \times 1.4(1.4 + 140\ (1 - 0.92)]} = 7.3$ cm$^{-1}$, and $\beta = (3 \times 12.6 \times 3.3)/(1 + 7.3 \times 3.3) - 2 = 3.0$. This calculation indicates that the actual fluence inside the bladder is 3-fold higher than calculated from the fiber output and the geometry. MC calculations led to higher values of β closer to the *in vivo* measurements.[12]

### 10.4.1.3 Cylindrical Surface Light Delivery

Cylindrical surface (**CS**) light delivery is used for PDT of a cylindrical lumen including the esophagus, trachea, bronchus, and vaginal wall. Typical optical fibers for endoscopic applications have diameters from 0.5 mm - 2.5 mm and diffuser lengths ranging from 0.5 cm - 7 cm. The **CS** light dose is specified as joules per unit of applicator length. The emitted radiance for commercial fibers depends on the distance along the fiber and the angle that a receiver makes with the fiber axis. Owing to these directionality effects, the calibration data provided with commercial fibers may exaggerate the uniformity of output, especially for short fibers. However, light scattering within an irradiated tissue should act to reduce the effects of fiber anisotropy. A special power meter is required to measure the output of an applicator longer than about 3 cm. Long fibers require careful handling because bending or kinking can lead to a permanent transmission loss.

**Example 10.3** PDT of Barrett's esophagus

High-grade dysplasia of the esophageal wall (Barrett's esophagus) is a potentially pre-cancerous disorder. Consider a PDT treatment with Photofrin® at standard conditions. The light is delivered *via* an optical fiber inserted in an endoscope having a 7 cm length of applicator. The fiber is centered in a balloon device that illuminates a 5 cm length of the esophageal wall. The balloon is inflated to maintain a uniform diameter of 10 mm. (a) Calculate the treatment time for a **CS** light dose of 200 J/cm and an output power of 1000 mW. (b) Estimate the equivalent surface dose.

(a) The **CS** dose calculation is based on the full length of the applicator: 200 J/cm = 1 W x $t$ /7 cm; $t$ = 1400 seconds.

(b) Consider a 1 cm length section of the esophageal wall. The circumference equals π times the diameter which is 3.14 cm. Therefore, the surface area of this section of the wall is 3.14 $cm^2$. The equivalent surface dose is 200 J/cm /3.14 $cm^2$ = 64 J/$cm^2$ . Note that this calculation requires an assumption about the diameter of the esophagus which is not the case for the nominal **CS** dose in units of J/cm.

#### 10.4.1.4 Cylindrical Insertion Light Delivery

Direct insertion of a PDT fiber into a tumor mass is termed cylindrical insertion (**CI**) light delivery. The cladding is removed from the active length of the fiber to facilitate insertion and improve the efficiency of energy transfer. The **CI** light dose is specified as joules per centimeter of unclad fiber. CI delivery permits relatively high power levels owing to good heat transfer from the fiber to the tissue. Care must be taken in measuring the power output in air because the unclad part of the fiber may overheat. A practical procedure is to measure the power output of the fiber at a low laser power level and then scale up the measurement to the laser power used for PDT.

**Example 10.4** PDT of a large tumor mass in the oral cavity

A large squamous cell tumor of the oral mucosa is treated with Photofrin® at standard conditions. **CI** delivery will be employed at a dose of 75 J/cm and a fiber with a 2 cm unclad length. Calculate the treatment time for 500 mW total fiber output.

The dosimetry relation leads to: 75 J/cm = (500/1000) W x $t$/2 cm; $t$ = 300 s.

### 10.4.2 Analytical Light Dosimetry Modeling

The dosimetry calculations in Section 10.4 do not provide information about effective depth of the PDT treatments. This key information can be estimated by analytical modeling. The relevant parameters are: (1) the energy fluence distribution within the tumor tissue, (2) the localized concentration of the PDT photosensitizer, and (3) the photochemical efficiency for initiation of tumor eradication. A unified modeling approach is described in Chapter 8. The overall modeling parameter $q^*$ should be constant for a given PDT drug and treatment conditions. A tentative value of $q^*$ for Photofrin® at standard conditions is 40 J/$cm^2$. Equation (8.16) expresses the necrosis depth $d_n$ in terms of the incident light dose and tissue-optic parameters. The optical penetration depth for diffuse flux $\delta_d$ is the most important quantity because the functions $F$ and $G$ appear inside the logarithmic terms and have a smaller effect on $d_n$. [For example, if (FG) increases from 5 to 50, $\text{Log}_e$ (FG) increases from 1.6 to to 3.9.] The forms of $F$ and $G$ calculated with the diffusion approximation are given in Table 8.2.

The applications of the analytical approach are illustrated for the hypothetical PDT treatments. Values of $\delta_d$ were either taken from the literature or calculated from the optical constants using Equation (5.15). All of the examples use $D$ = 0.055 cm for the optical diffusion constant at 630 nm and $n_r$ = 1.38 for the refractive index at an air-tissue interface. The tissue diffuse reflection coefficient calculated from the optical constants with Equation (5.17) taking $b$ = 3.11 (see Section 5. 4.3.) and $M$ was calculated with Equation (8.6) (see Section 8.2.1).

**Example 10.5** Modeling for PDT of skin cancer by FS light delivery

Estimate the necrosis depth for PDT of basal cell skin cancer with Photofrin® at standard conditions and a **FS** light dose of 30 J/cm². Assume $q*$ = 40 J/cm².

From Table 5.1: $\mu_a$ = 1.5 cm$^{-1}$, $\mu_s(1 - g)$ = 24 cm$^{-1}$ and $\delta_d$ = = 0.093 cm. The parameters in Equation (5.17) are $N'$ = 24/1.5 = 16 and $b$ = 3.11 leading to $R_d$ = 0.57 and Equation (8.6) gives $M$ = 6.5. Finally, substituting $(E_0 t)$ = 30 J/cm², $M$ = 6.5, $q*$ = 40 J/cm² and $\delta_d$ = 0.093 cm in Eq. (8.16) gives $d_n$ = 0.15 cm = 1.5 mm. This value of necrosis depth should be adequate for PDT of basal cell tumors which may be as thick as 1-2 mm.

Figure 8.1 shows plots of $d_n$ vs $(E_0 t)$ at 630 nm for several tissue types based on the optical constants in Table 5.1. The leveling of the response at high light dose shows that increasing the light dose provides only a small increase in necrosis depth.

**Example 10.6** Modeling for PDT of superficial bladder cancer by PS light delivery

Estimate the necrosis depth for PDT of multiple superficial bladder cancer with Photofrin® at standard conditions and a delivered **PS** light dose of 135 J/cm².

Assume that $q*$ = 40 J/cm². From Table 5.1: $\mu_a$ = 1.4 cm$^{-1}$, $\mu_s(1 - g)$ = 9.8 cm$^{-1}$ and $\delta_d$ = = 0.14 cm. The parameters in Equation (5.17) are $N'$ = 9.8/1.4 = 7 and $b$ = 3.11 leading to $R_d$ = 0.24 and Equation (8.6) gives $M$ = 4.5. Assuming the same form as **FS** delivery in Equation (8.16) gives $d_n$ = 0.07 cm = 0.7 mm. This calculation does not include the integrating sphere effect (see Section 8.2.2.1.) Assuming $\beta$ = 5 leads to $d_n$ = 0.3 cm = 3 mm.

**Example 10.7** Modeling for PDT of Barrett's esophagus

Estimate the necrosis depth for PDT of Barrett's esophagus cancer with Photofrin® at standard conditions and a delivered **CS** light dose of 200 J/cm. Assume that $q^* = 40$ J/cm$^2$. Assume that the diameter of the inflated esophagus is 15 mm and the esophageal mucosa has the same optical constants as the oral mucosa at 630 nm.

From Table 5.1: $\mu_a = 1.5$ cm$^{-1}$, $\mu_s(1 - g) = 5$ cm$^{-1}$ and $\delta_d = = 0.18$ cm. The parameters in Equation (5.17) are $N' = 5/1.5 = 3.3$ and $b = 3.11$ leading to $R_d = 0.14$ and Equation (8.6) gives $M = 3.9$. In applying Equation (8.16) the value of $d_n$ must be guessed in function G. A first approximation for $d_n << a = 0.75$ cm gives $G = 0.465$, $M' = 1.57$, and $F = 6.66$, leading to $d_n = 0.12$ cm. A second approximation inserting $d_n = 0.12$ cm in $G$ gives $d_n = 0.14$ cm.

**Example 10.8** Modeling of PDT for squamous cell carcinoma of the oral mucosa

Estimate the necrosis depth for PDT of a large squamous cell tumor of the oral mucosa with Photofrin® at standard conditions. **CI** delivery is employed at a dose of 75 J/cm and a 2 cm length unclad fiber length. Assume that $q^* = 40$ J/cm$^2$ and $D = 0.055$ cm.

From Table 5.1: $\mu_a = 1.5$ cm$^{-1}$, $\mu_s(1 - g) = 5$ cm$^{-1}$ and $\delta_d = = 0.18$ cm. The parameters in Equation (5.17) are $N' = 5/1.5 = 3.3$ and $b = 3.11$ leading to $R_d = 0.14$ and Equation (8.6) gives $M = 3.9$. In applying Equation (8.16) the value of $d_n$ must be guessed in function $G$. A first approximation for $d_n = 0.5$ cm cm gives $G = 0.465$ and $F = 34$ gives $d_n = 0.38$ cm. A second approximation inserting $d_n = 0.38$ cm in $G$ gives $d_n = 0.40$ cm. This calculation indicates that **CI** delivery initiates necrosis in a 2 cm x 0.8 cm cylindrical band around the inserted fiber.

### 10.4.2.1 Light Dosimetry Modeling for Second-Generation PDT Drugs

Examples 10.5 - 10.8 illustrate applications of the diffusion approximation to modeling of PDT light dosimetry for Photofrin®. The estimated value of $q^*$ used for Photofrin® at standard conditions must be modified for different values of the injection dose. Modeling for new PDT drugs requires corresponding values of $q^*$. The analysis can be simplified by noting that the scaled necrosis depth $d_n/\delta_d$ has a direct relationship to the scaled light dose $(E_0 t)/q^*$ with only a small dependence on $R_d$. The results in Figure 10.4 are calculated with Equation

(8.8) based on Monte Carlo simulation for **FS** delivery. Therefore, if the necrosis depth at a given light dose is known at one wavelength, it is possible to estimate $q^*$ at a different wavelength. and the light dose required to achieve the same necrosis depth at the new wavelength. The procedure is illustrated for a hypothetical animal model experiment using typical parameters for Lu-Tex.

---

**Example 10.9** Generalized light dosimetry modeling

Prostate tumors in an experimental animal model have the following optical constants at 630 nm: $\mu_a = 0.9$ cm$^{-1}$, $\mu_s(1 - g) = 12.3$ cm$^{-1}$; the corresponding values at 730 nm are 0.6 cm$^{-1}$ and 7.5 cm$^{-1}$. PDT is performed with a photosensitizer that has a 3 times higher absorption coefficient at 730 nm compared to 630 nm. An **FS** light dose of 300 J/cm$^2$ leads to a 5 mm depth of necrosis at 630 nm. Estimate the light dose required at 730 nm to achieve the same result. Assume $n_r = 1.38$ and $b = 3.11$.

The first step is to calculate $R_d$ and $\delta_d$ at each wavelength. From Equation (5.14), $\delta_d$ (630 nm) = 0.17 cm and $\delta_d$ (730 nm) = 0.26 cm . The values of $R_d$ from Equation (5.16 ) are 0.47 at 630 nm and 0.32 at 730 nm. . The value of $d_n/\delta_d$ at 630 nm is 0.5/0.17 =2.9. From Figure 10.4, the corresponding value of $(E_0t)/q^* = 4.0$, leading to $q^* = 300/4.0 = 75$ J/cm$^2$ at 630 nm. At 730 nm, $d_n/\delta_d = 0.5/0.26 = 1.9$ and $(E_0t)/q^* = 1.5$. At 730 nm $q^* = 75/3 = 25$ J/cm$^2$ owing to the stronger light absorption by the drug. Therefore $(E_0t) = 25 \times 1.5 = 38$ J/cm$^2$ at 730 nm.

The calculation shows that the 730 nm irradiation should require an approximately 8-fold lower light dose compared to 630 nm. This result is due in part to the higher absorption by Lu-Tex and in part to deeper light penetration at 730 nm compared to 630 nm.

---

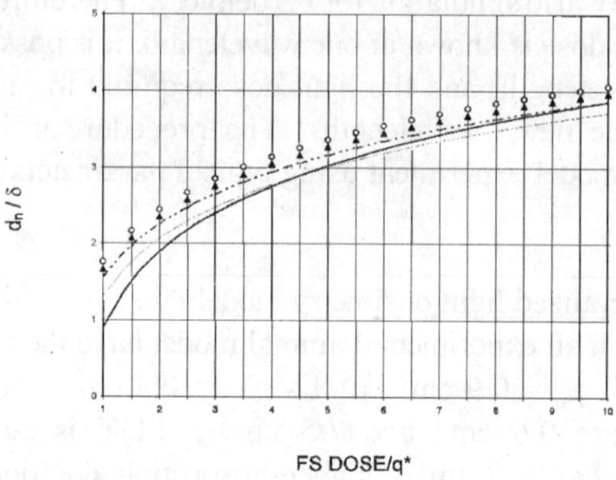


**Figure 10.4** Generalized PDT light dosimetry. Dependence of $d_n/\delta_d$ on $(E_0t)/q^*$ for different values of the diffuse reflection coefficient: —— $R_d$ = 0.1; - - - - $R_d$ = 0.2; –·–·– $R_d$ = 0.3; ···· $R_d$ = 0.4; ΔΔΔΔ $R_d$ = 0.5.

## 10.5 THERAPEUTIC RATIO

The *therapeutic ratio* (TPR) in radiotherapy is defined as the ratio of the normal tissue tolerance dose to the tumor lethal dose. The concept of TPR applies also for PDT when the normal and malignant tissues are exposed to the same treatment light. Factors leading to TPR > 1 are a higher intrinsic photosensitivity of the tumor tissue compared to normal tissue and a higher concentration of the drug in tumor tissue compared to normal tissue. The TPR may be estimated for PDT by assuming that $W^*$ is the same for the normal and tumor tissues, in which case the effective absorbed energy is proportional to the localized photosensitizer concentration ($C_s$). Equation (8.3) and Equation (8.15) lead to

$$TPR = \log_e(C_{st}/C_{sn}) = \mu_{eff}(d_{nt} - d_{nn}) \tag{10.1}$$

where $C_{st}$ and $C_{sn}$ are the drug concentrations in tumor and normal tissues, respectively, and $d_{nt}$ and $d_{nn}$ are the necrosis depths in tumor and normal tissues, respectively. Equation (10.1 ) assumes that both types of tissue have the same values of $\mu_{eff}$ A more complicated result is obtained for different values of $\mu_{eff}$ Equation (10.1) does not include the effects of drug photobleaching. Calculations based on Equation (8.19) show that photobleaching of irradiated normal tissues can increase the effective TPR and also protect the skin from post-PDT sunlight exposure.

**Example 10.10** Therapeutic ratio in PDT

A PDT drug has a 5 times higher concentration in tumor tissue compared to skin receiving the same light exposure. For $\mu_{eff}$ = 5 $cm^{-1}$ calculate the maximum depth of tumor tissue than can be treated to avoid skin tissue necrosis.

For $d_{nn}$ = 0, $d_{nt}$ = (1/5) x $Log_e$ 5 = 0.32 cm or 3 mm. This relatively short treatment depth can be increased by shielding normal tissues from the PDT light and selective light delivery with optical fibers.

## 10.6 DRUG PHOTOBLEACHING

Virtually all photosensitizing agents are altered by exposure to light. This effect is termed *photobleaching*. The fading of a dyed fabric is a familiar example of photobleaching. Photobleaching may be accompanied by the formation of colored products. For example, irradiation of PpIX and Photofrin® generate far-red absorbing photoproducts and new fluorescence bands. A significant contribution of colored photoproducts to the PDT response leads to complications beyond the scope of this treatment.

Photobleaching has a significant effect on light dosimetry. In the absence of photobleaching the PDT light dose and drug dose should be "reciprocal" quantities. For example, halving the drug dose should be compensated by doubling the light dose. However, clinical data for PDT of basal cell cancer show that the effective **FS** dose is about 7-fold higher at a Photofrin® dose of 1 mg./kg compared 2 mg/kg . A straightforward analysis shows that light dose-drug dose "non-reciprocity" is a consequence of photobleaching. In the absence of photobleaching, the local absorbed energy density depends only on the distribution of fluence rate within the tissue. The necrosis region begins closest to light entry and migrates deeper into the tissue as the irradiation proceeds. Photobleaching introduces an additional factor because the active photosensitizer is depleted faster in the frontal regions where the fluence is higher. . The net effect of photobleaching on the effective light dose is given by Equation (8.20) which predicts the dependence of the local energy absorption rate on depth into the tissue and the irradiation time. The corresponding necrosis depth can be derived by: (1) assuming **FS** delivery and substituting Equation (8.10) for $\Phi(\mathbf{r})$; (b) setting $W$ to $W^*$ which is energy density required for necrosis; (c) relating $W^*$ to $q^*$ by Equation (8.5). The resultant Equation (8.21) shows that effective necrosis depth $d_n$ is reduced by photobleaching. Numerical calculations give the photobleaching rate constant $K_s$. The estimated value $K_s$ = 2.5 (μg/g)/(J/$cm^2$) is obtained for Photofrin® at standard conditions, based on clinical data showing that reducing the drug dose from 2 mg/kg to 1 mg/kg requires an approximately 7-fold higher light dose. The results for other values of the Photofrin® dose are given in Table 10.3. Related calculations show that photobleaching lowers the necrosis depth $d_n$ for Photofrin® at standard conditions by $0.32\delta_f$ at any **FS** light dose. For a typical value of $\delta_f$ from 0.1-0.3 mm at 630 nm, the effect of Photofrin® photobleaching on necrosis depth is relatively small. The results in Section 8.5 assume the applicability of a *Type 1* photobleaching reaction, in which the rate of photobleaching is directly proportional to the absorbed fluence rate. However, similar results should be applicable for a *Type 2* photobleaching reaction. The reason for this is $^1O_2$ has a very short lifetime in tissues and must react close to the generation site, in which case the photobleaching reaction would be pseudo-first order. Elimination of a PDT drug from a tumor has similar effects on PDT as photobleaching. In this case the treatment time and the fluence rate are independent factors. Elimination during the PDT treatment is negligible for Photofrin® but it can be significant for a new drug with a short serum lifetime. Photobleaching also protects normal skin from sunlight exposure after PDT. Clinical evidence suggests that post-treatment sunlight exposure promotes the elimination of PpIX after PDT with ALA.

**Table 10.4 Effect of Photofrin® photobleaching on PDT light dose**

| *Drug dose (mg/kg)* | *Relative Light Dose* [a] |
|---|---|
| 2.0 | 1 |
| 1.75 | 1.2 |
| 1.5 | 1.6 |
| 1.25 | 2.3 |
| 1.0 | 6.7 |

[a] 630 nm irradiation at 48 hr post-injection.

## 10.7 PDT PHOTOTECHNOLOGY

The essential apparatus employed for PDT are a light source, light delivery system, and radiometric instrumentation. The properties of these devices are discussed in earlier chapters. Section 3.6 describes the operating principles, construction, and properties of lasers used for PDT. Optical fibers are the most widely used type of PDT light delivery system. This subject is described in Section 2.6.3. Radiometric measurements of light are described in Section 2.4.

This section summarizes some practical aspects of PDT phototechnology.

### 10.7.1 PDT Light Sources

A practical PDT light source should provide a uniform power output of at least 500 mW in the far-red or near-IR spectral regions. Other desirable features include stability, portability, and long lifetime. The properties of biomedical lasers are summarized in Table. 3.2. PDT treatments with Photofrin® are performed at 630 nm. A CW tunable dye laser pumped by 10-20 W argon-ion laser was the original PDT "workhorse". A dye laser has an approximately 100 nm tuning range with a given dye, typically Rhodamine 6G (575-665 nm) and DCM (620-740 nm). The output wavelength must be measured with a calibrated spectrometer prior to each PDT treatment. The argon-ion pumped dye laser has been superceded by more reliable "KTP-dye" laser. This system uses a solid state Nd-YAG laser as the primary source. The 1064 nm output is "doubled" to 532 nm by a passive potassium titanyl phosphate (KTP) crystal, and this output is used to pump the dye module. The typical "KTP-dye" dye laser output is 2 - 5W in the red or near-IR spectral regions. Other lasers that have been employed for pumping are the pulsed copper vapor laser (511 nm, 578 nm) and the pulsed XeCl excimer laser (308 nm). Non-dye lasers employed or proposed for PDT include the pulsed gold vapor laser (628 nm), the argon-ion pumped titanium sapphire laser (670 nm - 900 nm), and the Alexandrite laser (720 nm -790 nm). The PDT laser of the future may be the semiconductor diode laser. This device is inexpensive, very compact, and has a long lifetime. The gallium-aluminum-arsenide laser emits from 770 nm - 850 nm. Non-laser light sources have been used for PDT of superficial lesions. These include multiple LED sources, QTH projection lamps, and fluorescent lamps. A PDT light source is a medical device that requires regulatory approval. This approval by the Food and Drug Administration in the U.S. is specified for each PDT drug and the clinical application.

### 10.7.2 PDT Optical Fibers

The optical fibers used for PDT are usually of the step-index type with a fused-silica core and plastic cladding. Multi-mode fibers from 200 μm - 800 μm diameter are used for adequate power delivery. The radiance distribution of the exiting light depends on the geometry of the tip or "applicator". A flat- ended fiber emits a circular spot with a divergence angle in air of 20°- 30°. The irradiance profile is approximately gaussian (bell-shaped). A more uniform irradiance circular spot is obtained by fusing a "microlens" to the fiber tip. An internal reflection tip inclined at 45° projects a light beam inclined at 70°- 90° relative to the fiber axis. Fibers intended for direct insertion into tissue have a 0.5 cm - 3 cm length of the cladding removed at the tip. The tip of an endoscopic fiber is fabricated from a light-diffusing material such as an epoxy resin containing a white powder of high refractive index. Coupling a laser to an optical fiber is described in Section 2.6.5. The radiance entering the optical fiber cannot exceed the radiance exiting the laser. Efficient coupling is favored by a low laser beam divergence and a large acceptance angle by the optical fiber. PDT dye lasers normally satisfy this requirement. Semiconductor diode lasers have a large beam divergence and efficient coupling into an optical fiber requires a very thin active region to attain high output radiance.

NOTES

1. Bown SG, Rogowska AZ. (1999) New photosensitizers for photodynamic therapy in gastroenterology. *Can J Gastroenterol.* 13(5):389-92.

2. Bellnier DA, Greco WR, Loewen GM, Nava H, Oseroff AR, Pandey RK, Tsuchida T, Dougherty TJ. (2003) Population pharmacokinetics of the photodynamic therapy agent 2-[1-hexyloxyethyl]-2-devinyl pyropheophorbide-a in cancer patients. *Cancer Res.* 63(8):1806-13.

3. Bellnier DA, Dougherty TJ. (1996) A preliminary pharmacokinetic study of intravenous Photofrin in patients. *J Clin Laser Med Surg* 14(5):311-4.

4. In a three-compartment model the remaining drug concentration at time $t$ after the maximum uptake is expressed as: $C(t) = A \exp(-0.693/T_a) + B \exp(-0.693/T_b) + C \exp(-0.693/T_g)$ where $T$ is the half-time for each compartment.

5. Levy JG, Chan AH, Strong HA (1996) Clinical status of benzoporphyrin derivative. *SPIE Proceedings* Vol. 2625 pp.86-95

6. Boyle RW, Dolphin D. (1996) Structure and biodistribution relationships of photodynamic sensitizers. *Photochem Photobiol.* 64(3):469-85.

7. The lifetime of $^1O_2$ in water is approximately 4 ms. During this time a molecule diffuses approximately $L = (2Dt)$ where $t$ is the decay lifetime and $D$ is the diffusion constant. Assuming $D = 3 \times 10^{-5}$ cm$^2$/s for molecular oxygen in water gives $L = 15$ mm. Quenching of reactions of $^1O_2$ can further reduce the diffusion length.

8. Fingar VH. (1996) *J Clin Laser Med Surg.* (1996) Vascular effects of photodynamic therapy. 14(5):323-8.

9. Korbelik M. (1996) Induction of tumor immunity by photodynamic therapy. *J Clin Laser Med Surg.* 14(5):329-34.

10. Richter C (2000) Mitochondria as Targets for the Induction of Apoptosis in Photodynamic Therapy. Wyss P, Tadir Y, Tromberg BJ, Haller U (eds): *Photomedicine in Gynecology and Reproduction.* Basel, Karger, pp 157-168.

11. Kessel D, Luo Y. (1998) Mitochondrial photodamage and PDT-induced apoptosis. *J. Photochem Photobiol* B. 42(2):89-95.

12. van Staveren HJ, Keijzer M, Keesmaat T, Jansen H, Kirkel WJ, Beek JF, Star WM. (1996) Integrating sphere effect in whole-bladder wall photodynamic therapy: III. Fluence multiplication, optical penetration and light distribution with an eccentric source for human bladder optical properties. *Phys Med Biol.* 41(4):579-90.

# CHAPTER 11

# PHOTODYNAMIC THERAPY: CLINICAL ASPECTS

## Co-Author: B. H. Gerald Rogers, M.D.

Study protocols for PDT drugs are research projects with the objectives of acquiring information about toxicity, efficacy, and adverse reactions. A limited range of treatment variables are employed to permit statistical analysis of the results. HPD-derived drugs have been studied since 1978. Clinical trials In the U.S. are performed under an Investigational New Drug License (IND) authorized by the Food and Drug Administration. *Phase I* studies are directed towards evaluation of drug safety. Drug toxicity and dose-limiting side effects are determined for small numbers of patients with advanced cancers which have not responded to conventional treatments. The objectives of *Phase II* trials are to evaluate the response of cancers to PDT, establish effective treatment parameters, and identify adverse reactions. *Phase III* trials are randomized with respect to patient selection and compare the PDT results with conventional therapies. Table 11.1 lists the regulatory approvals of porfimer sodium (Photofrin®) as of 2003.

## 11.1 OVERVIEW OF A PDT PROCEDURE

The stages of a PDT procedure are as follows:

(1) *Drug administration*

PDT drugs are administered by intravenous injection, except for δ-aminolevulinic acid ( ALA) which is applied locally in a vehicle. The drug carrier, drug dose, and light dose vary with the specific agent. Photofrin® and most new PDT drugs photosensitize the patient's skin. Direct exposure to sunlight must be avoided for a variable period that depends on the drug. Conventional sun block agents are ineffective because they do not absorb visible light.

(2) *Tumor identification and visualization*

Selective uptake and retention of a PDT drug in tumors may be evaluated by fluorescence techniques. Photodynamic diagnosis (PDD) may be performed

**Table 11.1 Regulatory approvals of Photofrin®**

| Country | Approved Indications |
|---|---|
| Canada | Prophylactic treatment ofrecurrentpapillarybladdercancer (1993); palliative treatment of completely and partially obstructive esophageal cancers (1995), and lung cancer(1999) |
| Denmark | Treatment of lung and esophageal cancer (1999) |
| France | Treatment of recurrent lung and esophageal cancer (1996) |
| Finland | Treatment of lung and esophageal cancer (1999) |
| Germany | Treatment of early stage lung cancer (1997) |
| Iceland | Treatment of lung and esophageal cancer (1999) |
| Italy | Reduction of endobronchial obstruction or endobronchial mucosal lesions caused by non-small cell lung cancer or by metastases of other tumor cells to the lung, and palliative treatment of malignant dysphagia caused by esophageal carcinoma (2000) |
| Japan | Treatment of early stage lung cancer (1994), gastric and cervical cancers including cervical dysplasia and inoperable superficial esophageal and gastric cancers (1996) |
| The Netherlands | Treatment of superficial lung cancers, and palliative treatment of obstructive lung and esophageal cancers (1994) |
| United Kingdom | Treatment of lung and esophageal cancer (1998) |
| United States | Palliative treatment of totally or partially obstructing esophageal cancers that are unsuitable for Nd:YAG therapy (1995), curative treatment of microinvasive endobronchial non-small cell lung cancer (NSCLC) in patients, who are not indicated for surgery and radiotherapy (1998), and palliative treatment of late NSCLC (1998)<br>treatment of pre-cancerous lesions in Barrett's Esophagus (2003) |

shortly before applying the therapeutic light. Superficial tumors are visualized by illuminating with UV-A from a "black light" fluorescent lamp and viewing through amber goggles. Optical-electronic devices have been developed to record and analyze PDT drug fluorescence in superficial and interstitial sites.

(3) <u>*Light administration*</u>

The time delay between drug administration and the application of the activating light depends on the specific agent, ranging from 24-96 hours or longer for Photofrin® to a few hours for some new drugs. A laser delivering 1-3 watts at 630 nm is used for Photofrin®. Some new drugs may utilize non-laser light sources. Treatment times vary from a few minutes to about one hour depending on the specific drug, optical power level, and dimensions of treatment field. Superficial lesions are often treated without local anesthesia. Mild sedation may be employed for interstitial PDT.

(4) <u>*Post-irradiation responses*</u>

The observable tissue effects during or immediately after PDT with Photofrin® include loss of coloration or "blanching" and ecchymosis. Necrosis is evident within 24 hours and progresses for two to three weeks. Some new PDT drugs induce minimal necrosis. Tissue breakdown and healing typically requires about 6-8 weeks. The cosmetic results with Photofrin® for small superficial lesions are frequently excellent.

(5) <u>*Evaluation*</u>

The conventional grading scheme used for clinical trials is based on the tumor response at three months post-PDT: *complete response* (CR), biopsy-proven eradication of tumor; *partial response* (PR), 50% or greater reduction in tumor mass; *no response* (NR), less than 50% reduction in tumor mass; *progressive disease* (PR), tumor growth or metastasis. The Karnofsky Performance Status (KPS) is widely used to measure functional status after the procedure (Table 11.2).

(6) <u>*Re-treatment*</u>

Re-treatment may be employed for partial responses. Additional light treatments without drug administration are feasible with Photofrin® for periods ranging from 48 hours to many weeks. A follow-up PDT program with Photofrin® requires a delay of at least 3 months before another injection of the drug can be given.

**Table 11.2 Karnofsky performance status (KPS) scale**

| Rating | Activity | Living Situation |
|---|---|---|
| 100 | No restriction on activity based upon illness | Independent and able to work |
| 90 | Normal activity with minor signs or symptoms of illness | Independent and able to work |
| 80 | Normal activity with effort; some signs or symptoms of illness | Independent and able to work |
| 70 | Cares for self but unable to carry on normal activity or do active work | Independent; Unable to work |
| 60 | Requires occasional assistance; able to care for most needs | Independent; Unable to work |
| 50 | Requires considerable assistance and frequent medical care | Independent; Unable to work |
| 40 | Disabled; requires special care and assistance | Unable to care for self; unable to live independently |
| 30 | Severely disabled; hospitalization indicated; death not imminent | Unable to care for self; unable to live independently |
| 20 | Hospitalization indicated; very ill; active supportive treatment required | Unable to care for self; unable to live independently |
| 10 | Moribond; fatal processes proceeding rapidly | Unable to care for self; unable to live independently |

(7) *Other Treatments*

PDT has been employed after prior cancer treatments with radiotherapy and cytotoxic chemical agents. PDT has been used before and after surgery as an adjunctive procedure.

## 11.2 SUMMARY OF PDT RESULTS

PDT research and clinical results are reported in many publications including scientific and medical journals, conference proceedings and textbooks. A comprehensive review of the entire PDT literature is beyond the scope of this book. Instead, the current status of the more frequent PDT applications are reviewed with representative literature citations. Case histories of esophageal PDT treatments at the authors' institution are described to exemplify the clinical procedures.

### 11.2.1 PDT of Esophageal Cancer

Esophageal cancer represents 1.5% of all invasive cancers and the rate for adenocarcinoma is increasing annually. Most patients with esophageal cancer are diagnosed at an incurable stage with a 5-year survival rate of 5%-10%. Fifty percent of the patients are inoperable when the cancer is first detected. Dysphagia is the main symptom for advanced esophageal cancer. The alleviation of dysphagia and maintenance of dietary intake for these patients are the primary treatment goals. The palliation options for the advanced esophageal cancers are: surgery, radiotherapy, chemotherapy, and endoscopically-managed procedures such as dilation with polyvinyl wire-guided bougies and balloons, thermal ablation with bipolar electrocautery and the Nd:YAG laser, injection therapy with sclerosants, placement of self-expanding metal stents, and PDT.

Litle et al. reported on PDT treatment of 215 patients for bleeding or obstructing esophageal cancer[1]. PDT offered effective palliation in 85% of treatment courses Lightdale et al. reported on the PDT treatment of 110 patients with esophageal cancer. One month after PDT, objective tumor response based on measurements of an esophageal lumen diameter, was 33 % for both complete and partial responses. The median survival was 123 days for PDT, and palliation failure occurred after 34 days. Corti et al. reported the results of PDT for early esophageal cancer[2,3]. Twenty-three of 62 patients had complete responses with a median overall survival of 50 months. The complete response rate was increased to 82% (51 of 62) for patients who received radiotherapy in addition to PDT. Luketich et al. reported 91% improvement in dysphagia for 77 patients with a mean dysphagia-free interval of 80 days[4]. A recent study of pyropheophorbide-α (HPPH)-mediated PDT for esophageal cancer reported a lack of cutaneous photosensitivity[5].

### 11.2.2 PDT for Barrett's Esophagus

Barrett's esophagus is an important complication of gastroesophageal reflux disease (GERD). The normal squamous mucosa of the esophagus is replaced by metastastic columnar epithelium above the gastroesophageal junction.

Approximately 10 % of those with GERD develop Barrett's esophagus, and of these, 10 % will develop esophageal adenocarcinoma. The usual progression follows morphologically from low-grade dysplasia, to high-grade dysplasia, to early invasive carcinoma, and finally to advanced carcinoma. The diagnosis of Barrett's esophagus becomes crucial because of this high risk of cancer. Esophagectomy may be suggested when Barrett's esophagus with high-grade dysplasia is identified with multiple esophageal biopsies. PDT is an alternative treatment that leads to replacement of Barrett's mucosa with normal esophageal squamous epithelium. The PDT treatment follows the standard procedure with Photofrin®. In earlier clinical studies the laser light was delivered *via* endoscope using a cylindrical diffuser, resulting in strictures for over 50% of the patients[6]. In an effort to eliminate esophageal stricture after PDT due to the overlapping of treated regions, Overholt et al. developed a windowed centering esophageal balloon to deliver a uniform light dose[7]. The patients were given continuous omeprazole for 6 months after PDT. Complete elimination of Barrett's epithelium was achieved in 43 patients. Nd-YAG laser laser was required to ablate small residual areas of Barrett's mucosa in 35 patients. The formation of esophageal strictures has been a drawback of PDT. Limited data indicate that lower light dose does not reduce stricture formation. All post-PDT esophageal strictures reported by Dr. Overholt have responded to dilation. This result concurs with experience at this institution. A comparative study showed that oral administration of steroids after PDT did not reduce the incident of strictures, although the group receiving steroids had less discomfort[8].

**CASE HISTORY #1:**
**PDT for obstructing carcinoma of the esophagus**

This 71-year-old woman had received maximum-dose radiation plus chemotherapy prior to her referral for PDT. She was only able to swallow liquids. At endoscopy she was found to have an irregular obstruction just below the upper esophageal sphincter (Figure 11.1 left). She was injected with porfimer sodium (Photofrin) at a dose of 2 mg/kg. Probing determined the stricture was approximately 5 cm long. Two days after her injection she was treated with 300 joules/cm of 630 nm laser light utilizing a 5-cm long diffuser. Two days later at endoscopy the stricture was markedly softer. After careful balloon dilation it was possible to pass the 8.6 mm diameter endoscope through the stricture (Figure 11.1 right).

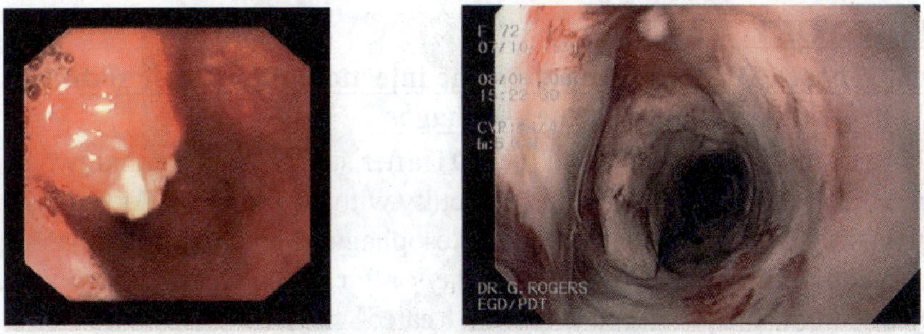


**Figure 11.1** Malignant Stricture of the Esophagus Located 17 cm from the Gingiva (left). The white area is necrotic tissue on the surface of the malignancy. A probe with a diameter of 2 mm could be passed through the stricture but the endoscope could not. Two days after PDT it is soft enough to be dilated to 12 mm diameter which allowed passage of the endoscope (right).

**COMMENT:**

This patient had received her maximum dose of radiation therapy so further treatment with that modality was impossible. At the time she was referred for PDT she was too debilitated to receive further chemotherapy. Placement of a self-expanding metal stent across the stricture was an option but the presence of a stent at this high level would almost certainly give her a continuous sensation of a foreign body in her upper esophagus. In this case PDT was the best treatment for this patient at this stage of her disease. A malignant stricture treated with PDT can usually be safely opened between 2 and 4 days after the light dose. PDT is a local form of therapy and therefore can be given in association with radiation or chemotherapy.

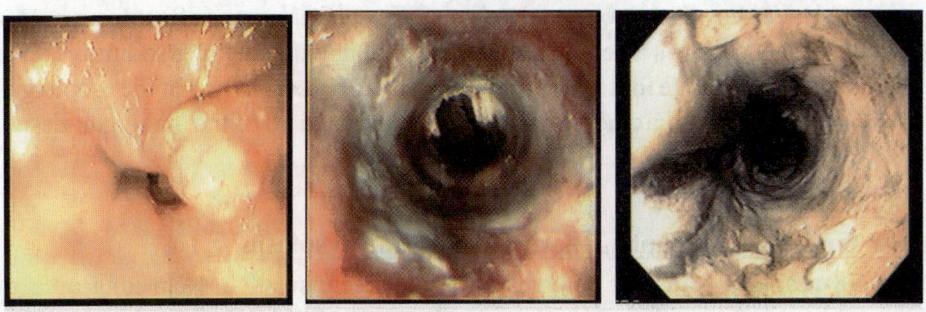

**Figure 11.2** Endoscopic view of obstructing adenocarcinoma of the distal esophagus (left). Same strictured area, three days after PDT (center). The malignant stricture was softened so it was possible to dilate up to 12 mm diameter. Results of repeat PDT 56 days after injection with porfimer sodium (right). This patient's malignant stricture was softened so it was possible to dilate it from 4 mm to 12 mm.

**CASE HISTORY #2:**

**Delayed repeat PDT treatment after one injection of porfimer sodium for a rapidly growing carcinoma of the esophagus.**

This 47-year-old man was referred for PDT after stent placement and two cycles of chemotherapy failed to relieve his difficulty with swallowing. At endoscopy a malignant stricture was found in the distal esophagus which prevented passage of the endoscope (Figure 11.2a). Three days after receiving the usual dose of porfimer sodium, the strictured area was treated with 200 joules/cm of 630 mm laser light utilizing a 2.5 cm-long diffuser. Three days later the malignant tissue was necrotic and could be easily debrided with the multibite forceps. After doing so it was easy to pass the endoscope all the way through the stricture . At endoscopy one month later the malignant stricture had closed down to 4 mm. Knowing that porfimer sodium is eliminated slowly from the body and is preferentially picked up by the neoplastic tissue, it was decided to retreat the patient. Accordingly he was treated with 400 joules/cm of 630 nanometer laser light 56 days after he had been injected with porfimer sodium. Two days later endoscopy showed a good effect. The necrotic malignant tissue was debrided and the stricture dilated up to 12 mm diameter without difficulty (Figure 11.3). Further chemotherapy was instituted at this time. Subsequently he ate without difficulty and gained six pounds.

**COMMENT:**

We now know that it requires about 12 weeks to eliminate porfimer sodium from the body to the point where sunburn is no longer a significant problem. The long half-life of this sensitizing agent can be used to the patient's advantage if recurrence of the malignant stricture occurs within that time frame. The case presented here showed that one can get a significant therapeutic effect from a laser light dose given 56 days after injection of the patient with porfimer sodium. I have also used this technique for treating patients with early carcinoma of the esophagus when it was not clear that the first light dose eliminated all of the malignant tissue.

**CASE HISTORY #3:**

**PDT for Barrett's esophagus with high-grade dysplasia**

This 78-year-old man was referred for PDT because of a 4-cm long area of Barrett's esophagus with focal high-grade dysplasia (Figure 11.4a). This condition is due to chronic gastroesophageal acid reflux and is prone to progress to adenocarcinoma. Ablation is thought to prevent malignant transformation. Ultrasound of the esophagus showed thickening of the muscular mucosa in the area of Barrett's esophagus but no evidence for invasive carcinoma. He met the inclusion criteria so he was entered into the Phase III trial of Photofrin (porfimer sodium) PDT for Barrett's esophagus with high-grade dysplasia initially sponsored by QLT Phototherapeutics of Vancouver, British Columbia. The patient was injected with 2 mg/kg of the sensitizing agent and treated two days later on November 13th, 1998 with 130 joules/cm from a 630 nanometer laser light source. The diffuser was held in the middle of the lumen by a centering balloon (Figure 11.4b). The centering balloon helps to evenly distribute the light dose over the area to be treated and prevents scatter to areas not needing treatment. Endoscopy two days after PDT the treated mucosa is necrotic (Figure 11.5a). Subsequently, the necrotic tissue sloughs and is gradually replaced by normal squamous mucosa as long as acid reflux is controlled with a proton pump inhibitor such as omeprazole. Follow-up endoscopy with biopsies every three months for one year then every six months for the remainder of the study showed complete disappearance of Barrett's esophagus and high-grade dysplasia (Figure 11.5b). At his last endoscopy on October 23rd, 2002, almost four years after his treatment, his condition remains stable with no evidence for recurrence of Barrett's or high-grade dysplasia. He was put on omeprazole 20 mg twice daily after PDT and will stay on it indefinitely.

**COMMENT:**

At the time of this patient's initial diagnosis of Barrett's esophagus with high-grade dysplasia, the standard treatment was esophagectomy. This elderly man had coronary artery disease which had been treated by coronary artery bypass grafts. His age and his heart disease made him a high-risk patient for esophagectomy. For this reason he was enrolled in the study. This patient was one of the 208 patients enrolled in the Phase III study. The study showed that high-grade dysplasia was completely eliminated in 77% of the patients treated with PDT. Also the progression to adenocarcinoma was reduced by half in the treated patients. Halfway through the study it was taken over by Axcan Pharma, also of Canada. The FDA approved porfimer sodium for the ablation of high-grade dysplasia and Barrett's esophagus on August 4th, 2003.

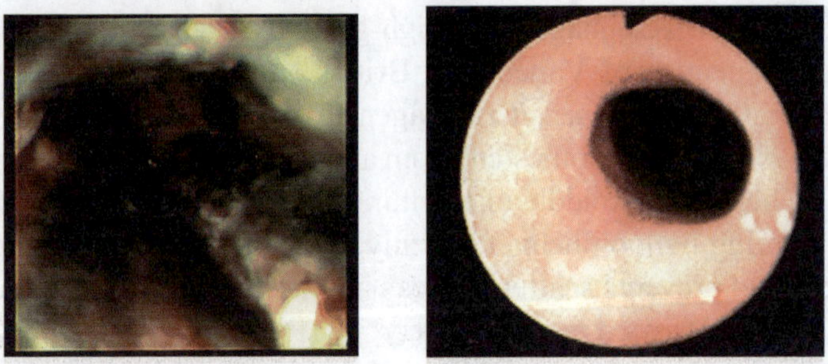

**Figure 11.4** Two long irregular tongues of salmon-colored mucosa invade the distal esophagus (left). Biopsies showed it to be Barrett's epithelium with high-grade dysplasia (carcinoma-in-situ). The centering balloon (right) is designed to help distribute the light dose over the area to be treated. The laser light diffuser is threaded into the center of the balloon. The walls of the balloon help to smooth out the surrounding mucosa.

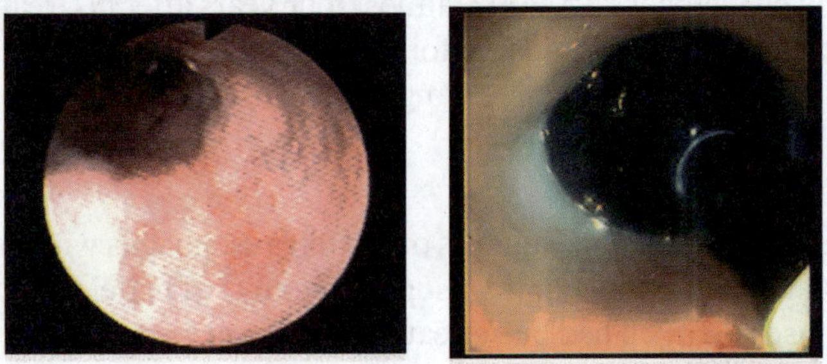

**Figure 11.5** The blackened necrotic mucosa of the distal esophagus two days after completion of PDT (left). The patient was started on a double dose of omeprazole at this time to control acid reflux. Regeneration of normal squamous mucosa after ablation of Barrett's mucosa with PDT. Biopsies showed no residual Barrett's and no high-grade dysplasia (right). Compare this photo with the pre-treatment photo in Figure 11.4.

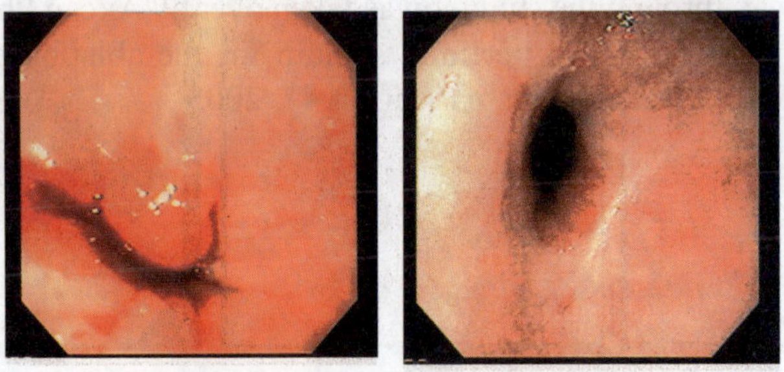

**Figure 11.6** One cm malignant polyp arising from Barrett's metaplastic esophageal mucosa (left). This early cancer was excised and the area treated with PDT. One year after excision of malignant polyp, treatment with PDT and omeprazole (right). Only a scar is seen where the malignant polyp was excised. Biopsies showed no residual Barrett's metaplasia.

**CASE HISTORY #4:**
**PDT for early carcinoma of the esophagus**

A 70-year-old man was found to have a 10-mm sessile polyp at the gastroesophageal junction associated with a 2-cm long segment of Barrett's esophagus (Figure 11.6a). Biopsies of the polyp showed well-differentiated adenocarcinoma. Endoscopic ultrasound showed no evidence for metastatic disease in the peri-esophageal tissue. PDT was recommended because of the patient's age, history of coronary artery disease, history of carotid endarterectomy and hypertension. The polyp was removed with electrosurgical snare technique. After a delay of five weeks to allow healing at the polypectomy site, the patient was given porfimer sodium at 2 mg/kg. Two days later a 630-nanomter light dose of 50 joules/cm was given to the area from which the polyp had been excised utilizing a 1-cm diffuser. Then the entire area of Barrett's esophagus was treated with a 150 joules/cm utilizing a 2.5-cm diffuser. Follow-up endoscopy the next day showed necrosis in the distal 3 cm of the esophageal mucosa. The patient developed difficulty swallowing and required five balloon dilations up to 15 mm during the two-month post PDT period. Subsequently, he had no further difficulty swallowing. Follow-up endoscopies showed no evidence for Barrett's esophagus and no residual adenocarcinoma. The patient is now over four years post treatment with no evidence for recurrence of Barrett's or his cancer. He was put on a proton pump inhibitor twice daily after PDT so that the esophageal mucosa would not regenerate in a highly acidic environment. Acid suppression after mucosal ablation promotes regeneration of normal squamous mucosa.

**COMMENT:**

This case illustrates the utility of PDT in treating early superficial carcinoma of the esophagus in an elderly patient with multiple medical problems. PDT is compatible with other modalities used for treating cancer of the esophagus. If PDT does fail it can be repeated or the patient could be treated with radiation, chemotherapy or surgery.

## 11.3 PDT OF SKIN CANCER

Various therapeutic modalities are used to treat skin cancer, including excisional surgery, radiotherapy, curettage and electrodesiccation, and cryotherapy. The conventional modalities have limitations in treating multiple lesions as well as difficult sites such as on the eye lid and the outer ear channel. Non-melanoma skin cancers have been treated with Photofrin® since 1978, including basal cell carcinoma (BCC), squamous cell carcinoma (SCC), Bowen's disease, and basal cell nevus syndrome (BCNS). In many studies Photofrin® was administered at standard conditions and irradiated at a light dose of 20-40 J/cm$^2$. Lower drug dose and higher light dose levels have been employed in some recent studies to reduce the duration and severity of skin photosensitization. The response rates for dermatological tumors vary widely.

At Roswell Park, more than 1400 superficial and nodular BCCs were treated at a dose of 1 mg/kg with 200 J/cm$^2$ and 630 nm red light, with a complete response rate of 92% and a clinical recurrence rate of 9% at 4 years.[9] Similar results were obtained with topical ALA.

Wilson et al. treated 37 patients, with 151 primary or recurrent BCC achieving an 88% CR rate[10]. In that study the Photofrin® dose was 1.0 mg/kg and the range of light dose was 70-290 J/cm$^2$. These results exemplify the non-reciprocity of light dose and drug dose characteristic of Photofrin®. Topical ALA has been used for PDT of BCC and SCC since the early 1990s. A 1996 review of Marcus et al. for superficial BCC shows CR rates of 88% -100%[11]. The PDT light was applied 3- 8 hours after 20% topical ALA. A more recent summary indicates CR rates of 79-100% for superficial lesions[12]. The range of light dose was 30-540 J/cm$^2$ for 20% ALA. Nodular and noduloluncerative BCCs were less responsive and required repeated treatments. Sclerodermiform BCCs were generally non-responsive.

PDT may be especially useful for patients with multiple recurring BCC including basal cell nevus syndrom (BCNS). PDT has not proven to be an acceptable treatment modality for malignant melanoma and AIDS-related Kaposi's sarcoma. PDT has been used to treat cutaneous T-cell lymphoma (*mycosis fungoides*, MF). Early results with systemic HPD were not continued. More promising results are reported for PDT of the patch/plaque stage of MF with topical ALA.[13,14] PDT is not suitable for management of advanced or inflammatory cutaneous carcinoma owing to the low response rates and induction of extensive cutaneous necrosis.

### 11.3.1 PDT of Metastatic Breast Cancer

Some of the earliest applications of PDT were for palliation of metastatic breast cancer on the chest wall. Eighteen different PDT studies are reported from 1979-1998 for HPD, Photofrin®, topical ALA and topical TPPS. Necrosis and other morbidity associated with Photofrin® was significantly reduced by reducing the drug dose to 0.8 mg/kg mg/kg and using a light dose to 135-170 J/cm$^2$ [15]. Sparing of normal tissue by photobleaching of Photofrin® is almost certainly the reason. A recent study for 9 patients with progressing chest wall recurrence, despite salvage surgery, radiation, and chemo-hormonal therapy, led to 89 % complete responses at 6 months post-PDT. Ongoing clinical trials indicate that Lu-Tex is a potentially useful drug for PDT of recurrent breast cancer[16]. A multiple LED source emitting at 732 nm was employed which permitted treatment of wide area fields. In a *Phase I* study employing a range of treatment variables, a complete response was achieved in 9 of 15 patients. Significant pain was experienced at the treatment sites but there was no necrosis. The typical duration of post-PDT skin photosensitization was less than a week and there was no detectable Lu-Tex in the patient's blood after 72 hours. An aluminum phthalocyanine preparation (Photosense®) was used in Russia in a preliminary study on PDT of pre-operative primary and recurrent breast cancer[17,18]. The drug was administered intravenously in saline at 0.5 mg/kg and irradiated at 672 nm using a semiconductor laser and a starting light dose of 150-200 J/cm$^2$ .

### 11.3.2 PDT of Pre-malignant and Proliferative Cutaneous Disease

*Actinic keratosis* (AK) is a rough, red or brown scaly patch on the surface of the skin. Approximately 40% of SCC begin as AK. PDT with ALA has led to excellent curative results. In 1992 Kennedy and Pottier reported that 90% of multiple AKs were eradicated by applying 20% topical ALA followed by a single exposure to filtered light from a slide projector[19]. This finding was confirmed by subsequent studies, with the best results achieved for Grade 1 (easily seen, slightly palpable) and Grade 2 (well developed, easily palpable) AKs and the poorest results for thick, hyperkeratotic lesions. ALA treatment of AK was approved in the U.S. in December 1999. The Levulan Kerastick® is a topical applicator for 20% ALA. After 18-18 hours the lesions are irradiated with a special blue fluorescent lamp. Clinical trials with this system led to 70% clearing with one treatment. Re-treatments after 8 weeks were effective for resistant lesions. PDT treatments of psoriatic plaque with HPD, BPD-MA, and ALA are experimental and not widely used.

### 11.3.3 PDT of Superficial Bladder Cancer

Superficial bladder cancer can occur as papillary tumor and carcinoma *in situ* (CIS). The lesions are not easily detected endoscopically and whole bladder wall therapy is required. CIS can lead to invasive cancer in 50-80% of the patients if intravesical therapy is not applied. Papillary bladder cancer is usually treated by transurethral resection (TUR) with a recurrence rate in the range of 40-70%. Following TUR intravesical chemotherapy with cytostatic agents or immunotherapy with bacillus Calmette-Guerin (BCG) may be performed. Complete response rates range between 60% and 79% and overall the average complete response is 72% in these patients.[20] PDT is an alternative treatment to cystectomy in superficial bladder cancer patients in whom transurethral resection and intravesical therapy with BCG have failed. Photofrin® has been employed at standard conditions. Smooth mucosal folds are provided by filling the bladder with isotonic saline solution to an average volume of 100 ml. The effective light dose within the bladder may be significantly higher than the delivered dose owing to the integrating sphere effect (see Section 8.2.2.1.). Typically, an applied dose of 15 J/cm$^2$ is amplified to 75 J/cm$^2$ by light that is back scattered and re-absorbed at the bladder wall. A modified PDT cystoscope described in Section 10.4.1.2. provides for *in situ* monitoring of the effective light dose. Kriegmair et al. treated 21 superficial bladder cancer patients who were candidates for cystectomy[21]. Intravenous injection of a porphorin was followed with illumination by 630 nm laser light. A complete response was achieved for 12 patients a mean cancer-free interval of 18 months; PR was achieved for 3 patients who were treated with transurethral resection and became cancer free; 6 patients did not respond to PDT and underwent cystectomy. Kriegmair et al. also employed intravesical administration of 5-aminolevulinic acid using a white light source. At a median follow-up of 18 months (range 3 to 25), 3 of the 7 patients with carcinoma in situ and 2 of the 4 patients with papillary tumors were free of disease[22].

Nseyo et al. treated 58 patients for resistant superficial transitional cell carcinoma (TCC) including Ta, T1, and refractory carcinoma in situ. Three months following treatment with Photofrin and 630 nm light, the complete response rates were 84% and 75% for residual resistant papillary TCC and refractory CIS respectively. At a median followup of 50 months, 59% of the responders were alive, with 31/34 still disease-free.[23] Manyak et al. report on Photofrin-mediated PDT using diffusion medium for isotropic light distribution. 29 patients had TCC carcinoma in situ and 5 had multiple papillary stage T(a) or T(1) lesions. At 3 months, a complete response was obtained in 14 of the 32 evaluable patients, a partial response in 4, and no response in 14. Four of the five patients with extensive papillary lesions did not respond.[24]

### 11.3.4 PDT of Gastric Cancer

PDT of gastric cancer with Photofrin® is performed in Japan where incidence is especially high. possibly owing to diet and genetic predisposition. The Health Organization of Japan has developed an intensive screening protocol, involving endoscopic ultrasound and biopsy, for diagnosis of gastric cancers. This screening allowed for the detection of gastric cancers in their early stages, and reduced the mortality rates. Surgery is the conventional treatment for early stage gastric cancer. PDT is an alternative for patients who refuse surgery or who have contraindications for surgery. After administration of 2 mg/kg Photofrin® the entire lesion is illuminated with a 630 nm laser light delivered by gastroscopy at 60 J/cm$^2$ or higher with about a 5 mm width of marginal mucosa. After PDT, an $H_2$-receptor antagonist is given for treating the sequential ulcer. Accurate light dosimetry is a major problem because of the anatomical shape of the stomach or invasive growth of cancer into the muscular layer. Also ongoing peristalsis must be taken into account. Kato et al. reported a complete response for 48 out or 58 patients diagnosed as stage I-III gastric cancer[25]. Mimura et al. treated 27 patients with early gastric carcinoma. A 100% complete response without recurrence was achieved for mucosal carcinoma 2 cm or less in diameter which declined to 20% for carcinoma invading more deeper than muscularis propria[26].

## 11.3.5 PDT of Bronchogenic Lung Cancer

The lung cancer mortality rate remains high although lung surgery, radiotherapy and chemotherapy have all improved survival. Early detection of lung cancer and complete resection of the primary lung cancer do not reduce the mortality in this disease because there is a high risk of developing another primary lung cancer and may lead to cardiovascular or pulmonary disorders. Endobronchial non-small cell lung cancer (NSCLC) in patients may have symptoms of hemoptysis (coughing up of blood) and dyspnea (shortness of breath). In later stages, NSCLC may lead to intrinsic (endobronchial) or extrinsic obstruction of the airways. Significant airway reopening is achieved with PDT in the majority of patients treated. The procedure requires flexible bronchoscopy, which is a minimally invasive technique. PDT may preserve normal lung tissue and reduce the risk of surgical resection in some patients. Photofrin® is administered at 2 mg/kg and the 630 nm laser light is delivered *via* a fiberoptic bronchoscope delivery system using either a microlens or a cylindrical diffuser. Power densities range from 200 to 400 mW/cm at light dose levels from 100 to 200 J/cm. Usually 48 hours after PDT, a second clean-up bronchoscopy is performed. Endo et al. reported that bronchogenic squamous cell carcinomas, visible with bronchoscope and less than 10 mm in diameter, were the good candidates for PDT.[27] Kato et al. treated 35 patients with mono-L-aspartyl chlorin e6 in an effort to reduce residual photosensitivity.Complete response was obtained in 82% of the patients[28]. Cortese et al. treated 21 patients with a total of 23 early stage lung cancers and followed the patients up to 116 months after PDT. A complete response of 71 % was achieved at the first follow-up bronchoscopy and lasted longer than 12 months for 52 % of these patients[29]. Patelli et al. treated 26 early-stage bronchogenic cancers (23 patients) and reported 16 complete responses and 10 partial responses[30].

### 11.3.6 PDT of Head and Neck Cancer

PDT of head and neck tumors with Photofrin® is a promising procedure which is currently in *Phase III trials* in the U.S. Oral cavity (mouth, tongue, and cheek) lip, pharynx, and larynx cancers are conservatively treated by radical surgery with reconstruction. Maximal dose radiation therapy and/or chemotherapy are applied following surgery. Local recurrences or residuals of the tumor are the more frequent causes of death rather than tumor metastasis. Rescue surgery has a high risk, causing cosmetic and functional tissue loss, and may not be suitable for patients with other medical problems. PDT is an alternative treatment for recurrent head and neck cancer. In early clinical trials at this institution MRI diagnosis was performed prior to PDT. The treatment plan employed a combination of **FS** and **CI** delivery modes depending on the location and dimensions of the lesions. Based on light dosimetry modeling, the typical light dose levels were 50-100 J/cm$^2$ for **FS** delivery and 100-200 J/cm for **CI** delivery. Twenty-six biopsy-proved squamous cell carcinoma (SCC) of the head and neck were treated with Photofrin® at 2 mg/kg. Complete responses were achieved for 77 % of the patients after 48-month follow-up[31]. Feyh reported on the PDT treatment of 83 patients with tumors of the face, oropharynx, and larynx. An overall complete response rate of 89% was confirmed by histologic diagnosis 13 - 71 months after PDT[32]. Biel treated primary and recurrent cancers of various types and staging in 87 patients[33]. Overall, a single PDT treatment led to a complete ? for all patients with CIS and T1 carcinomas. A complete response was achieved for 22 of 28 patients with laryngeal tumors, including CIS, recurrent leukoplakia, and T1 SCC after a mean follow-up of 30 months. PDT did not control severe and uncontrolled laryngeal tumors in 5 patients. Excellent PDT results were reported by Schweitzer for 20 patients squamous cell head and neck carcinoma[34]. Overall, PDT with Photofrin® led to excellent results for many types of head and neck carcinoma .

### 11.3.7 PDT of Gynecologic Cancer

Vaginal cancers have been treated primarily by radiotherapy, and vaginectomy has been considered as a radical treatment. However, vaginectomy is troublesome in elderly women with vaginal cancer and metastatic lesions. PDT can be an alternative for these patients. Cervical cancer is another common gynecologic cancer that can also be treated with PDT. Superficial cervical cancers

(CIS) are conventionally treated with conization (removing a cone-shaped area of the cervix with the cervical opening at the center of the large surface of the cone), cryotherapy, and Nd-YAG and $CO_2$ laser vaporization. These therapies have some complications such as scarring, severe cosmetic effects, and fertility problems. In an early study, Lele et al. reported on the PDT treatment of 21 patients with cancers of the perineum, vulva, cervix and vagina (26 lesions) and achieved a 43% complete response rate[35]. The irregular shape of gynecologic organs and areas of cervicovaginal junction made it difficult to control the light dosimetry. In early work, Lobraico et al. treated 38 cases of vulvar CIS with Photofrin®. A 71% complete response rate was observed at three months after PDT treatment[36]. However, there was no clear advantage over conventional treatments. PDT is an experimental adjunctive therapy to surgical debulking in patients with advanced recurrent ovarian cancer and other disseminated *intra*-peritoneal cancers.

Some results are reported for other PDT drugs. Antibody-bound phthalocyanine was employed for cytoreduction of ovarian cancer by *intra*-peritoneal PDT[37]. The mean survival for 31 PDT patients was 90% after 30 months and 40% after 60 month., compared to 78% after 30 months and 37% after 60 months in the non-PDT control group. A small scale study on PDT of vulvar *intra*-epithelial neoplasia III (VIN III) with ALA led to a complete response of the precancerous lesions in 3 out 5 patients[38]. In the same study, complete clearing of vulvar condyloma for four months was achieved in 3 of 7 patients. The patients experienced considerable pain during the irradiation which was reduced by employing a lower ALA concentration.

Approximately 40% of 700,000 hysterectomies performed annually in the U.S. are for treatment of excessive uterine bleeding (menorrhagia). Applications of PDT for endometrial ablation have been investigated as an alternative to hysterectomy. Recent work has emphasized ALA because this agent can be administered topically[39]. Histological examination showed selectivity with a lack of endometrial fibrosis or adhesions as a result of the phototherapy.

### 11.3.8 PDT of Brain Cancer

Malignant brain tumors have a very poor prognosis. The typical median survival ranges from approximately 20-30 weeks for patients after surgery followed by radiotherapy and drug treatments. Primary malignant brain tumors do not metastasize outside of the central nervous system. The objective of PDT is to improve local tumor control. The procedures require invasive surgery. In interstitial or "volume" illumination one or more optical fibers are inserted into the tumor either free hand or by stereotactic means. In "cavitary" illumination the PDT light is delivered *intra*-operatively after the tumor bulk has been removed.

This technique may be performed by **FS** illumination of the exposed tumor surface or by filling the tumor cavity with a light diffusing medium and placing the tip of the optical fiber within the medium. The diffusing medium is contained in an expandable balloon for more precise light dosimetry. The balloon technique was employed in a study for 56 patients with recurrent and with newly diagnosed gliomas who had failed previous surgery and radiotherapy[40]. Photofrin® was administered at 2 mg/kg and the PDT light was delivered after radical tumor resection or cyst drainage. The light dose at the applicator surface ranged from 8 - 110 J/cm$^2$. The median survival rates for 11 patients with newly diagnosed glioblastoma (GBM) was 36 weeks with 1 and 2 year survival of 36% and 0%, respectively; for 9 patients with malignant astrocytoma (MA) the survival was 48 weeks with 1 and 2 year survival of 44% and 33 %, respectively. Patients with recurrent gliomas had median survival rates of 30, 40, 56, and >174 weeks for GBM (32 patients), MA (14 patients), malignant mixed astrocytoma-oligodentroglioma (6 patients), and ependymoma (4 patients), respectively. A recent summary of results with PDT at different centers indicates median survivals of 24 months for newly diagnosed GBM and 7 - 9 months for recurrent GBM[41]. In an interesting recent development, the gadolinium texaphyrin derivative (Xcytrin® ) was used as an adjunctive agent in radiotherapy of GBM and also intrinsic pontine glioma, a usually fatal brain stem tumor in children. The selective localization of Gd-Tex in malignant tumors and its molecular paramagnetism permitting MRI diagnosis are important features. Gd-Tex is not an effective PDT agent. However, a combination of Lu-Tex and Gd-Tex may make it possible to employ follow-up PDT after radiotherapy.

## 11.4 CURRENT STATUS OF PDT TREATMENTS OF CANCER

Significant improvements in the efficacy of PDT and breadth of applications have occurred since the mid-1990s. The areas of recent advances include:

(1) <u>*Clinical studies*</u>

The results of Phase III clinical trials for major cancer diseases provide considerably more information about the effects of treatment variables and the expected clinical results.

(2) <u>*Improvements in light delivery and dosimetry*</u>

The development of new light applicator devices for PDT in the bladder, esophagus, brain, and female genital tract led to improved reliability and efficacy.

(3) *New PDT drugs*

The augmentation of Photofrin® with new PDT drugs permits a more selective matching of the drug to the specific malignancy and suggests the possibility of multi-drug PDT.

(4) *New PDT light sources*

The recent development of non-laser light sources will make PDT more accessible in community hospitals and clinics.

(5) *New developments in photodynamic diagnosis*

Computer-based optical-electronic devices based on the fluorescence properties of PDT drugs are being employed for tumor diagnosis and visualization.

(6) *Standardization of study protocols*

Prior experiences with Phase III clinical of Photofrin® have established a knowledge base that facilitates the granting of regulatory approvals for PDT studies with different cancers and new PDT drugs.

In recent years an enormous effort has been expended on the development and pre-clinical testing of new potential PDT drugs. This activity may be to the detriment of basic research on the action mechanism PDT and treatment planning for Photofrin® which is the most widely used PDT drug. The areas of recent progress include Monte Carlo modeling of light propagation in tissues and elucidation of the respective roles of vascular and cellular damage effects. More information about these and other basic aspects of PDT should help advance PDT from an essentially empirical procedure to one in which an individualized treatment plan is developed for each patient.

NOTES

1. Litle VR, Luketich JD, Christie NA, Buenaventura PO, Alvelo-Rivera M, McCaughan JS, Nguyen NT, Fernando HC. Photodynamic therapy as palliation for esophageal cancer: experience in 215 patients. *Ann Thorac Surg*. 2003 Nov;76(5):1687-92; discussion 1692-3.

2. Lightdale CJ, Heier SK, Marcon NE, McCaughan JS Jr, Gerdes H, Overholt BF, Sivak MV Jr, Stiegmann GV, Nava HR. Photodynamic therapy with porfimer sodium versus thermal ablation therapy with Nd:YAG laser for palliation of esophageal cancer: a multicenter randomized trial. *Gastrointest Endosc*. 1995 Dec;42(6):507-12.

3. Corti L, Skarlatos J, Boso C, Cardin F, Kosma L, Koukourakis MI, Giatromanolaki A, Norberto L, Shaffer M, Beroukas K. Outcome of patients receiving photodynamic therapy for early esophageal cancer.*Int J Radiat Oncol Biol Phys.* 2000 May 1;47(2):419-24.

4. Luketich JD, Christie NA, Buenaventura PO, Weigel TL, Keenan RJ, Nguyen NT. Endoscopic photodynamic therapy for obstructing esophageal cancer: 77 cases over a 2-year period. *Surg Endosc*. 2000 Jul;14(7):653-7.

5. Bellnier DA, Greco WR, Loewen GM, Nava H, Oseroff AR, Pandey RK, Tsuchida T, Dougherty TJ. Population pharmacokinetics of the photodynamic therapy agent 2-[1-hexyloxyethyl]-2-devinyl pyropheophorbide-a in cancer patients. *Cancer Res*. 2003 Apr 15;63(8):1806-13.

6. Overholt BF, Panjehpour M. Photodynamic therapy for Barrett's esophagus. *Gastrointest Endosc Clin N Am*. 1997 Apr;7(2):207-20

7. Overholt BF, Panjehpour M, Haydek JM. Photodynamic therapy for Barrett's esophagus: follow-up in 100 patients. *Gastrointest Endosc.* 1999 Jan;49(1):1-7.

8. Panjehpour M, Overholt BF, Haydek JM, Lee SG. Results of photodynamic therapy for ablation of dysplasia and early cancer in Barrett's esophagus and effect of oral steroids on stricture formation *Am J Gastroenterol.* 2000 Sep;95(9):2177-84.

9. Zeitouni, N.C., Shieh, S., Oseroff, A.R., 2001. Laser and Photodynamic therapy in the management of cutaneous malignancies. In: Ellerin, B. (Ed.), *Clinics in Dermatology.* Elsevier, New York, NY, pp. 328–339.

10. Wilson BD, Mang TS, Stoll H, Jones C, Cooper M, Dougherty TJ. Photodynamic therapy for the treatment of basal cell carcinoma. *Arch Dermatol.* 1992 Dec;128(12):1597-601.

11. Marcus SL, Sobel RS, Golub AL, Carroll RL, Lundahl S, Shulman DG. Photodynamic therapy (PDT) and photodiagnosis (PD) using endogenous photosensitization induced by 5-aminolevulinic acid (ALA): current clinical and development status. *J Clin Laser Med Surg.* 1996 Apr;14(2):59-66.

12. Kalka K, Merk H, Mukhtar H. Photodynamic therapy in dermatology *J Am Acad Dermatol.* 2000 Mar;42(3):389-413.

13. Paech V, Lorenzen T, Stoehr A, Lange K, Merz H, Meigel WN, Plettenberg A. Remission of a cutaneous Mycosis fungoides after topical 5-ALA sensitisation and photodynamic therapy in a patient with advanced HIV-infection. *Eur J Med Res.* 2002 Nov 25;7(11):477-9.

14. Orenstein A, Haik J, Tamir J, Winkler E, Trau H, Malik Z, Kostenich G Photodynamic therapy of cutaneous lymphoma using 5-aminolevulinic acid topical application *Dermatol Surg.* 2000 Aug;26(8):765-9; discussion 769-70.

15. Allison R, Mang T, Hewson G, Snider W, Dougherty D. Photodynamic therapy for chest wall progression from breast carcinoma is an underutilized treatment modality.*Cancer.* 2001 Jan 1;91(1):1-8.

16. M.F. Renschler, A.R. Yuen, T.J. Panella, T.J. Wieman, S. Dougherty, L. Esserman, M. Panjehpour, S.W. Taber, V.H. Fingar, E. Lowe, J.S. Engel, B. Lum, K.W. Woodburn, W.-F. Cheong and R.A. Miller, Photodynamic therapy trials with lutetium texaphyrin (Lu-Tex) in patients with locally recurrent breast cancer. *Proc SPIE Int Opt Eng* **3247** (1998), pp. 35–39.

17.Vakoulovskaya EG., Shental VV, Buidenok YV, Solomakho GI, Nadezhdina, TM Photodynamic therapy for breast cancer with photosense *Proc. SPIE* Vol. 4248, p. 46-48, Apr 2001

18. Vakoulovskaya EG, Shental_VV, Oumnova LV, Vorozhcsov GN_Photodynamic therapy of breast cancer with photosense *Proc. SPIE* Vol 4952, p. 149-151, Jun 2003

19. Kennedy JC, Pottier RH Endogenous protoporphyrin IX, a clinically useful photosensitizer for photodynamic therapy. *J Photochem Photobiol B.* 1992 Jul 30;14(4):275-92.

20.Bassi P. G (Bacillus of Calmette Guerin) therapy of high-risk superficial bladder cancer. *Surg Oncol.* 2002 Jun;11(1-2):77-83.

21. Kriegmair M, Waidelich R, Lumper W, Ehsan A, Baumgartner R, Hofstetter A. Integral photodynamic treatment of refractory superficial bladder cancer. *J Urol*. 1995 Oct;154(4):1339-41.

22. Waidelich R, Beyer W, Knuchel R, Stepp H, Baumgartner R, Schroder J, Hofstetter A, Kriegmair M. Whole bladder photodynamic therapy with 5-aminolevulinic acid using a white light source. *Urology*. 2003 Feb;61(2):332-7.

23. Nseyo UO, DeHaven J, Dougherty TJ, Potter WR, Merrill DL, Lundahl SL, Lamm DL.Photodynamic therapy (PDT) in the treatment of patients with resistant superficial bladder cancer: a long-term experience. *J Clin Laser Med Surg*. 1998 Feb;16(1):61-8.

24. Manyak MJ, Ogan K Photodynamic therapy for refractory superficial bladder cancer: long-term clinical outcomes of single treatment using intravesical diffusion medium *J Endourol*. 2003 Oct;17(8):633-9.

25. Kato H, Horai T, Furuse K, Fukuoka M, Suzuki S, Hiki Y, Ito Y, Mimura S, Tenjin Y, Hisazumi H, et al. Photodynamic therapy for cancers: a clinical trial of porfimer sodium in Japan *Jpn J Cancer Res*. 1993 Nov;84(11):1209-14.

26. Mimura S, Narahara H, Uehara H, Otani T, Okuda S. [Photodynamic therapy for gastric cancer] *Gan To Kagaku Ryoho*. 1996 Jan;23(1):41-6.

27. Endo C, Sagawa M, Sato M, Fujimura S. [Strategy of treatment for roentgenographically occult lung cancer] *Nippon Rinsho*. 2000 May;58(5):1112-6

28. Kato H, Furukawa K, Sato M, Okunaka T, Kusunoki Y, Kawahara M, Fukuoka M, Miyazawa T, Yana T, Matsui K, Shiraishi T, Horinouchi H. Phase II clinical study of photodynamic therapy using mono-L-aspartyl chlorin e6 and diode laser for early superficial squamous cell carcinoma of the lung. *Lung Cancer*. 2003 Oct;42(1):103-11.

29. Cortese DA, Edell ES, Kinsey JH. Photodynamic therapy for early stage squamous cell carcinoma of the lung. *Mayo Clin Proc*. 1997 Jul;72(7):595-602.

30.Patelli M, Lazzari Agli L, Poletti V, Falcone F. Photodynamic laser therapy for the treatment of early-stage bronchogenic carcinoma. *Monaldi Arch Chest Dis*. 1999 Aug;54(4):315-8.

31. Wenig BL, Kurtzman DM, Grossweiner LI, Mafee MF, Harris DM, Lobraico RV, Prycz RA, Appelbaum EL. Photodynamic therapy in the treatment of squamous cell carcinoma of the head and neck. *Arch Otolaryngol Head Neck Surg*. 1990 Nov;116(11):1267-70.

32. Feyh J. Photodynamic treatment for cancers of the head and neck. *J Photochem Photobiol B.* 1996 Nov;36(2):175-7.

33. Biel MA. Photodynamic therapy and the treatment of head and neck cancers. *J Clin Laser Med Surg.* 1996 Oct;14(5):239-44.

34. Schweitzer VG. PHOTOFRIN-mediated photodynamic therapy for treatment of early stage oral cavity and laryngeal malignancies. *Lasers Surg Med.* 2001;29(4):305-13.

35. Lele SB, Piver MS, Mang TS, Dougherty TJ, Tomczak MJ. Photodynamic therapy in gynecologic malignancies. *Gynecol Oncol.* 1989 Sep;34(3):350-2.

36. Lobraico RV, Grossweiner LI. Clinical experiences with photodynamic therapy for recurrent malignancies of the lower female genital tract. *J Gynecol Surg.* 1993 Spring;9(1):29-34.

37. Schmidt S. Antibody-targeted photodynamic therapy. *Hybridoma.* 1993 Oct;12(5):539-41.

38. Hillemanns P, Untch M, Dannecker C, Baumgartner R, Stepp H, Diebold J, Weingandt H, Prove F, Korell M. Photodynamic therapy of vulvar intraepithelial neoplasia using 5-aminolevulinic acid. *Int J Cancer.* 2000 Mar 1;85(5):649-53.

39. Wyss P, Caduff R, Tadir Y, Degen A, Wagnieres G, Schwarz V, Haller U, Fehr M. Photodynamic endometrial ablation: morphological study. *Lasers Surg_Med.* 2003;32(4):305-9.

40. Muller PJ, Wilson BC. Photodynamic therapy for malignant newly diagnosed supratentorial gliomas *J Clin Laser Med Surg.* 1996 Oct;14(5):263-70.

41.Radford IR. Gd-Tex Pharmacyclics Inc. *Curr Opin Investig Drugs.* 2000 Dec;1(4):524-8.

# CHAPTER 12

# PHOTOTHERAPY OF SKIN DISEASE: SCIENCE AND TECHNOLOGY

**Co-Author: James B. Grossweiner, M. D.**

UV radiations are used for phototherapy of skin diseases, either alone or in conjunction with a psoralen drug. Unlike the visible light used for PDT, the activating UV has both therapeutic and damaging effects on skin. The healing action of sunlight on skin has been known since ancient times. Although modern phototherapy relies on controlled artificial light sources, direct sunlight is still used in underdeveloped regions. Exposure of human skin to sunlight induces an extraordinarily large number of physiological responses as indicated by the partial listing in Table 12.1. The key role of wavelength is a consequence of the fundamental law of photobiology, which states that light must be absorbed in order to initiate a physical, chemical, or biological effect. Typical properties of some UV sources are indicated in Table 12.2. The ultraviolet spectrum is divided up into three ranges based on wavelength: UV-A (320-400 nm), UV-B (290-320 nm), and UV-C (200-290 nm). The UV-A spectrum has been subsequently divided up into UVA-1 (340-400 nm) and UVA-2 (320-340 nm). UV-C is potentially carcinogenic because DNA absorbs it. However, virtually no UV-C in sunlight reaches the surface of the earth owing to blocking by the ozone layer in the stratosphere and the earth's atmosphere. Human hazards from UV-C originate from unfiltered arc lamps and UV lasers. Sunlight is the most common source of UV-B and UV-A. UV-B has the capability of damaging normal skin and healing diseased skin. Induction of skin cancer is a major UV-B hazard. UV-A is less strongly absorbed in skin than UV-B. For a light skinned human, 100% of UV-C at 250 nm is absorbed in the outer layer of the skin, the epidermis. Approximately 5% of UV-B at 300 nm reaches the top of the second layer, namely the papillary dermis. On the other hand, about 50% of UV-A at 350 nm penetrates into the dermis. This deeper penetration into the skin helps make UV-A more effective in treating thicker skin lesions. Sunscreens effectively block UV-B induced erythema and sunburn. Many sunscreens also have partial UV-A protection. There are concerns that since sunscreens minimize sunburns but do not completely block UV-A, sunscreen users may actually have a higher incidence of skin cancer due to longer sun exposure times. This increased exposure to

**Table 12.1 Some UV effects of skin**

| Wavelength Band | Photosensitizer | Skin Response: Adverse | Non-Adverse | Therapeutic |
|---|---|---|---|---|
| UV-C | none | erythemna<br>skin cancer | melanogenesis (weak) | |
| UV-B | none | inflamation<br>immunosuppression<br>skin aging<br>polymorphous light eruption<br>skin cancer | tanning<br>vitamin D synthesis | psoriasis<br>eczema<br>pruritus<br>parapsoriasis<br>others |
| UV-B | sulfanilamide | photoallergy | | |
| UV-A | none | erythema<br>skin aging<br>skin cancer (?) | pigmentation | |
| UV-A | many drugs and chemicals | phototoxicity<br>photoallergy<br>induced porphyrias | tanning | |
| UV-A | psoralens | erythema (strong)<br>inflammation<br>phototoxicity<br>skin cancer | tanning | vitiligo<br>psoriasis<br>eczema<br>solar urticaria<br>graft-vs-host disease<br>mycosis fungoides |
| UV-A | endogenous porphyrins | porphyrias | | |

**Table 12.2 Spectral properties and MED for some optical radiation sources[a]**

| Radiance Source[b] | Distance to Skin (cm) | Irradiance (μW/cm$^2$ UV-C | UV-B | UV-A | VIS | MED(s) |
|---|---|---|---|---|---|---|
| F40 T12 daylight (5)[c] | 30 | 0 | 1 | 10 | 300 | — |
| F40 T12 sunlamp (6)[c] | 50 | 2 | 500 | 750 | 50 | 30 |
| F40 t12 PUVA (40)[c] | 20 | 0 | 10 | 10,000 | 600 | — |
| G15T8 germicidal (3)[d] | 50 | 200 | 3 | 5 | 10 | 30 |
| Hanovia Alpine[e] | 100 | 30 | 50 | 130 | 350 | 90 |
| 800 w xenon arc | 50 | 20 | 40 | 130 | 1,000 | 10 |
| Sunlight 20°f | | 0 | 250 | 4,500 | 50,000 | 1,000 |

[a]Abstracted from reference [1]; [b]Number of lamps in a planer array in parenthesis; [c]Fluorescent lamp; [d]Low pressure mercury arc; [e]Medium pressure "hot quartz" mercury arc; [f]Includes skylight.

UV-A light may increase skin cancer risk thru immunosuppression.[2] However, there is no good evidence that sunscreens themselves are carcinogenic. The *minimal erythema dose* (MED) is an indicator of individual sensitivity to ultraviolet light. This makes the MED useful for determining a starting light dosage for phototherapy. It is performed using a template of squares placed on the skin. These are exposed to incrementally higher doses of light. The MED is defined as the lowest exposure dose that results in detectable erythema with a sharply defined edge after 24 hours. Typical values of the MED for fair skinned individuals (skin types I and II) are 25 mJ/cm$^2$ for isolated UV-B and 30 J/cm$^2$ for isolated UV-A.[3] Overall, the entire UV-A band in sunlight contributes about 25% to MED. The damaging effects of UV-A are amplified many-fold in the presence of certain chemical agents, including natural pigments, clinical drugs, industrial products, cosmetics, and fragrances. Abnormal responses are classified either as phototoxicity or photoallergy depending on the pathophysiology. The available information about photosensitized toxicity is suggestive of *Type 2* reactions. The action mechanism in photoallergy is complicated and has been the subject of much research.

Photosensitization of skin to UV-A by psoralen derivatives is the basis of PUVA therapy. The PUVA concept originated in antiquity from the sunlight treatment of vitiligo with topical and ingested plant extracts containing psoralens. This history and subsequent research on photosensitization of nucleic acids by psoralens led to the modern concept of PUVA photochemotherapy. The progression from sunlight treatment of vitiligo to controlled PUVA therapy of psoriasis is a major accomplishment. The anti-psoriatic mechanism of PUVA has been investigated for more than four decades with only partial success. The lack of an animal model for psoriasis has been a major limitation. In the past few years, however, mouse models have been developed that may aid research into psoriasis.[4,5] *Extracorporeal photopheresis* (ECP) is an implementation of PUVA for treatment of cutaneous T-cell lymphomas (CTCL), autoimmune diseases, graft vs. host disease and prevention of transplant rejection. Instead of delivering the therapeutic light through the skin, some diseased lymphocytes are irradiated with UV-A ex *vivo* after being exposed to externally delivered 8-methoxypsoralen. The irradiated lymphocytes are then re-infused into the patient. The photosensitized T cells stimulate an enhanced immune response that has not been fully explained.

## 12.1 UV-B PHOTOTHERAPY

The acute responses of normal skin to a single UV-B exposure include erythema, hyperpigmentation, and skin thickening. The delayed phase of erythema appears after about 3-5 hours, reaches a maximum intensity between 12 and 24 hours, and fades after 72 hours. The minimal erythema dose MED depends on the light source, (spectral irradiance, collimation), factors related to the skin (skin type, age, previous light exposure, anatomic site, others) and physical parameters (elapsed time, angle of light incidence, the environment, others). UV-C at 254 nm is approximately 2-fold more erythemogenic than UV-B. UV-A at 365 nm is about 1000-fold less erythemogenic than UV-B. The MED for UV-B induced erythema depends on the product of the irradiance and the exposure time for a very wide range of the variables. This type of "reciprocity" is characteristic of a photochemical mechanism as opposed to a thermal process. The erythemal action spectrum is determined by plotting the reciprocal of the MED against wavelength. Measurements show decreasing erythemal effectiveness from 100% at 250 nm to 5% at 310 nm with a plateau from 280 to 290 nm (Fig 12.1). Scaling the erythema action spectrum to the spectral irradiance distribution of sunlight at the surface of the earth leads to an effective erythema spectrum for sunlight that peaks near 305 nm. Other light sources have different effective erythema spectra.

The initial erythemal reactions take place in the epidermis, where light absorbed by keratinocytes liberates transcellular vasoactive substances that diffuse into the dermis. The specific erythemogenic chromophores and the nature of the diffusing substances have not been completely identified although recent evidence suggests DNA is the chromophore.[6] The tanning reaction takes place in the epidermis. A very fast and short-lived tanning called immediate pigment darkening (IPD) begins during UV-A exposure and fades rapidly. This response may result from photochemical oxidation of existing melanin and melanosome redistribution.[7]

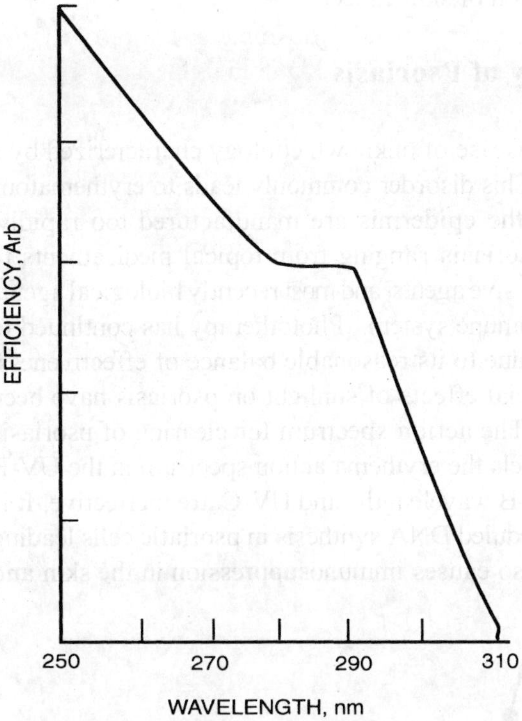


**Figure 12.1** Erythemal action spectrum for sunlight. Adapted from Hawk, J. L. M. and Parrish, J. A. *The Science of Photomedicine,* Plenum Press, New York., 1982. With permission.

Delayed tanning (DT) develops over a period of 3 days after UV-B exposure and may persist for weeks or months. DT requires UV-B dose levels above about 1.5 MEDs. The cytology of DT has been studied extensively. The process is initiated by light activation of melanocytes, followed by tyrosinase synthesis, melanosome formation, and transfer of melanosomes to keratinocytes. Skin thickening occurs 1-3 weeks after UV-B exposure. Evidence suggests that in fair to medium skin toned individuals, melanogenesis provides more photoprotection than stratum corneum thickening.[8] Langerhans cells, dermal dendrocytes, and macrophages mediate the immunological responses of skin to UV-B. It is theorized that, in addition to the direct effects on keratinocytes in the epidermis, UV-B damage to Langerhans cells and induction of suppressor T cells may cause disturbances in effective immune reactivity and immunosurveillance leading to skin cancer. UV-B can also cause the release of various immunosuppressive cytokines, such as Interleukin-10[9] and tumor necrosis factor (TNF).[10] More recently, mutations in tumor suppressor genes, such as p53 and PTCH among others, are considered to play a primary role in the generation of skin cancer.

### 12.1.1 UV-B Phototherapy of Psoriasis

Psoriasis is a chronic skin disease of unknown etiology characterized by a high rate of epidermal turnover. This disorder commonly leads to erythematous scaling plaques as the layers of the epidermis are manufactured too rapidly. Numerous treatments exist for psoriasis ranging from topical medicaments to systemic retinoids, immunosuppressive agents, and most recently biological agents that target specific steps of the immune system. Phototherapy has continued to have a place in psoriasis therapy due to its reasonable balance of effectiveness, cost and side effects. The beneficial effects of sunlight on psoriasis have been known for more than a century. The action spectrum for clearing of psoriasis begins near 295 nm[11] which parallels the erythema action spectrum in the UV-B region (Figure 12.2). Shorter UV-B wavelengths and UV-C are ineffective. It is postulated that UV-B inhibits scheduled DNA synthesis in psoriatic cells leading to more normal cell kinetics. It also causes immunosuppression in the skin and

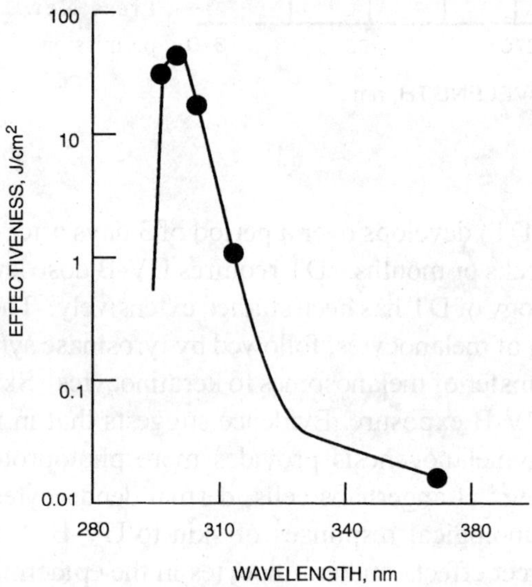


**Figure 12.2** Action spectrum for clearing psoriasis by UV radiations. The solid line is the action spectrum based on several studies. The closed circles are the reciprocal of the lowest daily dose that clears psoriasis. Adapted from Parish, J. A., *The Science of Photomedicine*, Plenum Press, New York, 1982. With permission.

induces apoptosis (which is a form of programmed cell suicide). The difference between the significant erythema response of normal skin below 295 nm and the absence of antipsoriatic action may result from increased light absorption and scattering by the abnormal stratum corneum and Malpighian layer. The dose of UV-B required for control of psoriasis above 295 nm is comparable to the MED for any light source. Average values of MED for UV-B according to skin type are given in Table 12.3. Erythemogenic UV-B phototherapy regimens start at 1 MED. Subsequent exposure doses are incrementally increased to maintain erythema. Typically, psoriasis clearing requires 25-30 treatments given 3-5 times a week. The remission induced by UV-B persists for about 9 weeks on the average.[12] Topical coal tar was combined with UV-B irradiation in the classical

**Table 12.3 MED for UV-B according to skin type**

| Skin Type | Criteria[a] | MED ($mJ/cm^2$)[b] |
|---|---|---|
| I | Always burn, never tan | 20-30 |
| II | Always burn, sometimes tan | 25-35 |
| III | Sometimes burn, always tan | 35-50 |
| IV | Never burn, always tan | 45-60 |
| V | Moderately pigmented | 60-100 |
| VI | Heavily pigmented | 100-120 |

[a]Reaction to first sun exposure of 3 MEDs; [b]Adapted from Coopman, S. A. and Stern, R. S., Phototherapy with ultraviolet B, in *Clinical Photomedicine*, Lim, H.W., and Soter, N.A. Eds., Marcel Dekker, New York, 1993, chap. 18. The entire field of cutaneous photomedicine is reviewed in this multi-author text, including the effects of light on normal skin, photosensitivity diseases, photoprotection, and phototherapy.

Goeckerman regimen originating in 1925. The procedure requires the application of crude coal tar for 24 hours, its removal with mineral oil, irradiation with a "hot quartz" mercury arc, and soaking in warm water with soap. Tar is known to have anti-psoriatic properties by decreasing DNA synthesis along with thinning of the epidermis. Presently, the Goeckerman regimen is not frequently performed, particularly since combining tar with UV-B only gives a marginal benefit at best.[13] Other topical adjunctive agents have been employed with UV-B to decrease the total dose required for clearing and prolong the symptom-free period, including inert lubricants, corticosteroids, and coal tar distillates. There is some evidence that the addition of anthralin, calcipotriene (a vitamin D3 analogue), or tazarotene (a retinoid prodrug) may improve the efficacy of UV-B treatments.[14,15,16,17] The potential long-term adverse reactions to UV-B phototherapy include accelerated skin aging and increased risk of non-melanoma skin cancer. UV-B has been used to treat other skin disorders including eczema, uremic pruritus, and several other inflammatory skin conditions such as pityriasis rosea[18] and pityriasis lichenoides.[19]

## 12.1.2 UV-B Phototherapy Light Sources

UV-B phototherapy is carried out with medium-pressure "hot quartz" mercury arcs (therapeutic sunlamp) and UV-B fluorescent lamps. A hot quartz lamp emits 10% of the power input as UV and 7% as visible light. The UV output is confined to the broadened mercury emission lines, with 37% in UV-B, 29% in UV- C, and 34% in UV-A. Approximately one-third to one-half of a human adult body is illuminated by a hot quartz lamp at a distance of 45 cm. The effective irradiance of a powerful 400-W therapy unit is about 0.4 mW/cm$^2$. The initial MED in a fair-skinned individual is attained after a minute at this distance. In comparison, the overhead sun requires about 15 minutes exposure to achieve 1 MED (This assumes a latitude of 30 degrees north at sea level.)[20] The irradiance of a hot quartz lamp varies with distance according to the "inverse-square" law. Hot quartz mercury lamps must be ventilated to remove heat and ozone, and the patient must wear UV protective goggles.

---

**Example 40** UV-B phototherapy with a hot quartz mercury arc

Estimate the initial exposure time required to achieve 1 MED with a hot quartz mercury arc for a patient with Type III skin at a distance of 20 cm. Assume the MED for Type I skin is 25 mJ/cm$^2$ and the lamp output at 45 cm distance is 0.4 mW/cm$^2$.

The approximate lamp output at 20 cm distance is: $(45/20)^2 \times 0.4 = 2.0$ mW/cm$^2$. The average MED is 40 mJ/cm$^2$ from Table 12.3. The exposure time is: $40/2 = 20$ s.

---

Metal halide lamps are an improved type of mercury arc lamp. The multiple emission lines of the metal atoms "fill in" the mercury lines providing a more uniform spectral distribution. Metal halide lamps can also be used as sources of UV-A with the addition of a filter to remove UV-C and UV-B. More uniform

UV-B irradiations are achieved with banks of fluorescent lamps. A typical UV-B fluorescent "sunlamp" with a thallium-activated calcium phosphate phosphor produces a single broad emission band from 270 to 390 nm peaking near 313 nm. A soda-lime glass envelope is used to filter out UV-C. The emission is approximately 60% UV-B and 40% UV-A. A typical 4 ft, 40-W FS40 sunlamp generates 4 W of UV-B. The actual output depends on the ambient temperature and the ballast. The erythemal efficiency of a bank of FS sunlamps is comparable to a hot quartz lamp at 50 cm. The advantages of fluorescent lamps over hot quartz include more uniform exposures of large treatment fields, faster startup, and less heat production. However, fluorescent lamps have a marked intensity decrease at the ends. This may be partially offset by having the patient stand on a stool or other support to elevate the lower legs which are often relatively more resistant to UV therapy. In the U.S. the Federal Bureau of Radiation Control sets standards for the spectral distributions of lamps to be used for UV-B phototherapy and PUVA therapy. Some fluorescent lamp designs employ phosphors which are claimed to be less erythemogenic than conventional broadband UV-B lamps. These include a "narrow-band" fluorescent UV-B lamp peaking at 310-313 nm, and UV-A/UV-B lamps for selective ultraviolet light phototherapy (SUP), which was popular in Europe. The treatment times are much longer for SUP compared to the conventional therapy. Narrow band UV-B (NB UV-B) has quickly grown in popularity. It has the advantage over broadband UV-B (BB UV-B) of emphasizing the wavelengths of UV-B most effective in clearing psoriasis. At the same time it minimizes shorter UV-B wavelengths that are highly erythemogenic but not anti-psoriatic. Due to the decreased amount of spectral energy produced by NB UV-B bulbs, increased numbers of fluorescent bulbs are used to help shorten treatment times.[21] Erythema from NB UV-B is longer lasting and more intense than BB UV-B.[22] Hence, suberythemogenic NB UV-B treatment protocols have been developed to minimize erythema while maintaining significantly more effectiveness than BB UV-B.[23] Topical combination therapies, such as with anthralin[24] and tazarotene,[25] are also being employed to help increase effectiveness and minimize ultraviolet exposure. NB UV-B is also being used for treatment of vitiligo,[26] atopic dermatitis,[27] lichen planus,[28] and cutaneous T-cell lymphoma.[29] Finally, spot treatment with 308-nm excimer laser has recently been used as an intense UVB light source for therapy of localized psoriatic plaques. It has the advantage of avoiding ultraviolet exposure to clinically uninvolved skin. At the same time, high UVB doses can be given to affected areas leading to fewer treatment sessions than with fluorescent light sources.[30,31]

**Case Study:**

A 48-year-old male presents with an exacerbation of his longstanding psoriasis. He is on no medications, has no history of photosensitivity, and is otherwise healthy. He states his previous treatments have included various topical medications including corticosteroids, calcipotriene, and tar. These medications initially helped, but their effect has diminished especially as the fall weather has approached. On examination there are erythematous well-defined scaling plaques over his scalp, trunk, and extremities. These plaques are thin and cover approximately 30% of his body. A clinical diagnosis of psoriasis vulgaris is made. Narrowband UVB treatment is suggested. He has type III skin, i.e. rarely burns and usually tans. The risks of ultraviolet therapy are discussed at length. He is given a pair of UV protective goggles, and the necessity for garment coverage of the genital area is emphasized due to its increased risk for UV induced squamous cell carcinomas. An MED is performed and is 400 $mJ/cm^2$. He is scheduled to receive treatments on Mondays, Wednesdays, and Fridays. Prior to therapy he applies mineral oil to his plaques of psoriasis and a high SPF sunscreen to his face and dorsal hands. The first treatment given is 200 $mJ/cm^2$. Subsequent treatments are increased by up to 15% depending on whether erythema or burning developed from the prior treatment. After 18 exposures his psoriasis has cleared except for a few plaques on his lower legs. He is satisfied with his progress and is switched to a maintenance regimen for two months in which he receives a once weekly treatment at the final UVB dose from the clearing phase. By the time his psoriasis starts to flare again it is summertime, and he maintains the plaques topically until the following fall.

## 12.2 UV-A PLUS PSORALEN PHOTOCHEMOTHERAPY

PUVA therapy is indicated for symptomatic control of severe psoriasis not responsive to other forms of therapy. The treatment objectives are to control the disease to less than 5% of the original extent, minimize the total UV-A dose, and reduce the risk of long-term adverse effects. Contraindications for PUVA include a history of photosensitivity diseases, such as lupus erythematosus, xeroderma pigmentosum, and albinism; ophthalmological disorders including cataracts and aphakia; and a history of melanoma or invasive squamous cell carcinoma. Caution and/or close follow-up must be used on patients with a history of basal cell carcinoma, severe photodamage, radiation therapy, arsenic therapy, hepatic or renal insufficiency, and cardiac disease. Caution must also be used if there is

a history of pemphigus, pemphigoid, or immunosuppression. Also, due to the need for prolonged standing and the high temperatures in the phototherapy unit (which can exceed 100 degrees F), any difficulty tolerating these conditions must be considered. PUVA is contraindicated during pregnancy and in nursing mothers.[32] Crystalline 8-MOP (methoxsalen) was employed for the original clinical trials of PUVA. The drug is administered orally in 10 mg capsules according to body weight (40 mg for 66-80 kg) and the light treatment is given 2-3 hours later. The light source consists of banks of special PUVA fluorescent lamps. The dosimetry parameters and results for the multicenter U.S. clinical trial on 1308 patients are summarized in Table 12.4.[33] Patients were treated two or three times per week using a variable light dose schedule scaled to skin type, accumulated dose, and the intermediate responses. The overall clearance rate was 88% in this trial and 3% of the patients failed to improve. In a 1981 European trial for 3175 patients, 90 - 100% clearance was reported for 89% of the patients and failure rate was 2.6%. A typical example of PUVA therapy is shown below.

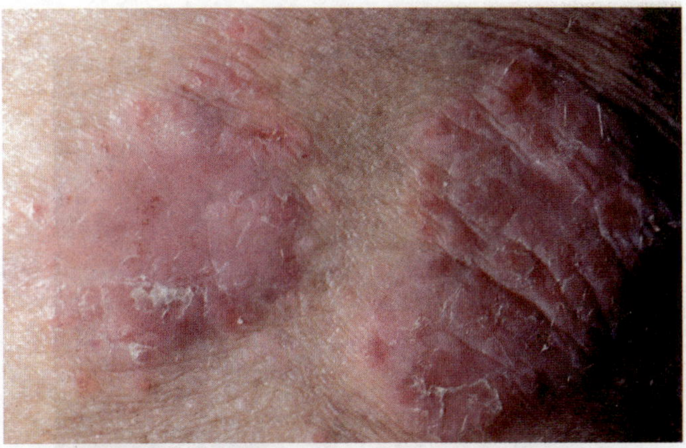

**Before PUVA Therapy (outer thigh skin)**

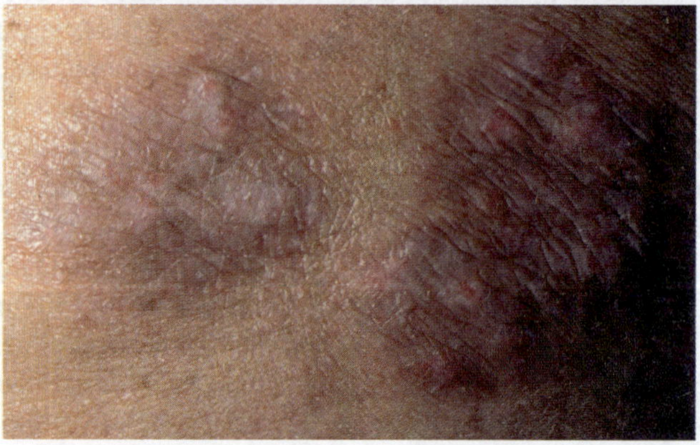

**After 10 PUVA Treatments**

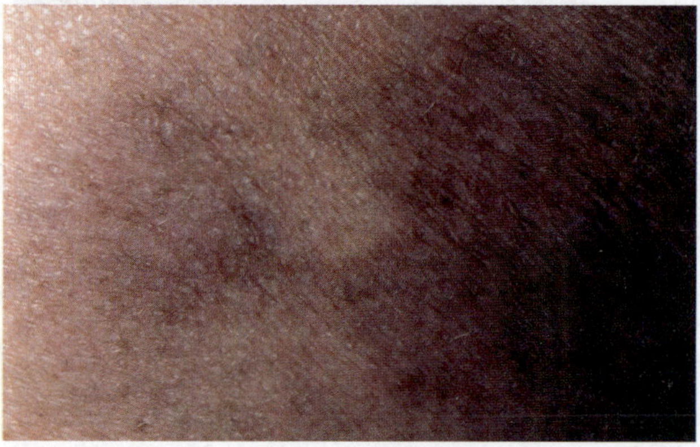

**After 23 PUVA**
**Treatments. (Note that typically the remaining hyper-pigmentation will gradually resolveover several months.)**

The skin should be protected from direct and indirect sunlight for at least 8 hours after ingestion of 8-MOP. The eyes must be completely shielded from UV-A light after ingestion of a psoralen drug and for the next 24 hours with UV protective wraparound goggles. Eye protection is required to prevent irreversible binding of 8-MOP to protein and DNA constituents of the lens which can photosensitize cataract formation. Sun exposure is undesirable on non-treatment days because pigmentation interferes with the therapy. An encapsulated liquid preparation of 8-MOP is available as Oxsoralen-Ultra®. The liquid preparation attains higher serum levels than crystalline 8-MOP in a shorter time and is effective at lower UV-A dosage levels. On the other hand, nausea and pruritus are increased with this formulation which might require a dosage reduction.[34] The recommended starting light dose levels are given in Table 12.4. An increasing schedule of light dose is employed during the clearing phase of PUVA according to skin type and response. A delayed erythemal response to PUVA usually appears 24-48 hours after exposure, peaks at 48-72 hours and persists for 7-14 days. Erythema is graded on a scale of 1-4, where 1 corresponds to minimally perceptible erythema and 4 corresponds to fiery erythema with edema and blistering. The erythemal dose-response curve for PUVA is much steeper than erythema induced by UV-B. A 2-fold increase of UV-A dose can increase the response from the minimum

**Table 12.4 Light dosimetry for PUVA therapy of psoraisis with 8-methoxypsoralen**

| Skin Type | First Light Dose[a] ($J/cm^2$) | Mean Clearing Light Dose[b] ($J/cm^2$) | Mean Total Light Dose[b] ($J/cm^2$) | Mean No. of Treatments[b] |
|---|---|---|---|---|
| I | 1.5 (0.5) | 10 | 190 | 24 |
| II | 2.5 (1.0) | 12 | 190 | 22 |
| III | 3.5 (1.5) | 14 | 250 | 24 |
| IV | 4.5 (2.0) | 16 | 300 | 24 |
| V | 5.5 (2.5) | 19 | 410 | 16 |
| VI | 6.5 (3.0) | 21 | 470 | 10 |

[a] The summary results in this table for crystalline methoxsalen were abstracted from reference[33]. The values of initial light dose in parentheses are recommended for the liquid formulation, Oxsoralen-Ultra; [b]Rounded to two significant figures.

phototoxic dose or MPD (that is, the dose of UV-A causing minimum erythema 48 hours later) to marked erythema with edema. By contrast, the equivalent responses for UV-B erythema would require a 5-fold dose increase. Accordingly, UV-A dose must be carefully monitored for each patient during the course of the treatment. The first change observed in the psoriatic lesions after PUVA therapy is a decrease in the amount of scaling, followed by flattening of the plaques, fading of erythema, and hyperpigmentation. The response to the therapy is graded on a scale of -1 (psoriasis worse) to 4 (clearing, complete flattening of plaques including borders). Pigmentation induced by PUVA is deeper and more persistent than UV-B tanning and lasts from several weeks to months. When control is achieved, maintenance treatments are given once per week, and then reduced to every other week, every 3 weeks, and every 4 weeks as tolerated. The short-term adverse effects of PUVA may include nausea and other reactions to the drug, pruritus, and phototoxic erythema. Photoaging, generalized PUVA lentigines (freckles), and a higher risk of non-melanoma skin cancer are long-term adverse effects. Squamous cell carcinoma risk, in particular, remains significantly elevated for many years after discontinuation of PUVA.[35] Also, a long term increased risk for melanoma from PUVA therapy has been found in a large cohort of patients followed since 1975 in the United States. This increased risk is found in fair skinned patients, especially those receiving numerous treatments. However, even those with a smaller overall exposure had an elevated risk.[36] On the other hand, a Swedish cohort of PUVA patients has not had an elevated risk of melanoma after more than 15 years of follow-up.[37] Skin cancer induction is particularly prevalent for patients with prior treatment with ionizing radiation, inorganic trivalent arsenic, and topical nitrogen mustards. It is not unusual for a patient with cutaneous T-Cell lymphoma to be treated with topical nitrogen mustard, PUVA, and electron beam therapy at various stages of their disease. Caution and close monitoring for skin cancers should be performed if PUVA is used in these situations.[38] PUVA has also been combined with systemic retinoids, such as acitretin, in a protocol called RePUVA.[39] This regimen is employed to decrease the dosages of both the retinoid and PUVA. It is hoped that reduced dosages will minimize the side effects of both the phototherapy and the retinoid. Retinoids and PUVA are also therapeutically synergistic, making RePUVA an option for resistant cases. In RePUVA, the retinoid is typically started 2 weeks prior to initiation of PUVA. As retinoids are very effective in clearing pustular psoriasis, RePUVA is often employed to treat the potentially life threatening generalized variant of pustular psoriasis called Von Zumbusch. Retinoids have also been combined with UV-B therapy for similar reasons also with good therapeutic results.

**Case Study**:

A 28 year old female with a history of psoriasis since age 8 had previously been controlled with UVB phototherapy. However, over the past several months her psoriasis has changed character with thicker and more widespread plaques covering approximately 40% of her body. On full skin examination, there are no lesions suspicious for skin cancer or atypia. She has no other medical problems and is not pregnant or nursing. She takes no medications for other reasons. PUVA is agreed upon after careful discussion of both short term and long term side effects, such as itching, UV burn risk, and carcinogenicity. Prior to beginning PUVA she has a baseline ophthalmological examination along with blood tests for liver function, antinuclear antibody, and a complete blood count. She is given both clear and tinted wrap-around UV protecting glasses. These will be used both indoors and outdoors for 24 hours after ingestion of methoxsalen (Oxsoralen-Ultra®). She is also instructed to avoid sunlight exposure on treatment days. Based on her weight of 115 pounds, she is prescribed a dosage of 30 milligrams of methoxsalen to be taken 1.5 hours before each treatment (115 lbs/2.2=52.3 kilograms, 52 x 0.6=31.2 milligram dose which is rounded to 30 milligrams). Treatments are scheduled on Mondays, Wednesdays, and Fridays. On her first phototherapy session she is instructed on use of the PUVA booth and is given a pair of tight fitting UV blocking goggles to be worn during the treatment. As she has type II skin, i.e. always burns, but sometimes tans, her initial dose of UVA is 1.0 J/cm$^2$. At each subsequent session her dosage is increased by 1.0 J/cm$^2$ as long as there is no erythema. After 6 sessions she begins to see improvement with decreased scaling of her plaques and mildly decreased thickness. After 20 sessions, all plaques except those on her lower legs have almost completely resolved leaving mild hyperpigmentation. There is a generalized bronze color of unaffected skin. At this point she is receiving 20 J/cm$^2$. She is now switched to once weekly PUVA treatments of 14 J/cm$^2$. Over the next 3 months the frequency of her treatments are gradually weaned from weekly to every 3 weeks.

### 12.2.1 PUVA Therapy of Other Cutaneous Diseases

Repigmentation of vitiligo was the earliest observed effect of sunlight and topical psoralen extracts. In clinical procedures, PUVA therapy with oral 8-MOP induced repigmentation in approximately 70% of vitiligo patients after 100-200 exposures. Head and neck lesions were most responsive, the trunk less well, and distal limbs not at all. Oral 4,5'8- trimethylpsoralen (trioxsalen, TMP) has been used in conjunction with PUVA lamps and sunlight exposures. PUVA has been used for numerous other skin diseases. A few of the more common conditions it is used for include mycosis fungoides, polymorphous light eruption, solar urticaria, atopic eczema, chronic cutaneous graft-vs-host disease, and palmoplantar dermatoses. For palmoplantar diseases, hand/foot phototherapy units which only expose these areas are employed. Treatment regimens for the above conditions may vary from those used for psoriasis.

### 12.2.2 PUVA Lamps

The requirements for a PUVA lamp are high output from 320-340 nm, low UV-B, and no UV-C. Fluorescent PUVA lamps generally use a cerium-activated calcium phosphate phosphor with 20% of the UV at 320 - 340 nm, 64% at 320 - 360 nm, and less than 3% UV-B. Commercial "black light" (BL) lamps used to excite fluorescence are not suitable for PUVA. The PUVA lamp output from 320 - 400 nm is approximately 10 mW/cm$^2$ compared to 5 mW/cm$^2$ for an ordinary fluorescent BL lamp and 5 - 6 mW/cm$^2$ for overhead sunlight. The light output drops significantly during the initial startup and PUVA lamps should be "broken-in" for 12 hours prior to use. A bank of PUVA lamps can deliver 30 J/cm$^2$ in 30 min. A phototherapy light box for institutional use typically consists of a walk-in unit with up to 48 bulbs. However, many units will combine UVA and UVB bulbs decreasing the total number of UVA bulbs which will lengthen treatment times accordingly. Safety concerns for these units include control of excessive heat, accurate treatment times, bulb shielding, and proper electrical wiring and grounding. Other recommended safety features are patient handholds, ability for the patient to stop and exit the box themselves, an emergency alarm for the patient to press, and a mechanism for the phototherapist to view the patient in the box, such as a window.

### 12.2.3 Psoralen Photochemistry

The detailed relationship of psoralen photochemistry and specific PUVA endpoints has not been fully explained. The linear furocoumarin derivatives, 8-MOP, 5-MOP, and TMP, are formed by fusion of a 2,3-furan bond and a 6,7-coumarin bond (Figure 12.3). The angular furocoumarin angelicin is formed by fusing the 2,3-furan bond to the 7,8-coumarin bond The absorption spectrum of 8-MOP has bands at 250 nm and 300 nm characteristic of all furocoumarins plus a broad band peaking near 305 nm extending to 400 nm (Figure 12.4). Optical excitation of 8-MOP follows the Jablonski energy level diagram for conjugated aromatic molecules (Figure 4.7). The $S_1$ fluorescent state of 1.9 ns lifetime in water populates a metastable $T_1$ state with approximately 0.06 quantum efficiency. Furocoumarins were originally characterized as "non-photodynamic" photosensitizers which do not require oxygen. In 1975 Poppe

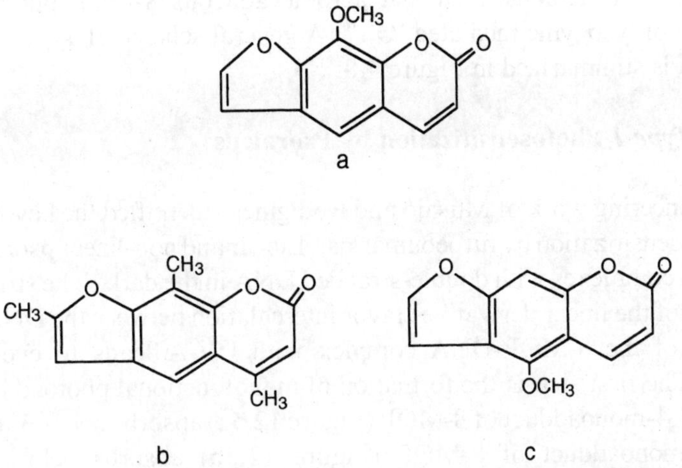


**Figure 12.3** Chemical structure of linear furocoumarin derivatives; (a) 8-methoxypsoralen, (b) 4,5',8-trimethoxypsoralen , (c) 5-methooxypsoralen

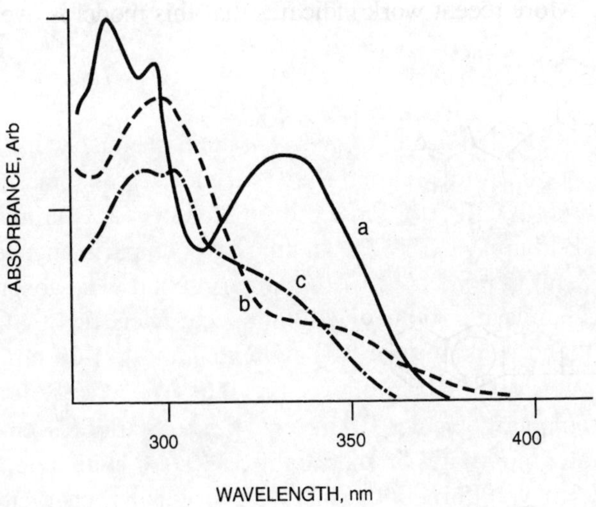


**Figure 12.4** Absorption spectra of psoralen derivitives in carbon tetrachloride; (a) psoralen, (b) 8-methoxypsoralen, (c) angelicin

and Grossweiner demonstrated that aerobic aqueous 8-MOP photosensitizes inactivation of lysozyme mediated $^1O_2$.[40] A general scheme of *Type 1* and *Type 2* pathways is summarized in Figure 4.8.

### 12.2.3.1 *Type 1* Photosensitization by Psoralens

The pioneering work of Musajo and Rodighiero identified the key features of DNA photosensitization by furocoumarins. Linear and non-linear psoralens form equilibrium complexes with double-stranded DNA in the dark. The structure and dimensions of the linear derivatives favor intercalation between the DNA strands. Irradiation of the 8-MOP-DNA complex with UV-A leads to cycloaddition products. The first step is the formation of monofunctional photoadducts. The non-linear 3,4-monoadduct of 8-MOP (Figure 12.5a) absorbs below 330 nm. The linear 4'5'-monoadduct of 8-MOP (Figure 12.5b) absorbs below 380 nm. Protracted irradiation at 360 nm converts the 4'5'-monoadduct to *inter*-strand cross- links (Figure 12.5c), while the 3,4-monoadducts accumulate. Non-linear furocoumarins such as angelicin form, at most, very low yields of cross-links. These early findings led to a proposed action mechanism for PUVA in which cross-links in psoriatic cells block scheduled DNA synthesis and inhibit proliferation. More recent work indicates that this model is oversimplified.

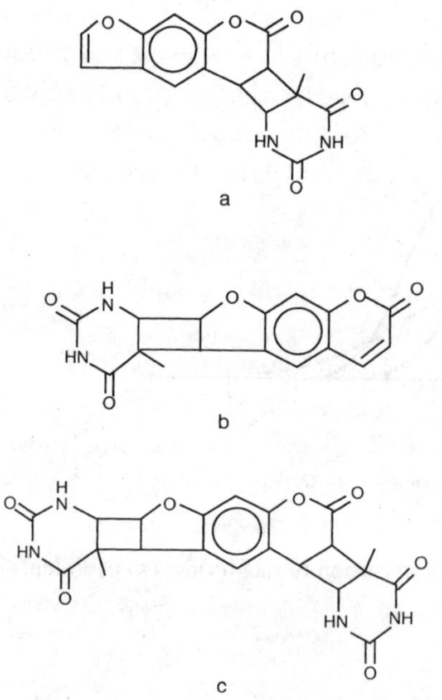


**Figure 12.5** Structure of psoralen photoaddition products with thymine: (a) 3,4-monocycloadduct, (b) 4', 5' -monocycloadduct, (c) 3, 4, 4', -5' cross-link

Monofunctional psoralens are weakly erythemogenic when applied topically and exposed to UV-A. Both monofunctional and bifunctional psoralens can induce hyperpigmentation in the approximate order: TMP > 8-MOP > 5-MOP > 5-methyl-angelicin > 4,5'-dimethyl-angelicin > 3-carbethoxypsoralen > angelicin. Measurements of the anti-psoriatic activity for topical application led to the same order of effectiveness. Lethality and mutation induction by psoralen photoadducts are mediated by repair mechanisms under genetic control. Studies on repair-deficient microorganisms indicate that the monoadducts are removed by error-free excision repair while repair of cross-links requires a slower, error-prone process. Both types of photoadducts may contribute to mutation induction owing to the much larger numbers of monoadducts. More recent studies with mammalian cells suggest that the mutation induction and repair in bacteria and yeast may not be the same as in humans. The photocarcinogenicity of 8-MOP has been the subject of concern. The hypothesis that error-prone repair of cross-links is responsible for cell transformation is supported by early findings that topical and intraperitoneal 3-carbethoxypsoralen (3-CP) are not carcinogenic in mice. However, other work indicates that the mono-functional psoralens, angelicin, 5-methylangelicin, and 4,5'-dimethylangelicin are as carcinogenic as 8-MOP.

### 12.2.3.2 *Type 2* Photosensitization by Psoralens

Generation of $^1O_2$ by aqueous 8-MOP has been confirmed and extended to other psoralens. Table 12.5 summarizes published values of the $^1O_2$ quantum yield and other photophysical properties of some psoralen derivatives. Natural psoralens are sparingly soluble in water which limits the yields of photoadducts with DNA. Synthetic AMT is highly soluble in water. *Type 2* processes mediated by $^1O_2$ are probably involved in the inflammatory responses to PUVA. Observations that topical application of antioxidants prior to PUVA therapy decreased the erythemal sensitivity support this conjecture. Psoralen triplet states are good electron acceptors. In addition to the *Type 1* photoaddition reactions with DNA, reduced psoralen radicals formed by selective photo-oxidation of other substrates may generate superoxide according to Equation (4.14) which can attack cellular targets. Membrane photosensitization by 8-MOP has been shown for liposomes and red blood cells. The photohemolysis reaction may not involve the direct attack of $^1O_2$ on red cell membranes. In one proposed mechanism, $^1O_2$ reacts with 8-MOP to form an oxygen-psoralen complex that reacts with the membrane. High concentrations of 3-CP and other psoralens can induce hemolysis of red cells in the dark. Photoconjugation of 8-MOP has been observed for many proteins and enzymes including epidermal soluble protein. The ability

**Table 12.5 Photophysical properties of psoralen derivatives**[a]

| Derivative[b] | Water Solubility (μg/ml) | $\Phi_{ST}$ [c] | $\Phi_{\Delta}$ [d] |
|---|---|---|---|
| Ps | 37 | 0.45 | 0.18 |
| 5-MOP | 5 | 0.03 | 0.03 |
| 8-MOP | 23 | 0.06 | 0.01 |
| TMP | 1 | – | 0.08 |
| DMA | 8 | – | 0.1 |
| 3-CP | 13 | 0.32 | 0.3 |
| Angelicin | 20 | 0.33 | ≤0.02 |

[a] Data from references[55,56,57,58]. [b] Ps, psoralen; 5-MOP,5-methoxypsoralen (bergapten); 8-MOP, 8-methoxypsoralen;TMP,4.5′8-trimethylpsoralen; DMA, 4.5′-dimethylangelicin; 3-CP, 3-carbethoxypsoralen; [c] Intersystem crossing efficiency in water, [d] singlet oxygen quantum yield in water.

of psoralens to photosensitize enzymes is specific to the given psoralen derivative and enzyme. The ocular hazard in PUVA therapy probably involves the formation of stable photo-conjugates with lens protein which acts as an *in-situ* photosensitizer. Concern has been for subsequent sunlight exposure to generate cataracts. However, a clinical study on 1,235 patients examined an average of 10 years apart could not demonstrate an increased risk for cataract formation with increased PUVA dose although the cohort overall had more cataracts than studies of those without psoriasis.[41]

#### 12.2.3.3 UVA1 Phototherapy

Another form of phototherapy being investigated recently is UVA1 which involves the spectrum from 340-400nm. In comparison to shorter wavelengths, UVA1 penetrates deeper into the skin with diminished erythema. This allows high intensity exposures. The mechanism of action appears to involve triggering apoptosis[42] and upregulating collagenase-1.[43] Due to the deep penetration into the skin, UVA1 is being used for treatment of both superficial and thicker lesions of mycosis fungoides (cutaneous T-cell lymphoma).[44] It is also being tried with other skin conditions including atopic dermatitis, urticaria pigmentosa, scleroderma,[45] and graft versus host disease.[46] Its use is presently limited by high cost of the phototherapy units and only short term knowledge of its side effects.

## 12.2.4 Extracorporeal Photopheresis

Mycosis fungoides (MF) is a form of cutaneous T-cell lymphoma (CTCL) that first manifests in the skin and progresses to involve lymph nodes, bone marrow, and internal organs. This frequently fatal form of lymphoma is uncommon, with an incidence rate of 0.36 per 100,000 people per year.**47** The successful clearing of the early eczematous or plaque stage MF with oral PUVA has been reported. The Sézary syndrome is an advanced stage of CTCL associated with the dissemination of malignant T cells with convoluted nuclei (Sézary cells) in the peripheral blood with a concomitant cutaneous erythroderma that is often disabling. The clinical course of the Sézary form of CTCL invariably progresses to death with a median survival of 30 months regardless of the treatment.[48] The extra-cutaneous stages of the disease are non-responsive to PUVA. Extracorporeal Photopheresis (ECP) is carried out with a bedside device that provides for leukapheresis and exposure of the separated leukocytes to UV-A within the same apparatus. The centrifugation step removes 90-95% of the red blood cells which would absorb the therapeutic UV-A. The maximum final hematocrit is less than 8%. In current procedures, the liquid form of 8-MOP is administered at 0.5 mg/kg 75-90 minutes prior to the procedure. The initial 8-MOP concentration in the blood is 100-200 ng/ml. Approximately 240 ml of buffy coat is mixed with 300 ml of the patient's plasma and 200 ml of saline plus heparin. The cells are pumped through a clear plastic plate between two banks of PUVA lamps for 1-2 hours starting from buffy coat collection. A typical 180 minute irradiation provides an average exposure of 1 - 2 J/cm$^2$. The treated cells and plasma are then reinfused into the patient for an overall treatment time of 3-4 hours.

A multicenter trial was reported in 1987 for 29 patients with exfoliative erythrodermic CTCL resistant to systemic chemotherapy, electron beam radiotherapy, and topical mechlorethamine, plus an additional 8 patients with localized plaque-type disease.[49] ECP was performed on two successive days every four weeks. Twenty-four of 29 erythrodermic patients and 3 of 8 plaque patients had improvement in cutaneous disease after a mean time of 22.4 weeks. Nine patients experienced a marked response with better than a 75% decrease in the extent of their skin lesions. An apparent decrease in peripheral blood involvement was observed, including a decrease of the absolute number of CD4$^+$ T cells and the number of atypical Sezary cells. Median survival of the original erythrodermic patients was 47.9 months from the start of ECP. ECP was well-tolerated by most patients. The more serious adverse reactions were episodes of hypotension during the initial leukapheresis step and the development of a spiking fever 4 to 12 hours after reinfusion of the treated cells in about 10% of the patients. Fever occurred most frequently in patients who had the most significant tumor regression and paralleled the release of tumor necrosis factor (NKF) and interferon gamma from macrophages and T cells, respectively. The best responders to ECP are patients with normal numbers of peripheral blood $CD8^+$ cytolytic and suppressor T cells, a normal CD4/CD8 ratio, minimal lymphadenopathy, and duration shorter than 2 years.[50] Also, patients without evidence of clonal sezary cells in the peripheral blood or with an especially large tumor burden are unlikely to respond as well to ECP. In a more recent application of ECP, 25% of the subjects began to show a response after 3.5 months of monthly treatments (Yale Dermatology, 2000). Adjunctive therapies for patients who are likely to be poor responders include methotrexate, electron beam therapy, and interferon. Total skin electron beam therapy used in addition to ECP was shown to significantly improve survival in a retrospective study of 44 patients.[51] Recently, in a few patients PUVA led to significant response rates when it was added to a combination of ECP, interferon, and bexarotene (a retinoid that selectively activates retinoid X receptors).[52] The photochemical efficiency of ECP is limited by the low solubility of 8-MOP in water. In a modified ECP procedure, a liquid form of 8-MOP (UVADEX®) is inserted directly into the white blood cell collection bag. This procedure improves reproducibility of the light dosimetry and exposure of the patient's system to the drug is very low. Another application for ECP is for prevention of acute rejection of allografts. In a randomized 1998 pilot study one group of 60 patients, recipients of primary cardiac transplants received the standard triple-drug immunosuppressive therapy (cyclosporine, azathioprine, and prednisone) and the other group received the standard therapy plus ECP. (Barr et al., 1998). A total of 24 ECP treatments were given during the first 6 months after transplantation. Twenty-seven of 33 patients in the ECP group had one or no rejection episodes compared to 14 in 27 of the standard therapy group. Finally, ECP has also been used for treatment of graft-versus-host disease and for systemic sclerosis along with other autoimmune disorders.[53]

#### 12.2.4.1 ECP Action Mechanism

The antineoplastic action mechanism of ECP is recently becoming evident. Photochemical alterations of the malignant cells appear to promote the immunological recognition and eradication of the remaining tumor cells in the skin, lymph nodes, spleen, and other sites of T-cell trafficking. It is unlikely that direct T-cell inactivation by psoralen-DNA photoadducts is responsible because only a small fraction of the T cells are treated. This conclusion is consistent with observations that leukapheresis without PUVA must be administered 2 to 3 times per week to produce improvement. Alterations of cell surface determinants may render malignant T cells more immunogenic. The dose-response relationship between damage to chromosomal DNA and lymphocyte membranes in ECP is clarified by work of Gasparro. Photosensitization studies on lymphocytes *in vitro* indicate that nuclear processes (thymidine incorporation, repair of photoadducts) were completely suppressed when the product of the 8-MOP concentration (in ng/ml) and the UV-A dose (in $J/cm^2$) reached about 50. This quantity may be taken as the "ECP dose". At an ECP dose less than 10, very few DNA photoadducts were formed and significant repair was observed. Non-repairable cell membrane damage (loss of dye exclusion) was induced at 300-400 accompanied by the formation of significant yields of photoadducts. The estimated ECP dose in photopheresis is 100-200. This dose level should inhibit nuclear processes while leaving the cell membranes intact but altered. These cells, when reinfused into the patient, could circulate for a sufficient time to induce changes in the immune system. It has recently been discovered that during ECP, monocytes, as they slowly pass and stick to the narrow (1mm) exposure plate are activated to become dendritic cells. The monocytes are relatively resistant to apoptosis. Newly formed immature dendritic cells phagocytize photoinduced apoptotic malignant T-cells which in turn leads to production of cytotoxic CD8 positive anti-tumor T-cells. These cytotoxic T-cells then may attack the remaining malignant T-cells on reinfusion into the patient. Due to the time delay needed for the production of dendritic cells from monocytes and for apoptosis of malignant T-cells, a new process called “transimmunization” has been developed. In transimmunization, a one day delay is added after ECP before reinfusion of the treated cells. This delay allows increased time for development of dendritic and apoptotic T-cells and for their direct exposure to each other. It is hoped that this increased time for formation and exposure will significantly increase the likelihood for production and expression of the cytotoxic antitumor cells. Cytokine release, such as with IL6, TNF$\alpha$, and IL1$\beta$, may also play a role in production of these antitumor cells.[54]

## NOTES

1. Kochevar, I. E., Pathak, M.A., and Parrish, J. A., Photophysics, photochemistry, and photobiology, in *Dermatology in General Medicine,* 3rd ed., Fitzpatrick, T. B., Eisen, A. Z., Wolff, K., Freedberg, I. M., and Austen, K. F., Eds., McGraw-Hill, New York, chap. 124, 1987.

2. Poon, TS et al., Prevention of immunosuppression by sunscreens in humans is unrelated to protection from erythema and dependent on protection from ultraviolet A in the face of constant ultraviolet B protection. *J. Invest Dermatol* Jul; 121(1): 184-90, 2003

3. *Fitzpatrick's Dermatology in General Medicine*, fifth ed., Freedberg, I. W. editor et al., McGraw-Hill, p. 1606, 1999.

4. Raychaudhuri SP, Dutt S, Raychaudhuri SK, Sanyal M, Farber EM. Severe Combined Immunodeficiency Mouse-Human Skin Chimeras: a Unique Animal Model for the Study of Psoriasis and Cutaneous Inflammation, *British Journal of Dermatology* May; 144 (5): 931-9, 2001.

5. Schon M.P. Animal Models of Psoriasis-What Can We Learn from Them? *Journal of Investigative Dermatology*, April; 112 (4): 405-10, 1999.

6. Young AR, Chadwick CA, Harrison GI, Nikaido O, Ramsden J, Potten CS, The Similarity of action spectra for thymine dimers in human epidermis and erythema suggests that DNA is the chromophore for erythema. *J. Invest Dermatol* Dec; 111 (6):982-8, 1998.

7. Routaboul C, Denis A, Vinche A. Immediate pigment darkening: description, kinetic and biological function. *Eur J. Dermatol,* Vol. 9, Issue 2, March 95-9, 1999.

8. Sheehan JM, Potten CS, Young AR. Tanning in human skin types II and III offers modest photoprotection against erythema. *Photochem Photobiol* Oct; 68 (4):588-92, 1998.

9. Niizeki H, Streilein J. W. Hapten-specific tolerance induced by acute, low-dose ultraviolet B radiation of skin is mediated via interleukin-10 *J. Investigative Dermatology* July; 109 (1): 25-30, 1997.

10. Hart PH, Grimbaldeston MA, Swift GJ, Sedgwick JD, Korner H, Finlay-Jones JJ., TNF modulates susceptibility to UVB-induced systemic immunomodulation in mice by effects on dermal mast cell prevalence. *European Journal Immunology* September; 28 (9): 2893-901, 1998.

11. Parrish JA, Jaenicke KF, Action Spectrum for Phototherapy of Psoriasis, *Journal of Investigative Dermatology* May; 76 (5): 359-62, 1981.

12. Larko O, Phototherapy of Psoriasis-Clinical Aspects and Risk Evaluation, *Acta Derm Venereol Suppl (Stockh)*, 103:1-42, 1982.

13. Stern RS, et al., Contribution of topical tar oil to ultraviolet B phototherapy for psoriasis, *J Am Acad Dermatol* 1986 May;14(5 Pt 1):742-7.

14. Carrozza P. et al., Clinical Efficacy of Narrow-Band UVB (311 Nm) Combined with dithranol in Psoriasis. An Open Pilot Study, *Dermatology* 200 (1): 35-9, 2000.

15. Schiener R, et al., Calcipotriol vs. Tazarotene as combination therapy with narrowband ultraviolet B (311 nm): efficacy in patients with severe psoriasis, *British Journal of Dermatology*, December; 143 (6): 1275-8, 2000.

16. Guenther L,Tazarotene Combination Treatments in Psoriasis, *Journal of the American Academy of Dermatology*, August; 43 (2 Part 3): S 36-42, 2000.

17. Molin L., Topical Calcipotriol Combined with Phototherapy for Psoriasis. The Results of Two Randomized Trials and a Review of the Literature. Calcipotriol-UVB Study Group, *Dermatology* 198 (4): 375-81, 1999.

18. Leenutaphong V, et al.,325 UVB Phototherapy for Pityriasis Rosea: a Bilateral Comparison Study. *Journal of the American Academy of Dermatology*, December; 33 (6): 996-9, 1995.

19. Gelmetti C, et al.,325 Pityriasis Lichenoides in Children: a Long-term Follow-up of Eighty-Nine Cases, *Journal of the American Academy of Dermatology*, September; 23 (3 Part 1): 473-8, 1990.

20. *Fitzpatrick's Dermatology in General Medicine*, fifth ed., Freedberg, I. W. editor et al., McGraw-Hill, p. 1601, 1999.

21. Karvonen J, Kokkonen EL, Ruotsalainen E, *311 nm UVB lamps in the treatment of psoriasis with the Ingram regimen*, Acta Derm Venereol, 69(1), 82-85, 1989.

22. Coven TR, Burack LH, Gilleaudeau R, Keogh M, Ozawa M, Krueger JG, Narrowband UV-B produces superior clinical and histopathological resolution of moderate-to-severe psoriasis in patients compared withn broadband UV-B, *Arch Dermatol*, Dec;133(12):1514-22, 1997.

23. Walters IB, Burack LH, Coven TR, Gilleaudeau P, Krueger JG, Suberythemogenic narrow-band UVB is markedly more effective than conventional UVB in treatment of psoriasis vulgaris, *J Am Acad Dermatol*, Jun;40(6 Pt 1):893-900, 1999.

24. Carrozza P, Hausermann P, Nestle FO, Burg G, Boni R, Clinical efficacy of narrow-band UVB (311 nm) combined with dithranol in psoriasis. An open pilot study, *Dermatology* 200(1):35-9, 2000.

25. Behrens S, Grundmann-Kollmann M, Schiener R, Peter RU, Kerscher M, Combination phototherapy of psoriasis with narrow-band UVB irradiation and topical tazarotene gel, *J Am Acad Dermatol* Mar;42(3):493-5, 2000.

26. Scherschun L, Kim JJ, Lim HW, Narrow-band ultraviolet B is a useful and well-tolerated treatment for vitiligo, *J Am Acad Dermatol* Jun;44(6):999-1003, 2001.

27. Reynolds NJ, Franklin V, Gray JC, Diffey BL, Farr PM, Narrow-band ultraviolet B and broad-band ultraviolet A phototherapy in adult atopic eczema: a randomised controlled trial, *Lancet* Jun 23;357(9273):2012-6, 2001.

28. Saricaoglu H et al., Narrowband UVB therapy in the treatment of lichen planus, *Photodermatol Photoimmunol Photomed* Oct; 19(5):265-7, 2003.

29. Clark C, Dawe RS, Evans AT, Lowe G, Ferguson J, Narrowband TL-01 phototherapy for patch-stage mycosis fungoides, *Arch Dermatol* Jun;136(6):748-52, 2000.

30. Feldman SR, Mellen BG, Housman TS, Fitzpatrick RE, Geronemus RG, Friedman PM, Vasily DB, Morison WL, Efficacy of the 308-nm excimer laser for treatment of psoriasis: results of a multicenter study, *J Am Acad Dermatol* Jun;46(6):900-6, 2002.

31. Asawanonda P, Anderson RR, Chang Y, Taylor CR, 308-nm excimer laser for the treatment of psoriasis: a dose-response study, *Arch Dermatol* May;136(5):619-24, 2000.

32. Contraindication information based on the "Guidelines of Care for Phototherapy and Photochemotherapy" from the American Academy of Dermatology and the package insert from the *Physician's Desk Reference*, Medical Economics Company, Inc., 2002.

33. Melski JW, Tanenbaum L, Parrish JA, Fitzpatrick TB, Bleich HL et al., Oral methoxsalen photochemotherapy for the treatment of psoriasis: a cooperative clinical trial, *J Invest Dermatol* Jun;68(6):328-35, 1977.

34. George SA and Ferguson J, Liquid Formulations of 8-methoxypsoralen (8-MOP) and 5-MOP: A Prospective Double-blind Crossover Assessment of Acute Non-

phototoxic Adverse Effects. *Photodermatol Photoimmunol Photomed* Feb; 9 (1): 33-5, 1992.

35. Nijsten TE and Stern RS. The increased risk of skin cancer is persistent after discontinuation of psoralen+ultraviolet A: a cohort study. *J Invest Dermatol* Aug: 121(2):ix-x, 2003.

36. Stern, RS and the PUVA Follow-up Study, The Risk of Melanoma in Association with Long-term Exposure to PUVA, *J Am Acad Dermatol*, 44:755-61, 2001.

37. Sigurgeirsson, LB, et al., PUVA and Cancer Risk: the Swedish Follow-up Study, *British Journal of Dermatology*, Jul;141(1):108-12, 1999.

38. Abel EA et al., Cutaneous Malignances and Metastatic Squamous Cell Carcinoma following Topical Therapies for Mycosis Fungoides *J Am Acad Dermatol,* Jun;14(6): 1029-1038, 1986.

39. Lebwohl, Mark et al., Consensus conference: Acitretin in combination with UVB or PUVA in the treatment of psoriasis, *J Am Acad Dermatol,* 45:544-53, 2001.

40. Poppe, W and Grossweiner, LI, Photodynamic sensitization by 8-methoxpsoralen via the singlet oxygen mechanism, *Photochem Photobiol,* 22, 217, 1975.

41. Stern, RS, Ocular lens findings in patients treated with PUVA. Photochemotherapy Follow-Up-Study, *J Invest Dermatol,* Oct;103(4):534-8, 1994.

42. Morita A, Werfel T, Stege H, Ahrens C, Karmann K, Grewe M, Grether-Beck S, Ruzicka T, Kapp A, Klotz LO, Sies H, Krutmann J, Evidence that singlet oxygen-induced human T helper cell apoptosis is the basic mechanism of ultraviolet-A radiation phototherapy, *J Exp Med*, Nov 17;186(10):1763-8, 1997.

43. Scharffetter K, Wlaschek M, Hogg A, Bolsen K, Schothorst A, Goerz G, Krieg T, Plewig G, UVA irradiation induces collagenase in human dermal fibroblasts in vitro and in vivo, *Arch Dermatol Res,* 283(8):506-11, 1991.

44. Zane C, Leali C, Airo P, De Panfilis G, Pinton PC, "High-dose" UVA1 therapy of widespread plaque-type, nodular, and erythrodermic mycosis fungoides, *J Am Acad Dermatol*, Apr;44(4):629-33, 2001.

45. Simon JC, Pfieger D, Schopf E, Recent advances in phototherapy, *Eur J Dermatol*, Dec;10(8):642-5, 2000.

46. Stander H, Schiller M, Schwarz T, UVA1 therapy for sclerodermic graft-versus-host disease of the skin, *J Am Acad Dermatol*, May;46(5):799-800, 2002.

47. Weinstock MA, Gardstein B, Twenty-year trends in the reported incidence of mycosis fungoides and associated mortality, *Am J Public Health*, Aug;89(8):1240-4, 1999.

48. Diamandidou E, Colome M, Fayad L, Duvic M, Kurzrock R, Prognostic factor analysis in mycosis fungoides/Sezary syndrome, *J Am Acad Dermatol*, Jun;40(6 Pt 1):914-24, 1999.

49. Edelson R, Berger C, Gasparro F, Jegasothy B, Heald P, Wintroub B, Vonderheid E, Knobler R, Wolff K, Plewig G, et al., Treatment of cutaneous T-cell lymphoma by extracorporeal photochemotherapy. Preliminary results, *N Engl J Med*, Feb 5;316(6):297-303, 1987.

50. Lehrer M et al., Extracorporeal Photopheresis, *eMedicine Journal*, Mar 25;2(3), 2001.

51. Wilson LD, Jones GW, Kim D, Rosenthal D, Christensen IR, Edelson RL, Heald PW, Kacinski BM, Experience with total skin electron beam therapy in combination with extracorporeal photopheresis in the management of patients with erythrodermic (T4) mycosis fungoides, *J Am Acad Dermatol*, Jul;43(1 Pt 1):54-60, 2000.

52. McGinnis KS et al., Psoralen plus long-wave UV-A (PUVA) and bexarotene therapy: An effective and synergistic combined adjunct to therapy for patients with advanced cutaneous T-cell lymphoma. *Arch Dermatol,* Jun;139(6):771-5, 2003.

53. Russo, M.D. and Mullen C., Cutaneous and noncutaneous disorders treated with Extracorporeal photopheresis, *International Journal of Dermatology,* 40:89-100, 2001.

54. Berger CL, Hanlon D, Kanada D, Girardi M, Edelson RL, Transimmunization, a novel approach for tumor immunotherapy, *Transfus Apheresis Sci*, Jun;26(3):205-16. 2002.

55. Ben-Hur, E. and Song, P.-S., Eds., The photochemistry and photobiology of furocoumarins (psoralens), in *Advances in Radiation Biology,* Vol. 11, 1982, Academic Press, New York, 1982, 131.

56. Blan, Q. A., and Grossweiner, L.I., Singlet Oxygen generation by furocoumarins effect of DNA and liposomes, *Photochem Photobiol,* 45, 177, 1987.

57. Potapenko, A. Ya., Mechanisms of photodynamic effects of furocoumarins, *J. Photochem. Photobiol. B: Biol.,* 9, 1, 1991.

58. Jones, S. G., Young, A. R., and Truscott, T. G., Singlet oxygen yields of furocoumarin and related molecules–the effect of excitation wavelength, *J. Photochem. Photobiol. B: Biol.*, 21, 223, 1993.

# CHAPTER 13

# PHOTOTHERAPY OF NEONATAL JAUNDICE

Neonatal jaundice is the most common problem encountered in newborns. 50% of term (> 37 weeks of gestation) and 80% of preterm babies are jaundiced within the first week. It is the most common cause for hospital admission in the first week of life, especially since the advent of early postpartum discharge programs. Unbound serum *bilirubin* (BR) or *hyperbilirubinemia* is the usual cause of neonatal jaundice. BR is a breakdown product of blood heme. Circulating free BR can enter almost all tissues of the body and is highly neurotoxic. Hyperbilirubinemia develops when the BR production exceeds the infant's capacity for its elimination. Normal plasma concentrations of BR are about 1.5 - 2.0 mg/dL (1mg/dL corresponds to 17 μmol/L). Jaundice is evident at 5.0 - 6.0 mg/dL. Total serum bilirubin (TSB) levels exceeding 10 mg/mL can induce delayed neurological damage evident in early childhood. *Kernicterus* is a severe form of hyperbilirubinemia in which very high serum BR concentrations induce deafness, cerebral palsy, or death. Hemolytic disease of the newborn is the classic cause of pathologic hyperbilirubinemia originating from maternal-fetal blood group incompatibility, genetic deficiencies in red blood cell metabolism, infection, and drugs. Total blood exchange was the only treatment for kernicterus prior to discovery of the phototherapy (PHB) in the late 1950's.[1] The phototherapy of neonatal jaundice originated from observations of J. Ward, a nursing sister who supervised a premature baby ward in Essex, U.K.. Sister Ward observed that the skin of jaundiced infants became bleached when exposed to sunlight. Controlled testing by Dr. R. J. Cremer and his associates demonstrated that sunlight reduces the serum BR levels in the infants. Cremer designed an irradiator consisting of eight blue fluorescent lamps and reported the successful reduction of both jaundice and unbound serum bilirubin. PHB is not without side effects, notwithstanding the widespread notion that visible light can only be beneficial. The eyes of the infant must be completely protected from the phototherapy light. The short-term adverse effects of PHB include temporary retardation of growth, increased insensible water loss, immune suppression, transient alterations in blood chemistry, and alteration of biologic rhythms. Clinical experience has not shown evidence of long-term adverse effects.

# 13.1 METABOLISM OF BILIRUBIN

The turnover rate of red blood cells is approximately three million cells per second. Humans utilize a multi-organ system for elimination of the excess heme (Figure 13.1). The membrane-bound enzyme heme oxygenase in the spleen opens the heme ring and removes the bridging α-carbon as CO, leading to the blue-green bile pigment biliverdin (BV) which has no known function (Figure 13.2). Biliverdin is reduced to BR by the enzyme biliverdin reductase. BR is insoluble
in water at physiologic pH and has an exceptionally high binding affinity for serum albumin (ALB). The BR-ALB complex cannot cross the blood-brain barrier. Excretion of the BR-ALB complex is mediated in the liver by the enzyme uridine diphosphate glucuronyl transferase (UDPGT), which conjugates BR with one or two molecules of glucouronic acid to form the mono- and di-glucuronides. The conjugated BR is excreted into the bile canaliculi of the liver, followed by transfer to the gall bladder and then to the intestine where it is excreted in feces. A small amount of BR is transferred to the kidneys and excreted in urine. Hyperbilirubinemia results when the formation and/or excretion of BR glucouronides becomes impaired. This conditions occurs most frequently in premature and newborn infants until UDPGT activity in the liver increases.

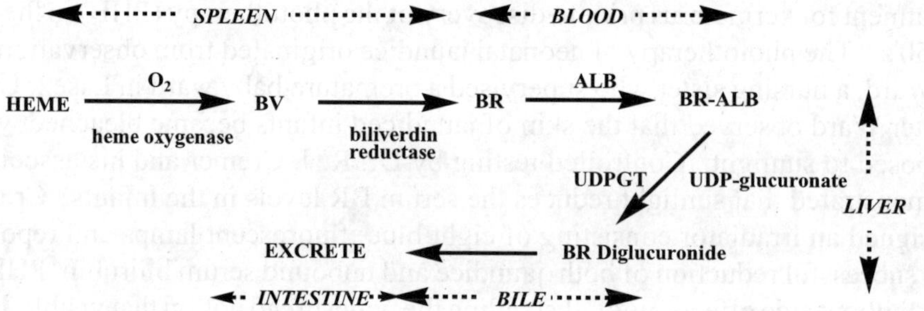


**Figure 13.1** Simplified biochemical pathway for degradation of heme

# 13.2 ACTION MECHANISM OF PHB

The interaction of light with BR is believed to take place mostly in extravascular tissues within 1 mm of the skin surface. Early workers postulated that PHB involves an *intra*molecular photosensitization, in which light absorption by BR generates $^1O_2$ which then oxidizes BR to biliverden. However, photooxidation products account for only a small fraction of the BR eliminated by PHB. There is now ample evidence that BR elimination is initiated primarily by anaerobic *photoisomerization* reactions. Two types of photoisomers have been identified. Four "configurational" photoisomers are formed *in vitro* by rotations of the BR molecule around the C=C bonds at C-4 and C-15, which are stabilized by *intra*molecular hydrogen bonds. Of these, only the 4Z,15E-photoisomer (referred to as *photobilirubin*) appears in infant's blood after phototherapy (Figure 13.2).[2] The photochemical conversion of bilirubin to photobilirubin is efficient *in vitro* with an initial quantum efficiency about 0.2 for blue light. However, this photoisomer also absorbs blue light and continuous irradiation establishes a *photoequilibrium* at a low level of conversion. Consequently, photobilirubin accumulates in blood of jaundiced infants and only a small amount is excreted BR binding to ALB does not inhibit the formation of photobilirubin. The other type of photoproducts are "structural" isomers referred to as *lumirubin* (Figure 13.2 ). The photochemical production of lumirubin is the order of 20-fold less efficient than the photobilirubin. However, this process is irreversible and provides the major pathway of BR removal. PHB also generates a low yield of colorless photooxidation products which are eliminated *via* the kidneys. According to the present model, light absorption in the skin of the infant converts unbound serum BR to albumin-bound photoisomers which are carried to the liver and eliminated in the intestine. BR photooxidation products are carried to the kidney and eliminated in urine.

## 13.3 PHB LIGHT SOURCES

The absorption spectrum of BR in a nonpolar solvent consists of a broad band extending from 350 nm - 500 nm peaking near 455 nm (Figure 13.3). Accordingly, PHB light sources are designed to emit the blue light most strongly absorbed by BR. The conventional light sources for PHB are fluorescent lamps providing the order of 1 $mW/m^2$ in the blue spectral region. In an early controlled study of Dr. T. R. C. Sisson, initiation of phototherapy in 2 - 3 day old infants with "blue" (B) fluorescent lamps led to a decrease of serum BR from an average of 10 mg/dL to 7 mg/dL after 1 - 2 days of treatment. The untreated control group reached BR levels of 13 mg/dL after 4 - 5 days of age, which decreased to 9 mg/dL after 3 - 4 additional days. Different types of fluorescent lamps have been used in phototherapy units, all of which have appreciable output from 420 to 470 nm.

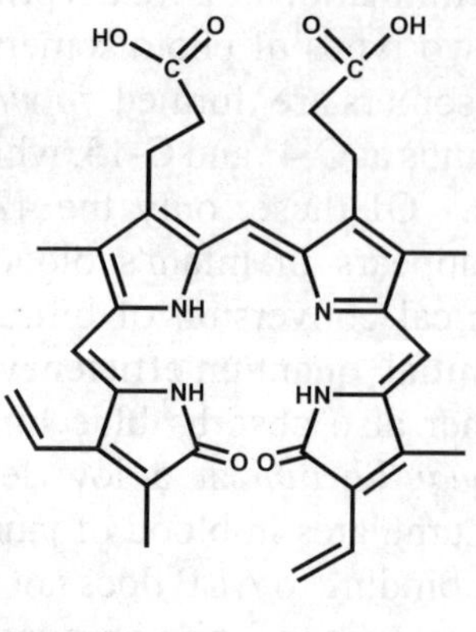

Biliverdin

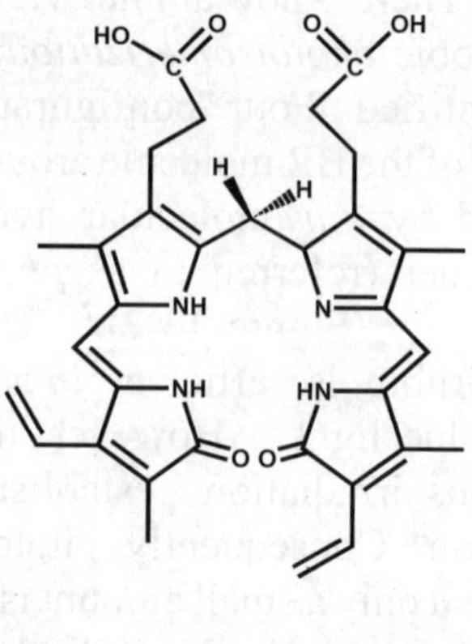

Bilirubin

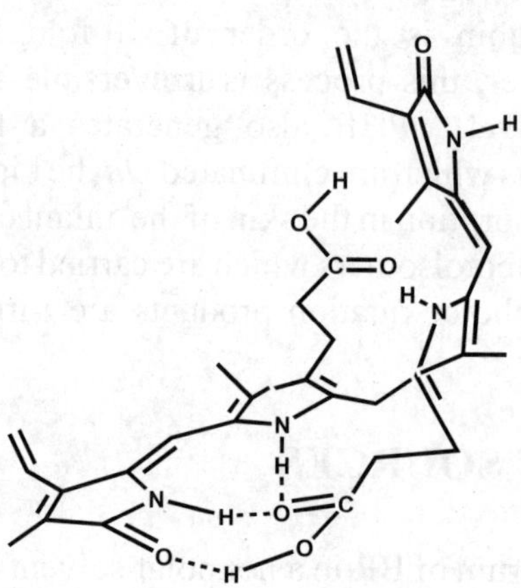

Photobilirubin

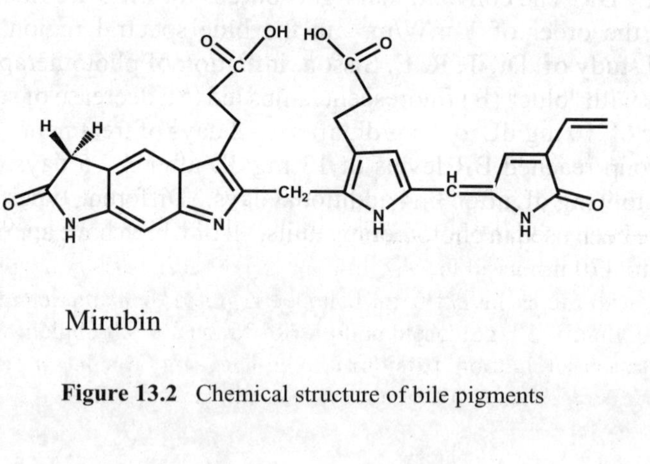

Mirubin

**Figure 13.2** Chemical structure of bile pigments

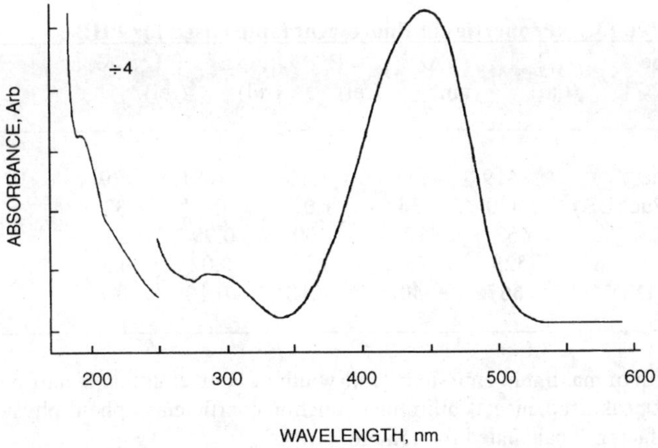


**Figure 13.3** Absorption spectrum of bilirubin in acetonitrile

The more common types are the "daylight" (DA) lamp with a broad emission from 400- 650 nm, "standard blue" (B) with a narrower band peaking near 450 nm, and "special blue" (BB) which is more intense than "standard blue" with a narrower bandwidth. Table 13.1 summarizes radiometry measurements of Pratesi *et al.* for several types of fluorescent lamps. The last column is a relative photophysical efficiency parameter, calculated by the present author, defined as $E_{max}\ \Delta\lambda\ \varepsilon_{BR}$, where $E_{max}$ is the peak irradiance of the lamp, $\Delta\lambda$ is the spectral width of the emission, and $\varepsilon_{BR}$ is the extinction coefficient of BR at the peak wavelength. This parameter suggests that either the "standard blue" (Type B) or "special blue" (BB) fluorescent lamps should be the most efficient light sources of this group. This photophysical calculation does not include the higher absorption of shorter wavelengths by the epidermis and the dependence of the photochemical quantum yields on wavelength. Alternative blanket-type irradiators have been available for about 10 years. In one type of arrangement, a flexible light-emitting panel is wrapped around the infant. A QTH lamp is used as the light source, delivered by multiple optical fibers to a phosphor emitting blue light. Representative systems include the Ohmeda BiliBlanket® and the Heathdyne Wallaby®. A different approach employs an illuminated blanket that delivers the therapeutic light to the infant's back, for example the Medela Bilibed®. These devices have the advantages that infants can be nursed close to their parents without the need for eye protection. A recent evaluation of 24 comparative studies does not indicate a clear superiority of either the conventional overhead lamps or the newer optical fiber devices, although combinations of the two modalities were more effective than the overhead lamp phototherapy alone[3].

**Table 13.1 Properties of fluorescent lamps used for PHB**

| Lamp Type | $\lambda_{max}$ [a] (nm) | $\Delta\lambda$ [b] (nm) | $E_{max}$ [c] (rel) | $\varepsilon$ [d] (rel) | $E_{max}\,\Delta\lambda\,\varepsilon$ [e] (rel) |
|---|---|---|---|---|---|
| Violet Blue | 419 | 33 | 1.46 | 0.81 | 39 |
| Special Blue (BB) | 440 | 34 | 1.01 | 0.97 | 33 |
| Blue (B) | 452 | 54 | 1.00 | 0.99 | 54 |
| Green (G) | 528 | 38 | 0.72 | 0.02 | 0.5 |
| Daylight (DL) | 667 | 40 | 0.18 | 0.10 | 0.7[f] |

[a] Wavelength of maximum emission; [b] full-width of lamp emission at half-peak irradiance; [c] peak irradiance; [d] bilirubin extinction coefficient; [e] photo-physical efficiency factor, [f] calculated for 455 nm.

## 13.4 CLINICAL ASPECTS OF PHB

The basic objective of PHB for healthy term newborns is to ensure that the TSB levels after 24 hours do not attain pathologic levels. There is no clear evidence for the superiority of different light sources and treatment protocols. Clinical studies comparing intermittent with continuous phototherapy have produced conflicting results. The clinical guidelines for management of non-pathologic hyperbilirubinemia recommended by the American Academy of Pediatrics are based on the belief that therapeutic intervention may carry significant risk relative to the uncertain risk of hyperbilirubinemia in this population[4]. The preliminary diagnosis requires visual observation of jaundice, determination of TSB levels, and acquisition of information that might indicate other causes of hemolytic disease. Recommendations for application of PHB depend on both the age of the infant and the TSB level. At 24 - 48 hours PHB should be considered for TSB > 12 mg/dL and is indicated for TSB in the range from 15 to 20 mg/dL. Higher levels of TSB require intensive phototherapy and possibly exchange transfusion. Term infants who are clinically jaundiced at < 24 hours old are not considered healthy and require further evaluation. Additional details are available in the pediatric literature.

## NOTES

1. There is no generally accepted abbreviation for the phototherapy of hyperbilirubinemia. PHB is used in this chapter.

2. The molecular structures employ the standard convention in which obvious C-H bonds are not shown and methyl and vinyl groups are indicated by single and double lines, respectively.

3. Tudehope D, Crawshaw A. Fibreoptic versus conventional phototherapy for treatment of neonatal jaundice. *J Paediatr Child Health.* Feb 1995;31(1):6-7.

4. Practice parameter: management of hyperbilirubinemia in the healthy term newborn. American Academy of Pediatrics. Provisional Committee for Quality Improvement and Subcommittee on Hyperbilirubinemia. *Pediatrics.* 1994 Oct;94(4 Pt 1):558-65.

# GLOSSARY

The vocabulary of phototherapy is derived from many different fields. Frequently employed abbreviations are listed at the end.

**ablation** a fast ejection of small amounts of matter from a material induced by pulsed laser radiation..

**absorption spectrum** a graph showing the dependence of absorptivity on wavelength or frequency.

**absorptivity** a parameter specifying the optical absorption by non-scattering layer defined as: $A = - \text{Log}_{10}\, T$, where $A$ is the absorptivity and $T$ is the transmission coefficient; see "optical density".

**acceptor** an impurity atom in a semiconductor that introduces additional positive charge carriers or "holes" in the valence band.

**achromat** a lens fabricated of two types of optical glass that corrects for spherical aberration.

**actinic rays** the ultraviolet wavelengths of sunlight shorter than 320 nm.

**actinometric units** a system of optical units based on the photon content of light.

**action spectrum** a graph showing the dependence of the biological effectiveness of a given radiation on the incident wavelength or frequency.

**active medium** the working substance in a laser responsible for light emission.

**active mode locking** a method for generating ultrashort laser pulses utilizing an amplitude or frequency modulating device driven at the round trip frequency of the optical cavity.

**albedo** ratio of the radiant power density reflected by a surface to the incident power density; the ratio of scattering coefficient of a medium to the sum of the absorption and scattering coefficient.

**amphipath (amphiphile)** a molecule with a hydrophobic head group and a long hydrophilic tail.

**anti-bonding molecular orbital** a molecular orbital that destabilizes a molecule when it is occupied by one or two electrons.

**atomic orbital** a quantum-mechanical wave function associated with single-electron states in an atom.

**aspheric lens** a lens in which at least one face is a surface of revolution

**astigmatism** an imaging defect in which off-axis rays are focused to an elliptical rather than a circular spot.

**asymmetry parameter** a parameter in radiative transfer theory defined as the mean cosine of the scattering angle for a single-particle.

**autofluorescence** the intrinsic fluorescence emitted by a biological tissue.

**avalanche photodiode** a semiconductor photodiode operated in reverse bias close to breakdown that amplifies electron-hole pairs generated by light absorption.

**average power** total energy delivered during an irradiation divided by the exposure duration.

**ball lens** a spherical, short-focal length lens useful for coupling lasers to optical fibers.

**band gap** the energy required to excite an electron in a crystal from the valence band to the conduction band.

**bandpass filter** an optical filter having high light transmission in one or several wavelength bands and very low transmission in other spectral regions.

**bandwidth** the spectral width associated with an electrical, optical, or acoustic signal.

**beam divergence** angular spread of a directed light beam.

**Beer's law** a postulated linear relationship between the absorptivity of a substance and the absorption coefficients of the constituents.

**B-field** the transverse magnetic field carried by electromagnetic radiation.

**bilirubin** a bile pigment formed by breakdown of heme.

**biostimulation** an experimental phototherapy in which very low power levels of laser light induce physiological responses.

**birefringence (double refraction)** a property of optically active crystals in which the refractive index varies with the direction of propagation and the polarization of the incident light.

**birefringent filter** an optical device fabricated from alternating layers of birefringent crystal plates and linearly polarizing films which transmits a series of widely-separated bands of narrow bandwidt.

**black body** a hypothetical object that absorbs all incident electromagnetic radiation.

**blaze** the uniform inclination angle of the elements in a reflection type of diffraction grating.

**bonding molecular orbital** a molecular orbital that stabilizes a molecule when it is occupied by one or two electrons.

**Brewster angle** the incidence angle at the interface between two media at which the reflected ray in the first medium is perpendicular to the refracted ray in the second medium.

**brightness** a measure of the visual response evoked by a light source; equivalent to radiance in radiometry and luminance in photometry.

**cavitation** production of electrically- charged species in an aqueous medium by absorption of ultrasound..

**chromatic aberration** an imaging defect in which different wavelengths are focused on different planes.

**chromophore** a light absorbing atom, molecule, or molecular unit.

**charge-injection device (CID)** a solid state imaging device for near-IR consisting of a two-dimensional metal-oxide-semiconductor (MOS) array in which readout is accomplished by injecting the stored charge into the substrate and monitoring the current flow.

**charge-coupled device (CCD)** a self-scanning semiconductor imaging device in which the light sensing element consists of a closely-spaced array of metal-oxide-semiconductor (MOS) capacitors.

**circular polarization** a type of light polarization in which the electric vector rotates at the wave frequency and the tip of the electric vector traces out a circle of constant radius.

***cis-trans* isomerization** a change in molecular structure resulting from rotation around a covalent double bond.

**cladding** the outer medium in an optical fiber surrounding a higher refractive index core.

**coherence** the existence of phase relationships between waves.

**coherence length** the maximum path difference at which light waves can form interference fringes.

**coherent bundle** an array of optical fibers capable of transmitting images.

**cold mirror** an interference filter that reflects ultraviolet and visible light and transmits infrared.

**color temperature** the temperature of black body radiation with approximately the same spectral distribution in the visible region as that of a luminous object.

**coma** an imaging defect leading to image blurring of off-axis rays caused by a variation of the focal length for different zones of the lens surface.

**Commission Internationale d'Eclairage (CIE)** the international committee that establishes the definition of optical units.

**compensated semiconductor** an extrinsic semiconductor in which the contributions of donors and acceptors to the electrical conductivity cancel out.

**conduction band** the highest energy band in a crystalline solid which is partially occupied by mobile electrons.

**continuous wave (CW)** the output of a laser operated in continuous rather than pulsed mode; arbitrarily defined as a continuous output for a period > 0.25 s.

**converging lens** a lens that focuses parallel rays to a real image.

**core** the inner medium of an optical fiber.

**cosine law** an optical principle stating that the irradiance of light reflected by a diffusing surface is proportional to the cosine of the angle between the normal to the surface and the direction of the incident rays; see lambertian surface.

**critical angle** the minimum angle of incidence for total internal reflection when light enters a medium of lower refractive index.

**critical micelle concentration (CMC)** the minimum detergent concentration at which micelles are formed.

**cross section** a concept in quantum physics in which the rate of a microscopic process is expressed as an area proportional to the probability of a specific event.

**curvature of field** an imaging defect resulting from imaging a flat object onto a curved surface.

**continuous wave (CW)** a mode of a laser operation in which the power output is constant in time.

**cylindrical lens** a lens fabricated with at least one cylindrical refracting surface.

**de Broglie wavelength** the wavelength of matter waves associated with a free particle.

**delayed fluoresence** molecular luminescence with the lifetime of a metastable state and the spectrum of fluorescence.

**delta-Eddington phase function** a light-scattering phase function for a single particle consisting of the superposition of an isotropic term, a term proportional to the cosine of the scattering angle, and a sharply-peaked solid-angle delta function in the forward direction.

**depletion region** a narrow zone in a semiconductor *p-n* junction depleted of mobile charge carriers.

**detectivity** a standard figure of merit of a photodetector specifying the minimum response that exceeds the intrinsic electrical noise power.

**diamagnetism** magnetic properties of a non-paramagnetic substance.

**dichroism** a molecular property in which the absorption of linear polarized light depends on the direction of polarization.

**dielectric polarization** a temporary electric dipole moment in a medium induced by an applied electric field.

**diffraction** bending of light waves by objects of dimensions comparable to the wavelength.

**diffraction grating** a device consisting of closely-spaced, parallel slits (transmission grating) or reflecting elements (reflection grating) used to separate polychromatic light into the constituent wavelengths.

**diffuse light** the type of light reflected by a rough surface or propagating in a turbid medium.

**diffuse reflection** light remitted at an external face of a turbid medium.

**diffusion current** flow of majority carriers across a semiconductor *p-n* junction that equalizes the Fermi energy on each side of the junction.

**diopter** the "power" of a lens or lens system defined as the reciprocal of the focal length in meters.

**dispersion** originally used for separation of a polychromatic light beam into the constituent wavelengths; also the variation of refractive index with wavelength or frequency.

**diverging lens** a lens that causes parallel rays of directed light to spread out.

**donor** an impurity atom in a semiconductor that introduces electrons into the conduction band.

**double refraction** see birefringence.

**doublet state** spectroscopic notation for an atom, molecule, ion, or free radical in which one electron spin is unpaired.

**drift current** flow of minority carriers across a semiconductor *p-n* junction.

**duty cycle** the fraction of elapsed time in which pulses are delivered.

**extracorporeal photophoresis (ECP)** a light treatment for cutaneous T-cell lymphoma in which a small blood sample is irradiated with UV-A outside the body and reinfused into the patient after oral administration of a psoralen drug.

**Eddington phase function** a light-scattering phase function for a single particle consisting of the superposition of an isotropic term and a term proportional to the cosine of the scattering angle.

**erythropoietic protoporphyria (EEP)** agenetic disease caused by the accumulation of protoporphyrin in red blood cells.

**E-field** electric field.

**electron-hole recombination** annihilation of conduction band electrons by valence band holes across the forbidden gap.

**electro-optic crystal** a crystal in which the refractive indices are altered by an external electric field.

**elliptical polarization** a type of light polarization in which the electric vector rotates at the wave frequency and the tip of the electric vector traces out an ellipse.

**emissivity** the ratio of the thermal radiation power emitted by an object to that of an ideal black body at same temperature and wavelength.

**electromagnetic radiation (EMR)** a transverse wave transporting electric and magnetic fields at the velocity of light.

**endoscope** a medical device used for visual examination and/or light delivery within a body cavity.

**energy** the capacity for doing work (J).

**energy band** closely-spaced electronic energy levels in a crystalline soli.

**energy fluence rate** the average radiant flux incident from all directions on a point in space or within a medium; also referred to as spherical irradiance and space irradiance.

**electron paramagnetic resonance (EPR)** an experimental technique that probes the magnetic energy levels of unpaired electrons in a material; the same as electron spin resonance (ESR).

**extended source** a source of radiation that can be resolved by the eye into a geometrical image.

**extraordinary ray (E-ray)** a ray of linearly polarized light propagated in a birefringent crystal for which the wave velocity depends on the propagation angle relative to the intrinsic optic axes of the crystal.

**excimer** a molecule formed between two atoms which is stable only in the excited state.

**excitation** a process that increases the internal energy of an atom or molecule.

**excitance** radiant flux emitted or reflected per unit area ($W/m^2$).

**excited singlet state** an electronically-excited state of an atom or molecule in which all electron spins are paired leading to zero total spin.

**exclusion principle** a fundamental principle of quantum physics stating that two electrons in a stable atom or molecule cannot have identical spatial motions and electron spin; also referred to as the Pauli principle.

**exitance** radiometric unit for radiant power reflected by or emitted by a surface.

**extrinsic semiconductor** a semiconductor in which additional electrical conductivity results from the presence of donor or acceptor impurities.

**Fabry-Perot** an optical resonator consisting of two parallel reflecting or partially reflecting surfaces separated by air or a dielectric medium.

**free electron model (FEM)** a molecular theory in which the electronic energy states are calculated for mobile electrons confined to a "box" corresponding to d the molecular backbone.

**Fermi energy** the electronic energy level in a crystalline solid in thermal equilibrium for which the probability of occupation by an electron equals 0.5; equivalent to the chemical potential of electrons in the solid.

**Fick's first law** A phenomenological relationship which states that the rate of conductive heat flow in a medium is proportional to the temperature gradient; also referred to as Fourier's law.

**first-order reaction** a reaction for which the time required for any given extent of completion, usually the half-life, is the same for any initial quantity of reactants.

**fluence** quantity of radiant energy incident on unit area in space or within a material.

**fluorescence** luminescence resulting from an allowed electronic transition in an atom or molecule.

**fluorescence efficiency** : the probability that absorption of one incident photon induces emission of a fluorescence photon.

**fluorophore** a molecule or molecular unit responsible for emission of a fluorescence.

**flux density** a generic term for radiant power per unit area in space or within a material; also referred to as fluence rate.

**focal length** the distance from the secondary nodal point of a lens or optical system to the primary focal point.

**focal point** the point in space at which light converges or from which light diverges or appears to diverge.

**F-number** the ratio of the focal length of an optical system to the diameter of the effective aperture; also referred to as "relative aperture".

**footcandle** a photometric unit of illuminance equal to one lumen per square foot.

**forbidden gap** a band of energy levels in a crystalline solid that cannot be occupied by electrons.

**four-level laser** a mechanism of laser action in which the two energy levels responsible for light emission lie between the pumped energy level and the ground state.

**free radical** a stable or unstable molecular species with an odd number of electrons.

**frequency domain** an experimental technique in which a band of sinusoidal input signals are applied to a system and the amplitude and phase of the output signals are measured as a function of the input frequency.

**frequency doubling** a non-linear interaction of laser radiation with an electro-optic crystal leading to an output at twice the input frequency.

**ferroelectric crystal** a crystal in which a permanent internal electric field exists below certain temperature referred to as the Curie temperature

**Fresnel lens** a lens constructed of concentric refracting grooves with the focusing properties of a high-aperture spherical lens.

**Fresnel relations** theoretical equations derived from electromagnetic theory which predict the dependence of the amplitude and polarization of light reflected and refracted at the interface between optical media of different refractive indices.

**front-surface mirror** a mirror in which the metallic reflecting surface is coated on top of a base material.

**gaussian beam** a light beam in which the irradiance profile transverse to the direction of propagation has the shape of a gaussian or bell-shaped function.

**geometrical optics** an approximate theory applicable when the dimensions of the optical elements are much larger than the wavelength in which light propagation is analyzed by in terms of light rays, instead of wave fronts as in physical optics.

**goniometer** a device that measures the angular direction of light rays.

**graded index fiber** an optical fiber in which the refractive index of the core decreases continuously with increasing distance from the fiber axis.

**gradient index lens (GRIN)** a lens fabricated from an optical material with a continuously varying refractive index.

**gyromagnetic ratio** the ratio of the magnetic moment to the angular momentum of a charged particle or a distributed charge.

**halflife** the time required for 50% completion of a rate process.

**halfwave plate** a retardation plate of half-wavelength thickness that rotates the plane of linearly polarized light.

**heat capacity** the absorbed energy required to cause a one degree temperature rise in a unit mass of material.

**Henyey-Greenstein phase function** a light-scattering phase function for a single particle which approximates the angular dependence of Mie scattering by a large particle relative to the wavelength.

**high-pass filter** an optical filter which transmits all wavelengths from a cut-off wavelength to the transmission limit of the material.

**hole** an unoccupied electronic energy level in the valence band of a semiconductor crystal with the transport properties of positively charged particle.

**holography** a technique for generating three-dimensional images by illuminating the object with coherent laser light.

**hot mirror** an optical filter that reflects infrared and transmits visible light and ultraviolet.

**hot molecule** a molecule which possesses higher vibrational energy than if it were in thermal equilibrium with its environment.

**horizontal transition** an electronic transition between two states having the same energy.

**hyperbilirubinemia** a metabolic disorder responsible for jaundice resulting from excess unbound bilirubin in the blood serum.

**hyperthermia (HT)** cell killing induced by temperature elevation.

**hydrated electron** a mobile electron in an aqueous medium stabilized by the induced dielectric polarization; the radical-anion of aqueous hydrogen atoms.

**internal conversion** a radiationless transition between equal vibrational energy states having different spin multiplicities.

**illuminance** photometric unit of luminous flux per units area (lux, $lm/m^2$, foot-candle).

**inner filtering** distortion of a fluorescence emission spectrum owing to light absorption by the medium.

**integrating sphere** a device use for a optical measurements consisting of a spherical shell with a highly reflecting inner surface and small ports for incident and emitted light.

**intensity** unit of flux in a given direction within a unit solid angle; (radiometric units, $W/m^2$; photometric units , cd, lm/sr).

**interference** interactions between coherent or partially coherent light waves.

**interference filter** an optical device consisting of alternate reflecting and dielectric layers designed to selectively reflect or transmit narrow wavelength bands.

**inter-modal dispersion** a property of optical fibers in which the optical signals of different linearly polarized modes arrive at different times

***inter*molecular energy transfer** non-radiative transfer of excitation energy from a donor molecule to an acceptor molecule.

**intimal hyperplasia** occlusion of an artery caused by abnormal cell growth in the smooth muscle cells of the intima.

**intrinsic semiconductor** a pure semiconductor crystal in which the thermal excitation across the forbidden generates equal number of electrons in the conduction band and holes in the valence band.

**inverse-square law** the principle of optics stating that the irradiance of light emitted by a point source decreases with the reciprocal of the second power of the distance between the receiver and source.

**ionization potential** the minimum photon energy required to eject an electron from an atom or molecule.

**infrared (IR)** the band of electromagnetic radiations having longer wavelengths than the visible spectrum and shorter than 1 mm.

**intersystem crossing (ISC)** a spontaneous electronic transition an atom or molecule leading to a change of electron spin direction; typically ,formation of a triplet state from an excited singlet state.

**irradiance** radiometric unit of flux per units area ($W/m^2$).

**isobestic point** a wavelength at which two molecular species have the same absorptivity.

**isotropic light** light waves that propagate with no preferred directionality.

**Jablonski diagram** a state-energy diagram for a typical highly-conjugated aromatic or heterocylic molecule depicting the transitions between the lower energy vibronic levels.

**Kerr cell** a device containing an optically active material that acquires birefringence in the presence of an applied electric field.

**lambertian surface** a rough reflecting surface for which the reflected intensity is proportional to the cosine of the angle from the normal to the surface and the radiance is constant in any direction.

**laser** a device that generates an intense, coherent, directional beam of light by stimulating electronic or molecular transitions to lower energe levels; acronym for Light Amplification by Stimulated Emission of Radiation.

**laser bioheat equation** the partial differential equation expressing conservation of thermal energy in a biological system.

**light emitting diode (LED)** a semiconductor diode in which an electrical current induces light emission.

**light ray** the trajectory of a point on the wave front of propagating light waves.

**light scattering** interaction of a light wave with a small particle or region that changes the direction and possibly the polarization without altering the wavelength.

**linear absorption coefficient** an optical constant proportional to the depletion of of flux density owing to absorption by the medium; the product of the particle concentration and the absorption cross section ($m^{-1}$).

**linear attenuation coefficient** an optical constant proportional to the depletion of of flux density owing to light scattering and absorption by the medium; the product of the particle concentration and the attenuation cross section ($m^{-1}$).

**linearity range** the range for which the response of a device is a linear function of the input.

**linear polarization** a type of light polarization in which the electric vector has a fixed orientation in space and oscillates with the wave frequency.

**linearly polarized mode** a propagation mode in a optical fiber corresponding to constant electric field pattern within the core.

**linear scattering coefficient** an optical constant proportional to the depletion of flux density owing to light scattering by the medium; the product of the particle concentration and the scattering cross section ($m^{-1}$) .

**linewidth** the bandwidth of a spectral line.

**liposome** an approximately spherical phospholipid vesicle with an interior aqueous phase suspended in an aqueous medium.

**lone-pair** valence electrons in an atom with opposite spins that do not participate in covalent bonding.

**longitudinal cavity mode** an axial standing wave pattern in an optical cavity.

**longitudinal wave** a wave motion in which the particle displacement is parallel to the direction of propagation.

**loop gain** the incremental increase of optical power in a laser for each round trip of light within the optical cavity.

**lumirubin** a structural photoisomer of bilirubin.

**lumen** photometric unit of luminous flux lm) .

**luminance** photometric unit of radiant flux in a given direction per unit solid angle amd per unit area perpendicular to the direction of propagation (cd/m$^2$).

**luminous energy** photometric unit of energy (lm-s, talbot).

**luminous excitance** luminous flux emitted or reflected per unit area (lm/m$^2$).

**luminous exposure** luminous energy arriving per unit area (lm-s/m$^2$).

**luminous flux** photometric unit of optical power (lm).

**luminescence** light emission excited by an external stimulus including light, x-rays, electrons, mechanical strain, and friction.

**magnetic dipole moment** a vector specifying the strength and direction of a magnetic dipole.

**magnification** the ratio of the size of an optical image to the size of the object.

**majority carrier** conduction band electrons in an *n*-type semiconductor and valence band "holes" in a *p*-type semiconductor.

**maser** acronym for microwave amplification by stimulated emission of radiation; a high-gain, low-noise device for amplification of microwave radiation.

**microchannel plate** a photodetector consisting of a planar array of electron-multiplying elements.

**minimum erythemal dose (MED)** the incident fluence leading to barely perceptible erythema after 24 hours.

**mean free path** the average distance traveled by particles or photons along the actual trajectory.

**micelle** an loosely-ordered spatial arrangement of amphipathic molecules in an aqueous medium.

**Mie scattering** the exact electromagnetic theory of light scattering by an insulating or conducting sphere.

**minority carrier** valence band "holes" an *n*-type semiconductor and conduction band electrons in a *p*-type semiconductor.

**molecular orbital (MO)** a quantum-mechanical wave function associated with a single-electron state of a molecule.

**mode** a standing-wave intensity pattern within an optical cavity.

**mode-locking** coherent addition of longitudinal laser cavity modes of constant phase differences leading to a sequence of uniformly-spaced, high- power, short-duration pulses.

**monochromatic radiation** radiation with a very narrow spectral bandwidth.

**monochromator** an optical device that separates approximately monochromatic wavelengths from incident broad-spectrum radiation.

**Monte Carlo** a statistical technique in which random numbers are used to predict the outcome of a succession of microscopic events; used in tissue optics to predict the trajectories of photons.

**metal-oxide-semiconductor** (MOS) a light-sensitive capacitor used in a charge-coupled device (CCD).

**magnetic resonance imaging (MRI)** a tomographic imaging technique based on the differential relaxation time constants of tissue protons after alignment by a strong external magnetic field.

**multimode fiber** an optical fiber that transmits more than one linearly polarized mode.

**multiple scattering** a mode of light propagation in which photons are scattered more than once before being absorbed or escaping from the medium.

**mycosis fungoides (MF)** alternative designation of cutaneous T-cell lymphoma (CTCL).

**numerical aperture (N.A.)** the light gathering efficiency of a lens or optical fiber defined as the sine of the half-acceptance angle.

**negative temperature** see population inversion.

**Newton's law of cooling** the assumption that the rate of surface heat transfer at a phase boundary is proportional to the temperature difference between the two phases.

**Nicol prism** a split and cemented calcite crystal that acts as a linear polarizer.

**normalized frequency parameter** a standard characterization of the number of modes propagated by an optical fiber; also referred to as the "V-number".

***n*-type semiconductor** a semiconductor in which conduction band electrons are the majority charge carriers.

**optical activity** the property of crystals and liquids to rotate the plane of linearly polarized light; the direction of rotation is described as right-handed or dextrorotary and left-handed or levorotary.

**optic axis** one or two unique directions of light propagation in a birefringent crystal at which double refraction does not take place.

**optical axis** the centerline of an optical system on which light rays are not deviated.

**optical cavity** a device that confines light rays within a region defined by surface reflectors.

**optical density** a parameter specifying the optical absorption by a scattering or non-scattering layer defined as: O.D. $= - \text{Log}_{10} T$, where O.D. is the optical density and $T$ is the transmission coefficient .

**optical fiber** a dielectric filament in which light is propagated by total internal reflection (TIR).

**ordinary ray (O-ray)** a ray of linear polarized light propagated in a birefringent crystal for which the phase velocity is independent of the angle relative to the optic axes and which obeys Snell's law.

**optical penetration depth** the straight-line distance within a medium in which the energy fluence rate is attenuated by a factor of 03679 ($1/e$).

**optical rotary dispersion( ORD)** a wavelength dependence of the rotation angle of linearly polarized light.

**paramagnetism** magnetic properties of a substance that contains permanent magnet dipoles owing to unpaired electron spins.

**paraxial ray** a light ray that lies close to and almost parallel to the optical axis of an optical system.

**passive mode-locking** a technique used for laser mode-locking in which a saturable absorber is present in the optical cavity.

**photodynamic therapy (PDT)** a phototherapy for malignant tumors utilizing the combined action of light and a photosensitizing drug in which molecular oxygen is required in the initial photochemical reaction.

**peak power** a parameter used to characterize the power delivered in a single pulse for repetitively-pulsed irradiations.

**phase function** used in radiative transfer theory to specify the angular dependence of light scattering by a single particle.

**phase shift** the shortest difference in the time at the same position, or shortest difference in position at the same time, in which two sinusoidal waves attain their maximum amplitudes; frequently described in units of the wavelength.

**phosphor** a substance that continues to emit luminescence after the excitation is terminated.

**phosphorescence** a relatively long lifetime luminescence compared to the corresponding fluorescence; also referred to as afterglow.

**photoallergy** an allergic response initiated by exposure of the skin to light, frequently associated with chemicals and drugs.

**photobilirubin** a configurational photoisomer of bilirubin.

**photochemotherapy** a phototherapy utilizing the combined action of light and an exogenous chemical drug; usually employed to describe PUVA.

**photoconductive detector** a photodiode in which light increases the electrical conductivity.

**photodiode array** a linear arrangement of semiconductor photodiodes.

**photodynamic action** a biological photosensitization that requires the presence of molecular oxygen in the initial photochemical step; frequently termed a *Type 2* photosensitiztion..

**photoelectric detector** a device that converts optical radiation to an electrical signal.

**photoelectric effect** electron ejection from metals induced by ultraviolet.

**photoequilibrium** a chemical equilibrium established by light absorption in two, *inter*convertible species.

**photohemolysis** light-induced release of hemoglobin from red blood cells.

**photoionization** ejection of electrons from atoms or molecules induced by light; also used for electron ejection by x-rays and gamma-rays.

**photoisomerization** a photochemical reaction leading to product molecules formed from the same atoms having different structures; structural photoisomers have different spatial arrangements of the atoms; configurational photoisomers are *inter*convertable by molecular rotations.

**photolytic bond cleavage** a photochemical reaction leading to splitting of a covalent chemical bond.

**photometric units** a system of optical units applicable only for visible light based on the visual response of the standard, light-adapted human eye.

**photomultiplier** a type of photoelectric detector in which the primary photocurrent emitted by a photocathode is amplified within the device by impinging on a succession of electron-multiplying elements referred to as dynodes..

**photon** a hypothetical weightless particle moving with the speed of light and carrying electromagnetic momentum and energy.

**photoredox reaction** a photochemical reaction leading to transfer or an electron or hydrogen atom from a donor molecule to an acceptor molecule.

**photosensitization** a light-activated process in which the presence of a photosensitizing agent, usually in small amounts, leads to photophysical, bphotochemical, or photoiological processes that do not take place without the photosensitizer.

**phototoxicity** a morbidity initiated by exposure of the skin to light caused by the presence of a chemical agent or drug.

**photovoltaic detector** a device in which light generates a potential difference.

**physical optics** a theory of optics applicable when the dimensions of the optical elements are comparable to the wavelength in which light propagation is analyzed by in terms of wave fronts instead of light rays as in geometrical physical optics.

**piezoelectric crystal** a crystal in which mechanical strain induces an internal electric dipole moment and an electric field at the external faces.

***pi* molecular orbital** a type of molecular orbital which the electron density is localized in two interior lobes (binding) or four exterior lobes (anti-bonding) above and below the molecular axis.

**plasma** a gaseous mixture of electrons, excited atoms, and ions generated by an electrical discharge.

**plume** aerosol created by vaporization of tissue that may contain toxic agents.

***p-n* junction** a semiconductor element consisting of a layer of *p*-type material in contact with a layer of *n*-type material.

**Pockel's cell** a laser Q-switch in which the fast removal of a voltage applied to an electro-optic crystal increases the transmission of the optical cavity for linearly polarized light.

**population inversion** a non-equilibrium condition in which the occupation of an upper energy level of an atom or molecule is higher than that of a lower energy level.

**principal plane** a plane perpendicular to the optical axis of a lens which acts as the plane of refraction.

**probability amplitude** a solution to Schroedinger's wave equation; also referred to as a wave function.

**probability density** the magnitude of the square of the probability amplitude; interpreted as the probability that a measurement will find the particle(s) in a volume element at the given values of the space and time coordinates.

***p*-type semiconductor** a semiconductor in which "holes" in the valence band are the majority charge carriers.

**point source** a source of radiation that cannot be resolved by the eye into a geometrical image.

**pumping** a method for achieving the population inversion in a laser.

**psoralen-plus-UVA (PUVA)** a photochemotherapy for skin diseases utilizing the combined action of an oral psoralen drug and UV-A..

**pyroelectric crystal** a ferroelectric crystal in which the magnitude of the intrinsic electric dipole moment is sensitive to temperature.

**Q-switch** a fast shutter used with a pulsed laser which blocks the optical path until population inversion has been achieved.

**tungsten-quartz-halogen lamp (QTH)** a tungsten filament lamp with a fused-silica envelope containing a halogen vapor that permits higher temperature operation than conventional incandescent lamps.

**quantum theory** a general theory of particles and radiations in which it is postulated that their physical properties are characterized by discrete values of the relevant variables.

**quantum yield** the number of product molecules in a photophysical or photochemical system generated by the absorption of one photon; also referred to as "quantum efficiency".

**quarter wave plate** a retardation plate of quarter-wavelength thickness which converts linearly polarized light to elliptical polarized or circular polarized light.

**radian** a unit of angle in a plane equal to the angle subtended at the center of a circle by an arc equal in length to the radius of that circle; one radian equals 57.3 degrees of angle.

**radiance** SI unit of radiant flux in a given direction per unit solid angle amd per unit area perpendicular to the direction of propagation ($W/m^2 \cdot sr$).

**radiant exposure** radiant energy arriving per unit area ($J/m^2$).

**radiant flux** radiometric unit of optical power (W).

**radiation pressure** the force per unit area exerted by electromagnetic adiation on the surface reflecting and/or absorbing body.

**radiative lifetime** the lifetime of an electronically-excited state for which emission of fluorescence is the only decay process.

**radiative transfer** a general theory of light propagation in random media based on scattering and absorption of photons by constituent particles.

**radiometric units** a system of optical units based on radiant power (W) and radiant energy (J).

**ray** the trajectory of a point on the wave front of a propagating wave.

**Rayleigh range** a measure of the distance over which a laser beam remains highly collimated.

**Rayleigh scattering** an early theory of light scattering applicable for small particles relative to the wavelength.

**ray matrices** 2 x 2 matrices used for analytical ray tracing in optical systems.

**ray tracing** graphical construction of optical imaging by mapping the ray trajectories.

**real image** an optical image that can be projected on a screen.

**rectifier** an electrical device with a higher resistance to current flow in one direction than the opposite direction.

**reduced scattering coefficient** in radiative transfer theory, the product of the linear scattering coefficient and (1 - g) where g is the asymmetry parameter.

**reflection coefficient** the fraction of incident optical power reflected by a surface; also referred to as "reflectivity".

**refractive index** ratio of the light velocity in vacuum at a given wavelength to that in a medium.

**relative refractive index** ratio of the refractive indices of two media at a given wavelength.

**relaxation** any process by which a system not in thermal equilibrium with the environment releases its excess energy.

**remission** exiting of light from an illuminated object.

**resolving power** the smallest object distance that can be distinguished in an image; also used for the limit of wavelength separation by a monochromator or spectrograph.

**resonance energy transfer** transfer of energy between two atoms or molecule in which energy is conserved.

**response time constant** a parameter that characterizes the time required for the response of a system to attain the maximum value ; usually taken as 0.6321 (1 - 1/*e*) of the maximum response.

**responsivity** the strength of the electrical signal generated by a photodetector per unit incident optical power.

**retardation plate** a birefringent crystal plate containing an optic axis with a thickness equal to a multiple of a full-wavelength, half-wavelength or quarter-wavelength.

**scattering angle** in scattering theory, the angles between the directions of the incident and scattered objects in an appropriate coordinate system.

**second-order process** a process for which the instantaneous rate of response is proportional to the square of the concentration of one reactant or the product of the concentrations of two reactants.

**selection rule** a quantum mechanical relation which must be satisfied for an allowed (high-probability) optical transition in an atom or molecule.

**self-quenching** an interaction of an excited molecule with an unexcited molecule of the same type which deactivates the excited molecule.

**sensitized fluorescence** an energy transfer process in which light absorption by a donor molecule leads to fluorescence of an acceptor molecule.

**single-mode fiber** an optical fiber that transmits only one linearly polarized mode.

**SI units** a complete set of physical units established by international agreement

***sigma* molecular orbital** a type of molecular orbital which the electron density is localized in one interior lobe (binding) or two exterior lobes (anti-bonding) centered on the molecular axis.

**singlet molecular oxygen** the lowest-energy electronically excited state of molecular oxygen which is a singlet state; frequently termed "singlet oxygen".

**singlet-singlet energy transfer** *inter*molecular transfer of energy from the excited singlet state of a donor molecule to the singlet ground state of an acceptor molecule leading to the excited singlet state of the acceptor.

**singlet state** spectroscopic notation for an atom, molecule, or ion in which all electron spins are paired leading to a total spin of zero.

**Snell's law** relationship between the angle of incidence and the angle of refraction at the interface between two optical media.

**solid angle** angular coordinates describing directionality in three-dimensional space.

**soliton** a wave that propagates over long distances with a constant or oscillatory pulse shape.

**Soret band** an intense absorption band near 360-400 nm in the absorption spectra of porphyrin molecules.

**space charge** a time-independent distribution of electric charge.

**spatial coherence** a property of waves occupying a region of space that have the same frequency, direction, and polarization.

**spectral irradiance** wavelength distribution of the irradiance emitted by a light source.

**spectral radiance** wavelength distribution of the radiance emitted by a light source.

**spectrograph** a spectroscopic instrument that records the spectral distribution of entering light.

**spectrometer** a spectroscopic instrument that measures the wavelength of entering light.

**spectrophotometer** a spectroscopic instrument that measures and records the optical absorption or emission spectra of inserted samples utilizing photoelectric detection.

**spectroradiometer** an instrument used for measuring radiant power or energy; a "power meter".

**specular reflection** reflection of light by a highly polished surface.

**spherical aberration** an imaging defect in which parallel rays are not focused at a point.

**spherical lens** a lens with at least one spherical refracting surface.

**spin** the intrinsic angular momentum associated with many fundamental particles including electrons, protons, and neutrons.

**spin conservation** a physical or chemical process in which the total spin of the reactants equals the total spin of the products.

**spontaneous emission** radiative decay of an excited atom or molecule which takes place without external stimulus.

**Stefan-Boltzmann law** the relationship stating that the total thermal radiation emitted by a black body is proportional to the fourth power of the absolute temperature.

**step-index fiber** an optical fiber in which the core and cladding have constant refractive indices.

**steradian** a unit of angular measure in three dimensions equal to the solid angle subtended at the center of a sphere by a spherical surface equal in area to the square of the radius of that sphere.

**stimulated emission** a process in which absorption of electromagnetic radiation by an excited atom or molecule induces radiative transitions to a lower energy states.

**streak camera** a high-speed motion picture camera used to record very fast events.

**Stokes' shift** the long-wavelength shift of a fluorescence band relative to the corresponding absorption band.

**substrate** the major part of a photosensitized system as distinguished from the photosensitizer and the external environment.

**SUP , selective ultraviolet light phototherapy** a type of fluorescent lamp used for phototherapy of skin diseases which emits UV-A and UV-B.

**superoxide** the one-electron reduction product of molecular oxygen; the radical-anion of the perhydroxyl free radical.

**target theory** a model for biological radiation damage in which the probability of an observed response depends on the number of "hits" on one or more "targets".

**temporal coherence** waves that have the same phase and velocity.

**tomograph** an imaging technique in which a narrow planar region of the object is scanned in multiple directions by a co-linear source and detector.

**TPR**, therapeutic ratio: the ratio of the photodynamic therapy light dose required for normal tissue necrosis to that required for tumor necrosis.

**thermal conductivity** the proportionality constant between the rate of heat flow and the temperature gradient in Fick's law.

**thermal detector** a device which measures the power of optical radiation by means of its heating effects; exemplified by a thermopile and a pyroelectric crystal.

**thermal diffusivity** the ratio of the thermal conductivity to the product of the density and specific heat.

**thermal penetration depth** a measure of surface heat penetration into a tissue perfused by blood flow; see Equation (112).

**thermal relaxation time constant** a measure of the time required for a heat pulse to diffuse a given distance; see Equation (111).

**thermal waves** highly-damped, dispersive waves corresponding to conductive heat flow.

**thermister** a solid state device that changes its electrical resistance with temperature.

**thermopile** an array of dissimilar metal junctions used to measure radiant power by the heating effect.

**three-level laser** a type of laser action in which the laser light is emitted by radiative transitions from an indirectly pumped intermediate energy level to the ground state.

**threshold process** a process in which the observed response does not start until a threshold or "conditioning" dose has accumulated.

**throughput** the fraction of incident power transmitted by an optical device.

**time domain** an experimental technique in which a pulsed input signal is applied to a system and the response is measured in real time.

**tissue optics** the theory of light propagation and utilization in biological tissues.

**TIR** total internal reflection: reflection of light at an interface with a medium of higher refractive index.

**transmission coefficient** fraction of incident optical power transmitted at the rear surface of a planar layer.

**transport albedo** ratio of the scattering cross section of a particle to the sum of the absorption and scattering cross sections.

**transverse electromagnetic mode** a standing wave pattern of electromagnetic radiation in the plane perpendicular to the longitudinal axis in an optical cavity.

**transverse wave** a wave motion in which the displacement is perpendicular to the direction of propagation.

**triplet state** spectroscopic notation for an atom, molecule, or ion in which two electron spins are unpaired leading to a total electron spin of h/2 p .

**triplet-triplet energy transfer** intermolecular energy transfer in which a donor triplet state excites the singlet ground state of the acceptor to the acceptor triplet state.

**turbidity** cloudy appearance of a substance such as opal glass caused by the presence of light-scattering particles.

**Type 1 photosensitization** a photosensitized chemical reaction in which the initial interactions of the excited photosensitizer are with the substrate.

**Type 2 photosensitization** a photosensitized chemical reaction in which the initial interactions of the excited photosensitizer are with molecular oxygen leading to the production of singlet molecular oxygen.

**ultrasound** acoustic waves with frequencies higher than can be perceived by the human ear; typically 1-10 MHz for medical applications.

**ultraviolet radiation (UV)** the band of electromagnetic radiations having shorter wavelengths than the visible spectrum.

**valence** the number of chemical bonds made by an atom in a compound.

**valence band** an energy band in a crystalline solid fully occupied by electrons.

**vibronic levels** quantum states of a molecule associated with electronic and vibrational energies.

**vidicon** a television picture tube in which the optical image projected on a thin photoconductive layer is scanned by a moving electron beam.

**virtual image** a perceived optical image located at the origin of divergent rays.

**V number** see normalized frequency parameter.

**vr , vibrational relaxation** transfer of excess molecular vibrational energy to the external medium as heat.

**wave** a moving oscillatory disturbance.

**wave frequency** the repetition rate of the amplitude oscillations in wave motion.

**wavefront** a surface of a propagating wave containing points of constant phase.

**wave function** see probability amplitude.

**wavelength** the distance between two successive points on a periodic wave that have the same phase.

**wave velocity** the velocity of propagation of a wavefront.

**Wien's displacement law** the inverse relationship between the peak wavelength of black body radiation and the absolute temperature; see Equation (40).

**white light** polychromatic light perceived without hue.

**work function** the minimum photon energy required to eject electrons from a metal.

**Zener diode** a heavily doped p-n junction in which a high current flow is initiated by a small increase of reverse bias.

**zero order** a rate process in which the response does not depend on the amounts of the constituents.

# INDEX